AF598057

Green Energy and Technology

For further volumes:
http://www.springer.com/series/8059

Anoop Singh · Deepak Pant
Stig Irving Olsen
Editors

Life Cycle Assessment of Renewable Energy Sources

Editors
Anoop Singh
New Delhi
India

Stig Irving Olsen
Bagsværd
Denmark

Deepak Pant
PRODEM
Mol
Belgium

ISSN 1865-3529 ISSN 1865-3537 (electronic)
ISBN 978-1-4471-5363-4 ISBN 978-1-4471-5364-1 (eBook)
DOI 10.1007/978-1-4471-5364-1
Springer London Heidelberg New York Dordrecht

Library of Congress Control Number: 2013946320

Printed on acid-free paper

Springer is part of Springer Science+Business Media (www.springer.com)

Foreword

With the advent of modern civilisation and continuously growing human population, there is constant increase in the demand for the energy world over for livelihood and recreational purposes. The major sources of conventional energy derived through petroleum resources and coal reserves are depleting, which have raised the concerns and led to growing global interest in developing alternative sources of energy. National governments also see energy independence as a kind of security for the country. There have been intensive efforts all over the world to explore and exploit the alternative energy sources, such as solar energy, wind energy, bioenergy, etc. Bioenergy largely relies on biomass-based processes for the development of liquid and gaseous fuels, which have often been termed as first generation (ethanol from corn and other starchy sources), second generation (bioethanol from lignocellulosic feedstocks and biodiesel from vegetable oils), third generation (biofuels derived from algae) and fourth generation (biohydrogen). Biofuels derived from renewable materials offer much promise. In addition to serve as alternative source of energy, they also offer potential benefits on environmental impact in comparison to fossil fuels.

For the development of technologically and economically feasible renewable energy process, not only one requires substantial basic R&D data, but must also develop suitable models and integrate them with scale-up data. Yet another important aspect in this regard is life cycle assessment (LCA) study, which should be accomplished for a complete economic, environmental and social sustainability scenario development. LCA studies could involve the production and use of a product or the development of a service or product. In either cases, environmental and economic scenarios must be given due consideration.

The book on 'Life Cycle Assessment of Renewable Energy Sources' provides state-of-the-art information on the LCA studies and scenarios for the renewable energy. The editors have put together a host of highly relevant topics, ranging from the importance of LCA for renewable energy sources, key issues for bio-based renewable energy sources LCA, LCA for the production of biogas, bioethanol, biodiesel from different feedstocks, LCA for wind energy, solar energy, hydropower and comparison of different LCA studies. These aspects have been dealt by the peers.

LCA should involve the elements of life cycle inventory, life cycle impact assessment and interpretation. All these have been achieved in this book by describing the specialty processes and pioneering works. The editors have brought together a pool of expertise to present the state-of-the-art information, which have presented in-depth analysis of the knowledge on various aspects.

Overall, the information provided in this book is highly scientific, updated and would be beneficial for the researchers and practitioner equally; this will be also useful for those entering into this area.

Ashok Pandey
National Institute for Interdisciplinary Science and Technology, CSIR
Trivandrum, India
Editor-in-Chief, Bioresource Technology (Elsevier)

Preface

In recent years, a lot of emphasis has been given to renewable, sustainable and environment friendly energy sources in order to offset the dependence of mankind on conventional and non-renewable sources of energy most of which are fossil-based. However, the plethora of options available today makes it difficult for the users, policy makers as well as the researchers in this area to identify the right source for a specific situation as the usage and implementation depends on a variety of factors such as availability, ease of transportation, maintenance and end-of-life options. Energy and environment are closely interlinked and therefore any alternative energy option brings with it a certain impact on the environment. Several terms such as 'cradle to grave', 'cradle to cradle', 'cradle to gate' are used in this regard to denote the impacts at each stage of a product's life-cycle. This has led to a lack of understanding among the practitioners in this field and often leads to complicated situations where no agreement can be found over one single source of renewable energy. The integrated assessment of all environmental impacts from cradle to grave is the basis for many decisions relating to achieving improved products and services. The assessment tool most widely used for this is the environmental Life Cycle Assessment (LCA).

This book is intended to have three roles and to serve three associated audiences namely, the students and research community who will benefit from the lucid explanation of the LCA aspects of different bioenergy systems, the policy makers who will find it easier to identify the pros and cons of one type of bioenergy systems against another and finally the industries involved as it will give them a feeling about the current loopholes and ways to fix them. New developments in LCA methodology from all over the world have been discussed and, where possible, complemented with real life examples by the renowned experts in the field. Integration of all the recent developments into a new, consistent methodology for each type of renewable energy system has been the main aim for this book. Though we have tried to be very objective in our choice of topics to be covered in this book, some not so common themes might have been missed but which may become important in future which we will try to cover in the second edition of the book. "Importance of Life Cycle Assessment of Renewable Energy Sources" gives an overview of LCA for renewable energy sources, "Key Issues in Conducting Life Cycle Assessment of Bio-Based Renewable Energy Sources" –"Sustainability of (H_2 + CH_4) by Anaerobic Digestion via EROI Approach and LCA Evaluations" discusses the LCA of different types of biofuel systems. "Life-Cycle Assessment of Wind Energy" explores the LCA of wind energy and "Comparing Various Indicators for the LCA of Residential Photovoltaic Systems" deals with photovoltaic systems. "Hydropower Life-Cycle Inventories: Methodological Considerations and Results Based on a Brazilian Experience" explain the LCA aspect of hydropower while "A comparison of Life Cycle Assessment Studies of Different Biofuels" compares the LCA approaches for different renewable energy sources.

A major advantage of this book is that it also provides advice on which procedures should be followed to achieve adequate, relevant and accepted results. Furthermore, the distinction between detailed and simplified LCA makes this book more broadly applicable, while guidance is provided as to which additional information can be relevant for specialised applications.

We sincerely hope that this book will contribute to the necessary transition to environmentally benign and sustainable energy production and consumption.

Anoop Singh
Deepak Pant
Stig Irving Olsen

Acknowledgments

We, the editors would like to thank several people who helped us as we continued to work on this book. We begin by thanking Anthony Doyle from Springer who first approached us with the proposal for the book and helped us crystallise our ideas on the topic. Afterwards Quinn Grace and Christine Velarde took over and provided excellent support with all the administrative work. We also thank all the authors who kindly agreed to provide the chapters and worked with us throughout the process. We are also grateful to the reviewers who took time out of their busy schedule to critically review the chapters of this book and provided very valuable suggestions for their improvements. This book is a labour of love for us since we spent a lot of our weekends and free time on working on it. For this reason alone, our families deserved to be thanked for bearing with us all this while. Anoop and Stig like to thank the management of Technical University of Denmark (DTU), Lyngby, Denmark for their support. Deepak would like to thank the management of VITO especially Dr. Karolien Vanbroekhoven, Programme Manager at Separation and Conversion Technology unit for her unflinching support and encouragement towards this endeavour.

Contents

Importance of Life Cycle Assessment of Renewable Energy Sources 1
Anoop Singh, Stig Irving Olsen and Deepak Pant

Key Issues in Conducting Life Cycle Assessment of Bio-Based Renewable Energy Sources 13
Edi Iswanto Wiloso and Reinout Heijungs

The Application of Life Cycle Assessment on Agricultural Production Systems with Reference to Lignocellulosic Biogas and Bioethanol Production as Transport Fuels 37
Nicholas E. Korres

Life-Cycle Assessment of Biomethane from Lignocellulosic Biomass 79
Abdul-Sattar Nizami and Iqbal Mohammed Ismail

Life Cycle Assessment of Biodiesel from Palm Oil 95
Keat Teong Lee and Cynthia Ofori-Boateng

Environmental Sustainability Assessment of Ethanol from Cassava and Sugarcane Molasses in a Life Cycle Perspective 131
Shabbir H. Gheewala

Comparison of Algal Biodiesel Production Pathways Using Life Cycle Assessment Tool 145
Anoop Singh and Stig Irving Olsen

Sustainability of (H_2 + CH_4) by Anaerobic Digestion via EROI Approach and LCA Evaluations 169
B. Ruggeri, S. Sanfilippo and T. Tommasi

Life-Cycle Assessment of Wind Energy 195
E. Martínez Cámara, E. Jiménez Macías and J. Blanco Fernández

Comparing Various Indicators for the LCA of Residential Photovoltaic Systems 211
Ruben Laleman, Johan Albrecht and Jo Dewulf

Hydropower Life-Cycle Inventories: Methodological Considerations and Results Based on a Brazilian Experience 241
Gil Anderi da Silva, Flávio de Miranda Ribeiro and Luiz Alexandre Kulay

A Comparison of Life Cycle Assessment Studies of Different Biofuels 269
Dheeraj Rathore, Deepak Pant and Anoop Singh

Index 291

About the Editors

Dr. Anoop Singh, is a Scientist at Department of Scientific and Industrial Research (DSIR), Ministry of Science and Technology, Government of India. He completed his Doctoral degree in Botany in 2004 from AAIDU, Allahabad, India and his Master's degree in Environmental Sciences in 2001 from GBPUAT, Pantnagar, India. Before joining DSIR, he worked at Technical University of Denmark, Denmark; University College Cork, Ireland; The Energy and Resources Institute (TERI), New Delhi, India; Indian Agricultural Research Institute (IARI), New Delhi, India; Banaras Hindu University, Varanasi, India; and VBS Purvanchal University, Jaunpur, India. He has visited several European countries and participated in several international conferences. He has published more than 50 research articles in scientific journals (>1,000 citations, h-index 17) and is a member of several scientific communities. He is serving as Editorial board member for a number of journals besides being a reviewer. His research interests are focused on sustainable agriculture, waste management through agriculture, the utilization of industrial, agricultural, and household waste for eco-friendly energy production, renewable energy, and their life cycle assessment (LCA).

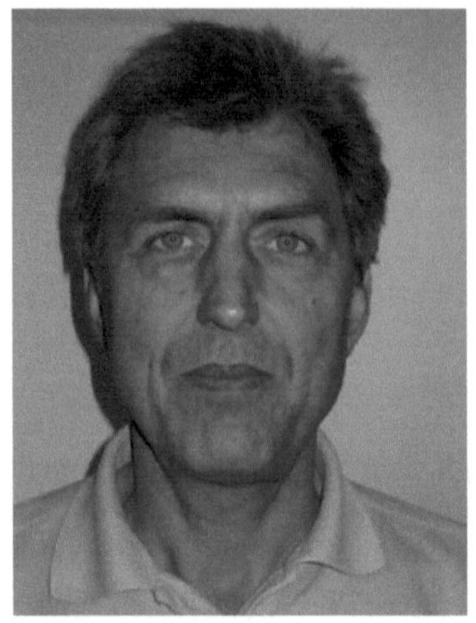

Dr. Stig Irving Olsen is an Associate Professor in sustainable production at the section for Quantitative Sustainability Assessment, Department of Management Engineering at the Technical University of Denmark. He obtained his Ph.D. in LCA from Technical University of Denmark in 1997 and a Master of Science in Biology from University of Copenhagen in 1988. Since his Ph.D., his main research area has been in methodology development in LCA, particularly in the life cycle impact assessment of human health impact. During the last years his research has focused more on application of LCA in several technology areas, including renewable energy and nanotechnology.

Dr. Deepak Pant is a Research Scientist at the Flemish Institute for Technological Research (VITO), Belgium, currently working on bioenergy, specifically, the design and optimisation of bioelectrochemical systems for energy recovery from wastewaters and microbial electrosynthesis for production of value added chemicals and fuels through electrochemically driven bioprocesses. He has a Ph.D. in Environmental Biotechnology (2007) from TERI University, New Delhi (India) and has 28 peer-reviewed publications with 950 citations (h-Index 14) and 10 book chapters to his credit. He is involved in several European projects on biomass, biowaste, wastewater treatment and feasibility studies. His research experience lies in industrial wastewater treatment, wasteland reclamation and restoration, biofertilisers, sustainable agriculture, biofuels and bioenergy and life cycle analysis (LCA).

Contributors

Johan Albrecht Faculty of Economics and Business Administration, Ghent University, Tweekerkenstraat 2, 9000 Ghent, Belgium

Jo Dewulf Research Group ENVOC, Ghent University, Coupure Links 653, 9000 Ghent, Belgium

J. Blanco Fernández Department of Mechanical Engineering, University of La Rioja, Logroño, La Rioja, Spain

Shabbir H. Gheewala The Joint Graduate School of Energy and Environment, King Mongkut's University of Technology Thonburi, Bangkok, Thailand; Center for Energy Technology and Environment, Ministry of Education, Bangkok, Thailand

Reinout Heijungs Institute of Environmental Sciences (CML), Leiden University, P.O. Box 9518, 2300 RA Leiden, The Netherlands

Nicholas E. Korres 26 Grigoroviou street, Patisia, 11141 Athens, Greece

Luiz Alexandre Kulay University of São Paulo, São Paulo, Brazil

Ruben Laleman Faculty of Economics and Business Administration, Ghent University, Tweekerkenstraat 2, 9000 Ghent, Belgium

Keat Teong Lee School of Chemical Engineering, Universiti Sains Malaysia, Seri Ampangan, Nibong Tebal, 14300 SPS, Pulau Penang, Malaysia

E. Jiménez Macías Department of Electrical Engineering, University of La Rioja, Logroño, La Rioja, Spain

E. Martínez Cámara Department of Mechanical Engineering, University of La Rioja, Logroño, La Rioja, Spain

Flávio de Miranda Ribeiro Pollution Prevention Group—University of Sao Paulo (GP2-USP), Sao Paulo, Brazil

Abdul-Sattar Nizami Center of Excellent in Environmental Studies (CEES), King Abdulaziz University (KAU), P.O. Box 80216, 21589 Jeddah, Saudi Arabia

Iqbal Mohammed Ismail Center of Excellent in Environmental Studies (CEES), King Abdulaziz University (KAU), P.O. Box 80216, 21589 Jeddah, Saudi Arabia

Cynthia Ofori-Boateng School of Chemical Engineering, Universiti Sains Malaysia, Seri Ampangan, Nibong Tebal, 14300 SPS, Pulau Penang, Malaysia

Stig Irving Olsen Quantitative Sustainability Assessment, Department of Management Engineering, Technical University of Denmark, Kongens Lyngby, Denmark

Deepak Pant Separation and Conversion Technology, Flemish Institute for Technological Research (VITO), Mol, Belgium

Dheeraj Rathore Department of Conservation Biology, School of Biological Science, College of Natural and Mathematical Sciences, University of Dodoma, Dodoma, Tanzania

B. Ruggeri Department of Applied Science and Technology, Politecnico di Torino, Corso Duca degli Abruzzi 24, 10129 Turin, Italy

S. Sanfilippo Department of Applied Science and Technology, Politecnico di Torino, Corso Duca degli Abruzzi 24, 10129 Turin, Italy

Gil Anderi da Silva Brazilian Life-Cycle Association, Coordinator of Pollution Prevention Group—University of Sao Paulo (GP2-USP), Sao Paulo, Brazil

Anoop Singh Quantitative Sustainability Assessment, Department of Management Engineering, Technical University of Denmark, Kongens Lyngby, Denmark; Department of Scientific and Industrial Research (DSIR), Ministry of Science and Technology, Government of India, Technology Bhawan, New Mehrauli Road, New Delhi 110016, India

T. Tommasi Center for Space Human Robotics, Istituto Italiano di Tecnologia, Corso Trento 21, 10129 Turin, Italy

Edi Iswanto Wiloso Institute of Environmental Sciences (CML), Leiden University, P.O. Box 9518, 2300 RA Leiden, The Netherlands

Importance of Life Cycle Assessment of Renewable Energy Sources

Anoop Singh, Stig Irving Olsen and Deepak Pant

Abstract The increasing demand for sustainable renewable energy sources to reduce the pollution and dependency on conventional energy resources creates a path to assess the various energy sources for their sustainability. One renewable energy source might be very attractive for heat production and not so attractive for electricity and transport purposes. The commercial-scale production of these energy sources requires careful consideration of several issues that can be broadly categorized as raw material production, technology, by-products, etc. The life cycle assessment (LCA) is a tool that can be used effectively in evaluating various renewable energy sources for their sustainability and can help policy makers choose the best energy source for specific purpose. Choice of allocation method is very important in assessing the sustainability of energy source as different allocation methods respond in present differently. The present chapter is an effort to highlight the importance of LCA of renewable energy sources.

1 Introduction

Progressive depletion of conventional fossil fuels with increasing energy consumption and greenhouse gas (GHG) emissions has led to a move toward renewable and sustainable energy sources (Singh et al. 2011, 2012; Nigam and

A. Singh (✉) · S. I. Olsen
Department of Scientific and Industrial Research (DSIR), Ministry of Science and Technology, Technology Bhawan, New Mehrauli Road, New Delhi 110016, India
e-mail: apsinghenv@gmail.com

S. I. Olsen
e-mail: siol@dtu.dk

D. Pant
Separation and Conversion Technology, Flemish Institute for Technological Research (VITO), Mol, Belgium
e-mail: deepak.pant@vito.bepantonline@gmail.com

A. Singh et al. (eds.), *Life Cycle Assessment of Renewable Energy Sources*,
Green Energy and Technology, DOI: 10.1007/978-1-4471-5364-1_1,

Singh 2011). The production of sustainable energy based on renewable sources is a challenging task for replacing the fossil-based fuels to get cleaner environment and also to reduce the dependency on other countries and uncertainty of fuel price (Singh and Olsen 2012, 2011; Pant et al. 2012). A worrying statistic is that the global production of oil and gas is approaching its maximum and the world is now finding one new barrel of oil for every four it consumes (Aleklett and Campbell 2003). All these serious concerns related to energy security, environment, and sustainability have led to a move toward alternative, renewable, sustainable, efficient, and cost-effective energy sources with lesser emissions (Prasad et al. 2007a, b; Singh and Olsen 2012).

The life cycle assessment (LCA) of renewable energy sources is the key to observe their sustainability. There is a need to conduct LCA of renewable energy production system on the basis of their local conditions, as one energy source cannot be sustainable for all geographical locations, due to variations in resources availability, climate, environmental, economical and social conditions, policies, etc. Therefore, LCA can be used as a tool to assess the sustainability of various energy sources for different locations. LCA techniques allow detailed analysis of material and energy fluxes on regional and global scales. This includes indirect inputs to the production process and associated wastes and emissions, and the downstream fate of products in the future (Singh et al. 2011). LCA studies vary in their definition of the various criteria, such as, scope and goal, system boundaries, reference system, allocation method. LCA studies of renewable energy sources calculate the environmental impact and can relate the results against sustainability criteria. The present chapter is an effort to highlight the importance of LCA of renewable energy sources to get a more holistic perspective of their environmental sustainability.

2 Renewable Energy Sources

The most common renewable energy sources are presented in the Fig. 1. Each renewable energy source is performing differently; one could be best option for one location/purpose/season and could not perform with that efficiency at another location/purpose/season. The solar energy sources are best in remote or under developed areas having bright sunshine (Jayakumar 2009). Windmills are best suited near sea shore, as there winds are enough strong to get decent production of energy. Similarly, tidal, hydroelectric, geothermal, and ocean thermal energies have their importance. Among the renewable energy sources, biofuels are the most popular renewable energy source because of the availability of raw material (biomass), everywhere and round the year and also due to its suitability in transport vehicles and industries. The detailed description of different biofuels is published by Nigam and Singh (2011).

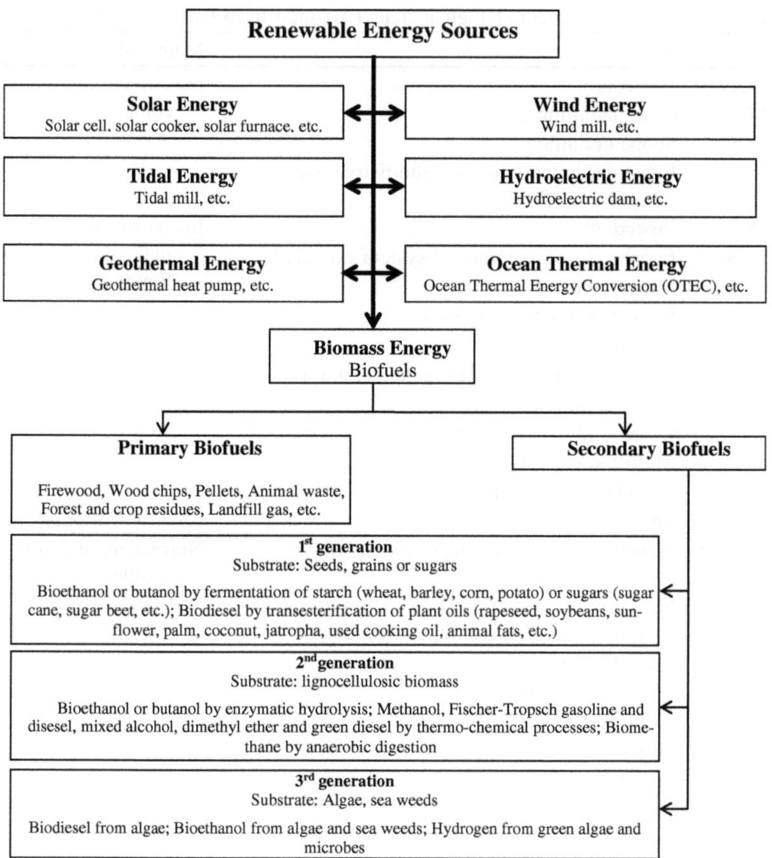

Fig. 1 The most important renewable energy sources

3 Life Cycle Assessment

ISO 14040 defined LCA as the "compilation and evaluation of the inputs, outputs and potential environmental impacts of a product system throughout its life cycle" (ISO 2006). Thus, LCA is a tool to assess the environmental impacts and resources used throughout a product's life cycle and consider all attributes or aspects of natural environment, human health, and resources (Korres et al. 2010) and can be defined as a method for analyzing and assessing environmental impacts of a material, product, or service along its entire life cycle (ISO 2005). LCA analyzes the environmental burden of products at all stages in their life cycle (from the cradle to the grave) from the extraction of resources, through the production of materials, product parts and the product itself, and the use of the product to the management after it is discarded, either by reuse, by recycling, or by final disposal (Guinée 2004).

Table 1 Overview of LCA methodological steps (Adapted from Guinée 2004)

Phase	Steps	Main result
Goal and scope definition	Procedure	Functional unit, alternatives compared
	Goal definition	
	Scope definition	
	Function, functional unit, alternative and reference flows	
Inventory analysis	Procedure	Inventory table, other indication (e.g., missing flows)
	Economy—environmental system boundary	
	Flow diagram	
	Format and data categories	
	Data quality	
	Data collection and relating data to unit processes	
	Data validation	
	Cutoff and data estimation	
	Multifunctionality and allocation	
	Calculation method	
Impact assessment	Procedures	Environmental profile
	Selection of impact categories	Normalized environmental profile
	Selection of characterization methods: category indicators, characterization models	Weighting profile
	Classification	
	Characterization	
	Normalization	
	Grouping	
	Weighting	
Interpretation	Procedure	Well-balanced conclusion and recommendations
	Consistency check	
	Completeness check	
	Contribution analysis	
	Perturbation analysis	
	Sensitivity and uncertainty analysis	
	Conclusions and recommendations	

Various steps involved in the LCA methodology are listed in Table 1. The complete life cycle of the renewable energy sources includes each and every step from raw material production and extraction, processing, transportation, manufacturing, storage, distribution, and utilization. Each of these can have an impact (harmful or beneficial) of different environmental, economical, and social dimensions. It is therefore of crucial importance to assess the complete fuel chains from different perspectives in order to achieve sustainable biofuels (Markevičius et al. 2010).

The environmental burden covers all types of impacts on the environment, including extraction of different types of resources, emission of hazardous substances, and different types of land use. Reinhard and Zah (2011) distinguished the two main approaches of LCA, i.e., the attributional and the consequential approach: both approaches differ with respect to system delimitation and the use of average versus marginal data. Attributional LCA describes the environmentally relevant physical flows to and from a life cycle and its subsystems, while consequential LCA describes how environmentally relevant flows will change in response to possible decisions. Marginal data are represented by the product, resource, supplier, or technology, which are the most sensitive to changes in demand, and economic value criteria are used to identify the marginal products (Ekvall and Weidema 2004).

Attributional LCA is limited to a single full life cycle from cradle to grave, and consequential LCA is not limited to one life cycle, but uses system enlargement to include the life cycles of the products affected by a change in the multifunctional processes will often be handled through allocation, physical flows in the central life cycle. In attributional LCA multifunctional processes will often be handled through allocation, while in consequential LCA, allocation will generally be avoided through the system expansion. Additionally, marginal data are used, whereas average data are applied in attributional LCA (Ekvall and Weidema 2004; Reinhard and Zah 2011).

Various scientists have employed LCA on renewable energy production systems (Reinhard and Zah 2011; Biswas et al. 2011; Ribeiro and Silva 2010; Gabrielle and Gagnaire 2008; Gnansounou et al. 2009; Kiwjaroun et al. 2009; Martínez et al. 2009; Suri et al. 2007; Laleman et al. 2011; Zah et al. 2007), and some useful results considering the factors (e.g., biomass, technologies, use, system boundary, allocation, reference system) affecting the outcome of the analysis have been obtained (Singh et al. 2010).

4 Importance of Life Cycle Assessment

The purpose of LCA is to compile and evaluate the environmental consequences of different options for fulfilling a certain function (Guinée 2004), and it is a universally accepted approach of determining the environmental consequences of a particular product over its entire production cycle (Pant et al. 2011). The LCA methodology can be useful to acquire a comprehensive knowledge of the environmental impacts generated by industrial products during their whole life cycle (de Eicker et al. 2010). LCA can play a useful role in public and private environmental management in relation to products as this may involve both an environmental comparison between existing products and the development of new products (Guinée 2004). LCA has been the method of choice in recent years for various kinds of new technologies for bioenergy and carbon sequestration.

The "holistic" nature of LCA depicts both its major strength and, at the same time, its limitation. The broad scope of analyzing the complete life cycle of a product can only be achieved at the expense of simplifying other aspects (Guinée 2004). LCA of renewable energy production system requires a careful design regarding the goal and scope definition, choice of functional unit, reference system, system boundaries and appropriate inventory establishment and allocation of emissions in products and by-products (Singh and Olsen 2012). Larson (2006) describes four input parameters to cause the greatest variation and uncertainties in LCA results of energy production, namely climate-active plant species (species with ability or otherwise to adapt to climate change); assumptions about N_2O emissions; the allocation method for co-product credits; and soil carbon dynamics.

In general, LCA is in fact developed for impacts with an input–output character, and extractions from the environment and emissions to the environment can both be well linked to a functional unit (Udo de Haes and Heijungs 2007). LCA regards all processes as linear, both in the economy and in the environment. The LCA model focuses on physical characteristics of the industrial activities and other economic processes; the attributional LCA does not include market mechanisms or secondary effects on technological development (Guinée 2004).

The results of LCA study are as much science based as possible and aim to enlighten stakeholders in a production–consumption chain, thus contributing to rational decision-making. LCA study can also be of use inside a company; by implementing an LCA study on a product, the processes of the product system can be identified, which largely appear to contribute to its total environmental burden. This may help to direct environmental management of the company, for instance to support its investment decisions or to influence its supply management (Udo de Haes and Heijungs 2007). The main applications of LCA are analyses of the origins of problems related to a particular product; comparing improvement variants of a given product; designing new products; choosing between a number of comparable products. Similar applications can be distinguished at a strategic level, dealing with government policies and business strategies for renewable and sustainable energy source. The way an LCA project is implemented depends on the intended use of the LCA results (Guinée 2004). This reasoning can be predominantly true for decisions in the energy sector. In year 2010, EPA applied the consequential LCA approach in its regulation for US renewable fuel standards under the 2007 US Energy Independence and Security Act (RFS2, as opposed to renewable fuel standards under the 2005 U.S. Energy Policy Act, RFS1) (EPA 2010; Wang et al. 2011).

5 LCA and Sustainability of Renewable Energy Sources

The general principles of sustainable biofuel production are relatively easy to define (as shown in Fig. 2). However, it is quite challenging to derive a sound framework that is able to characterize environmental, economical, and social impacts in an adequate way. World Commission on Environment and Development defined the term "sustainability" as "the development that meets the needs of the present without compromising the ability of future generations to meet their own needs" (UNCED 1992). The methodologies to address LCA and sustainability are advancing although the availability of practical data remains an issue (Black et al. 2011). Sustainable development can be defined as the fulfillment

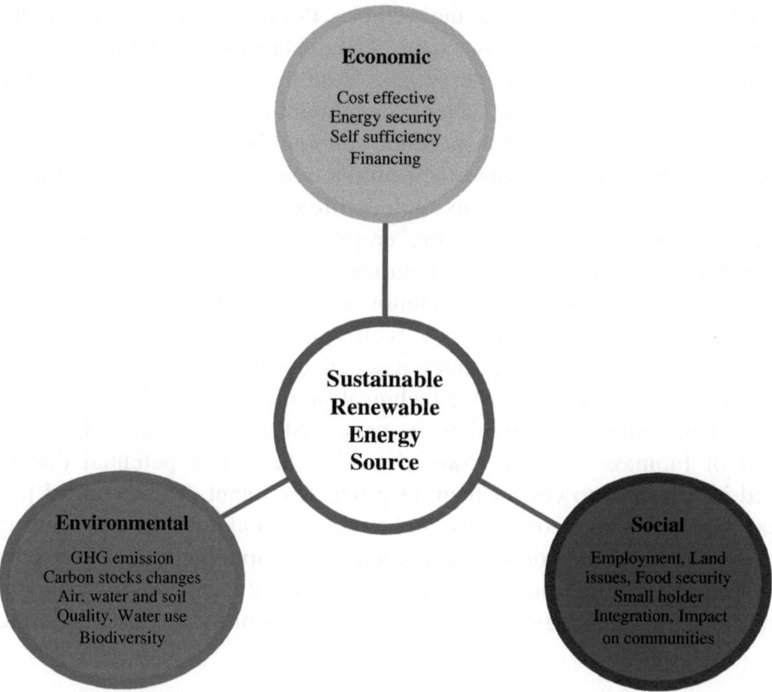

Fig. 2 Economic, social, and environmental aspects of sustainable renewable energy sources (Adapted from IEA 2011; Singh and Olsen 2012)

through the optimal use of any available source within a production system. Energy conversion, utilization, and access underlie many of the great challenges associated with sustainability, environmental quality, security, and poverty (Korres et al. 2010, 2011). Sustainability assessment of products or technologies is normally seen as encompassing impacts in three dimensions, i.e., social, environmental, and economic (Elkington 1998). These three dimensions form the backbone of sustainability standards. To replace the fossil fuels with biofuels, there is a need to maximize the environmental and social value of biofuels that is also important for the future of biofuels industry and market potential depends on being cost competitive with fossil fuels (Fig. 2). The environmental dimension comprises amongst others the GHG emissions, global ecological performance, conservation of energy resources, rational life cycle water use, effect on soil quality, conservation of biodiversity, use of chemicals, and the practice of slash and burn and the socioeconomic dimensions includes competition with food and feed, contribution to local well being, impact on communities and the quality of working conditions. These three interrelated goals must stay in balance for biofuels to remain sustainable.

Environmental impacts occur in all stages of the energy production system: the transformation of the land needed, production and application of chemicals and other input, cultivation of energy crops, production of the biofuel, transportation to the gauging station, and use in the vehicle. Pollutants are generated in many different steps of the production chain. The sustainability of renewable energy production depends on the net energy gain fixed in the output that depends on the production process parameters, such as the amount of energy-intensive inputs and the energy input for harvest, transport and running the processing facilities (Haye and Hardtke 2009), emissions and their production cost. The most used indicators to measure the energy sustainability include life cycle energy balance, quantity of fossil energy substituted per hectare, co-product energy allocation, life cycle carbon balance, and changes in soil utilization (Silva Lora et al. 2011). Gnansounou et al. (2009) stated that monitoring reduction in GHG emissions and estimations of substitution efficiency with respect to fossil fuels is subject to significant uncertainty and inaccuracy associated with the LCA approach.

The schematic illustration of the technical biomass potential and constraints to the sustainable biomass potentials is presented in the Fig. 3. The technical potential of biomass is much lower than the theoretical potential due to cost involved in transport to collect them at production plant. The technical potential also has several social, economical, and environmental constraints, resulting only in a part of the technical potential that could be suitable for sustainable renewable energy production. Gnansounou (2011) suggested that due to the multidimensional impact of renewable energy sources, the sustainability impact assessment of

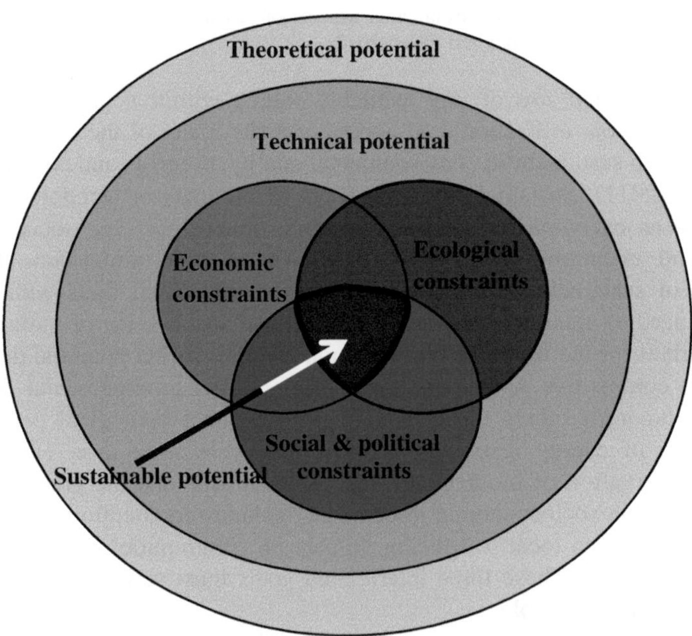

Fig. 3 Schematic illustration of the technical potential and constraints to the sustainable biomass potentials (Adapted from Steubing et al. 2010)

policies is as relevant as the sustainability assessment of production pathways and regulatory impact assessment.

Sustainability evaluation of biofuels is a multicriterial problem; Silva Lora et al. (2011) suggests the following main indicators for a sustainable energy production:

- To be carbon neutral.
- Not to affect the quality, quantity, and rational use of available natural resources.
- Not to affect the biodiversity.
- Not to have undesirable social consequences.
- To contribute to the societal economic development and equity.

The major factors that will determine the impacts of renewable energy production system include their contribution to land use change, the feedstock/input used, technology adopted, scale of production, use of by-products (if any), wholesale trade and retail of energy product and by-product, and emissions after end use of produced energy. Yan and Lin (2009) revealed that the interactions among various sustainability issues make the assessment of biofuel development difficult and complicated. The complexity during the whole renewable energy production chain generates significantly different results due to the differences in input data, methodologies applied, and local geographical conditions.

In order to ensure net societal benefits of biofuels production, governments, researchers, and companies will need to work together to carry out comprehensive assessments, map suitable and unsuitable areas, and define and apply standards relevant to the different circumstances of each country (Phalan 2009). The length and complexity of the supply chains make the sustainability issue very challenging. The main aim is to improve the performance of the strategies by enhancing positive effects, mitigating negative ones, and avoiding the transfer of negative impacts to future generations (Gnansounou 2011). The science of LCA is being stretched to its limits as policy makers consider direct and indirect effects of biofuels on global land and water resources, global ecosystems, air quality, public health, and social justice (Sheehan 2009). The sustainable renewable energy production is directed by environmental impacts (direct and indirect), economic viability including societal and political acceptance.

6 Conclusions

The increasing demand for renewable energy challenges societies to find out sustainable and renewable energy source. LCA is a tool which can be used effectively in assessing the sustainability of renewable energy sources. The collection of actual data for such study is a quite challenging task, as these data sets have very high variations with the temporal and spatial variation. The sustainability basically depends on three pillars of social, economical, and environmental performance of the renewable energy source. The social, economical, and environmental constraints reduce the potential of sustainable renewable energy sources.

References

Aleklett K, Campbell CJ (2003) The peak and decline of world oil and gas production. Miner Energ 18:35–42

Biswas WK, Barton L, Carter D (2011) Biodiesel production in a semiarid environment: a life cycle assessment approach. Environ Sci Technol 45:3069–3074

Black MJ, Whittaker C, Hosseinid SA, Diaz-Chaveza R, Woodsa J, Murphy RJ (2011) Life cycle assessment and sustainability methodologies for assessing industrial crops, processes and end products. Industrial Crops Products 34(2):1332–1339

de Eicker MO, Hischier R, Kulay LA, Lehmann M, Zah R, Hurni H (2010) The applicability of non-local LCI data for LCA. Environ Impact Assess Rev 30:192–199

Ekvall T, Weidema BP (2004) System boundaries and input data in consequential life cycle inventory analysis. Int J Life Cycle Assess 9(3):161–171

Elkington J (1998) Cannibals with forks: the triple bottom line of 21st century business. New Society Publishers, Canada

EPA (U.S. Environmental Protection Agency) (2010) Regulations of fuels and fuels additives: changes to renewable fuel standard program; final rule. Fed Reg 5(58):14670–14904

Gabrielle B, Gagnaire N (2008) Life-cycle assessment of straw use in bioethanol production: a case study based on biophysical modeling. Biomass Bioenerg 32:431–441

Gnansounou E (2011) Assessing the sustainability of biofuels: a logic based model. Energy 36:2089–2096

Gnansounou E, Dauriat A, Villegas J, Panichelli L (2009) Life cycle assessment of biofuels: energy and greenhouse gas balances. Bioresour Technol 100:4919–4930

Guinée JB (ed) (2004) Handbook on life cycle assessment operational guide to the ISO standards. Kluwer Academic Publishers, New York

Haye S, Hardtke CS (2009) The roundtable on sustainable biofuels: plant scientist input needed. Trends Plant Sci 14(8):409–412

IEA (2011) Technology roadmap biofuels for transport. International Energy Agency, France

ISO (2005) 14040-Environmental management—life cycle assessment—requirements and guidelines. International Standard Organisation, p 54

ISO (2006) ISO 14044: environmental management—life cycle assessment—requirements and guidelines. ISO 14044:2006(E), International Standards Organization

Jayakumar P (2009) Solar energy resources assessment handbook. Prepared for Asian and Pacific centre for transfer of technology (APCTT) of the united nations—economic and social commission for Asia and the Pacific (ESCAP), p 117. http://recap.apctt.org/Docs/SOLAR.pdf. Accessed Jan 2013

Kiwjaroun C, Tubtimdee C, Piumsomboon P (2009) LCA studies comparing biodiesel synthesized by conventional and supercritical methanol methods. J Cleaner Prod 17:143–153

Korres NE, Singh A, Nizami AS, Murphy JD (2010) Is grass biomethane a sustainable transport biofuel? Biofuels, Bioproducts Biorefining 4:310–325

Korres NE, Thamsiriroj T, Smyth BM, Nizami AS, Singh A, Murphy JD (2011) Grass biomethane for agriculture and energy. In: Lichtfouse E (ed) Genetics, biofuels and local farming systems, sustainable agriculture reviews. Springer Science + Business Media B.V., p 5–49

Laleman R, Albrecht J, Dewulf J (2011) Life cycle analysis to estimate the environmental impact of residential photovoltaic systems in regions with a low solar irradiation. Renew Sustain Energ Rev 15(1):267–281

Larson ED (2006) A review of life-cycle analysis studies on liquid biofuel systems for the transport sector. Energy Sus Dev 10(2):109–126

Markevičius A, Katinas V, Perednis E, Tamasauskiene M (2010) Trends and sustainability criteria of the production and use of liquid biofuels. Renew Sustain Energy Rev 14:3226–3231

Martínez E, Sanz F, Pellegrini S, Jiménez E, Blanco J (2009) Life-cycle assessment of a 2-MW rated power wind turbine: CML method. Int J Life Cycle Assess 14(1):52–63

Nigam PS, Singh A (2011) Production of liquid biofuels from renewable resources. Prog Energy Combust Sci 37:52–68

Pant D, Singh A, Bogaert GV, Diels L, Vanbroekhoven K (2011) An introduction to the life cycle assessment (LCA) of bioelectrochemical systems (BES) for sustainable energy and product generation: relevance and key aspects. Renew Sustain Energ Rev 15:1305–1313

Pant D, Singh A, Bogaert GV, Olsen SI, Nigam PS, Diels L, Vanbroekhoven K (2012) Bioelectrochemical systems (BES) for sustainable energy production and product recovery from organic wastes and industrial wastewaters. RSC Adv 2:1248–1263

Phalan B (2009) The social and environmental impacts of biofuels in Asia: an overview. Appl Energy 86:S21–S29

Prasad S, Singh A, Joshi HC (2007a) Ethanol as an alternative fuel from agricultural, industrial and urban residues. Resour Conserv Recycl 50:1–39

Prasad S, Singh A, Jain N, Joshi HC (2007b) Ethanol production from sweet sorghum syrup for utilization as automotive fuel in India. Energ Fuels 21(4):2415–2420

Reinhard J, Zah R (2011) Consequential life cycle assessment of the environmental impacts of an increased rapemethylester (RME) production in Switzerland. Biomass Bioenerg 35:2361–2373

Ribeiro FM, Silva GA (2010) Life-cycle inventory for hydroelectric generation: a Brazilian case study. J Cleaner Prod 18(1):44–54

Sheehan JJ (2009) Biofuels and the conundrum of sustainability. Curr Opinion Biotech 20:318–324

Silva Lora EE, Escobar Palacio JC, Rocha MH, Grillo Renó ML, Venturini OJ, del Olmo OA (2011) Issues to consider, existing tools and constraints in biofuels sustainability assessments. Energ 36:2097–2110

Singh A, Olsen SI (2011) Critical analysis of biochemical conversion, sustainability and life cycle assessment of algal biofuels. Applied Energ 88:3548–3555

Singh A, Olsen SI (2012) Key issues in life cycle assessment of biofuels. In: Gopalakrishnan K et al (eds) Sustainable bioenergy and bioproducts. Green energy and technology. Springer-Verlag London Limited, London, pp 213–228

Singh A, Pant D, Korres NE, Nizami AS, Prasad S, Murphy JD (2010) Key issues in life cycle assessment of ethanol production from lignocellulosic biomass: challenges and perspectives. Bioresourc Technol 101(13):5003–5012

Singh A, Olsen SI, Nigam PS (2011) A viable technology to generate third generation biofuel. J Chem Tech Biotech 86:1349–1353

Singh A, Pant D, Olsen SI, Nigam PS (2012) Key issues to consider in microalgae based biodiesel. Energ Edu Sci Technol Part A: Energ Sci Res 29(1):687–700

Steubing B, Zah R, Waeger P, Ludwig C (2010) Bioenergy in Switzerland: assessing the domestic sustainable biomass potential. Renew Sustain Energy Rev 14:2256–2265

Suri M, Huld TA, Dunlop ED, Ossenbrink HA (2007) Potential of solar electricity generation in the European union member states and candidate countries. Solar Energ 81(10):1295–1305

Udo de Haes HA, Heijungs R (2007) Life-cycle assessment for energy analysis and management. Appl Energy 84:817–827

UNCED (1992) Report of the united nation conference on environment and development, vol 1, Chap. 7

Wang M, Huo H, Arora S (2011) Methods of dealing with co-products of biofuels in life-cycle analysis and consequent results within the U.S. context. Energy Policy 39(10):5726–5736

Yan J, Lin T (2009) Bio-fuels in Asia. Appl Energy 86:S1–S10

Zah R, Hischier R, Gauch M, Lehmann M, Boni H, Wager P (2007) Life cycle assessment of energy products: environmental impact assessment of biofuels. Bern: Bundesamt fur Energie, Bundesamt fur Umwelt, Bundesamt fur Landwirtschaft, p 20

Key Issues in Conducting Life Cycle Assessment of Bio-based Renewable Energy Sources

Edi Iswanto Wiloso and Reinout Heijungs

Abstract Although there is an ISO-standardized method for conducting life cycle assessment (LCA) studies, its application to renewable energy sources, in particular to bio-based renewable energy (bioenergy) involving agricultural chains, is not straight forward. There are theoretical and practical issues in goal and scope definition, functional unit, inventory analysis, and impact assessment. The debate between attributional LCA and consequential LCA is, for bioenergy, even more crucial than for ordinary products, especially when it comes to either direct or indirect land-use change. Data are often highly variable, and system boundaries are quite arbitrary. For bioenergy from biomass residues, allocation and recycling provide complications. The treatment of biogenic carbon is of particular interest. The choice of impact categories and the necessity of a regionalized impact assessment are another problem. This chapter provides a systematic overview of these topics.

1 Introduction

Our economy has long been dependent on non-renewable energy carriers, especially on fossil energy. The high dependence on non-renewable energy sources developed over a relatively short period of time. From the middle of the nineteenth century, there was a rapid increase in the use of fossil fuels. These non-renewables replaced wood and soon became the basis of an exponential growth in energy use associated with a number of novel energy-demanding activities (Sørensen 2002). Early man was only capable of causing environmental disturbance on a local scale; however, man has currently achieved a technological level, enabling him to

E. I. Wiloso (✉) · R. Heijungs
Institute of Environmental Sciences (CML), Leiden University, P.O. Box 9518, 2300 RA Leiden, The Netherlands
e-mail: wiloso@cml.leidenuniv.nl; ediiswanto@yahoo.com

A. Singh et al. (eds.), *Life Cycle Assessment of Renewable Energy Sources*, Green Energy and Technology, DOI: 10.1007/978-1-4471-5364-1_2,

convert energy at rates that are responsible for climate change over extended areas. With 81 % of recent global energy use originating from fossil fuels, 6 % from nuclear, and 13 % from renewable energy (IEA-Bioenergy 2009), it is understandable that human societies have recently begun to reconsider the use of renewable sources. In light of this development, we are now, along with other environmental impacts, facing two major problems: depletion of fossil resources and an increase in anthropogenic levels of carbon dioxide.

Alternative options that are available to reduce our dependence on non-renewable sources and simultaneously mitigate climate change are already in development. The use of bio-based renewable energy (bioenergy) is now deemed to be one of the most promising renewable energy alternatives. Reasons typically given for why bioenergy should be promoted are diverse. Bioenergy is considered carbon neutral, it is made from renewable resources, it stimulates the agricultural sector, and it may be produced domestically in many countries, hence diminishing political and economic dependency on other countries (Guinée et al. 2009). However, criticisms have also developed against biofuels, particularly on their role in the food price spikes and the nature of land-use change. A specific example of this case is the maize to bioethanol for transportation fuel in the United States that induced land-use impact, direct and indirect (Harvey and Pilgrim 2011). WRI (2005) indicated that land use (18.2 %) and agriculture's (13.5 %) contribution to greenhouse gas emissions (GHGs, including N_2O and CH_4 in addition to CO_2) are globally estimated to be at least twice the amount of the total emissions from global transport (13.5 %). This assessment indicates the importance of the potential contribution of the land-use aspect to the overall environmental burden of bioenergy systems. Major activities related to these land-use-related impacts are deforestation that releases carbon dioxide from burning or decomposing biomass and oxidizing uncovered humus. In addition to other impact categories such as biodiversity loss and soil quality degradation, all these emissions may negate any GHG benefits of biofuel systems for decades to centuries (Tilman et al. 2009). In this regard, these same authors proposed that biofuels should receive policy support as substitutes for fossil energy only when they make a positive impact on four important objectives: energy security, GHG emissions, biodiversity, and the sustainability of the food supply.

Bioenergy is presently the largest global contributor (77 %) to renewable energy and has contributed significantly to the production of heat, electricity, and fuels for transport (IEA-Bioenergy 2009). Therefore, in the following parts of this chapter, discussion will be focused on bioenergy as the dominant fraction of renewable energy. The main feedstocks for bioenergy are biomass residues from forestry, agriculture, and municipal waste. Only a small portion of sugar, grain, and vegetable oil are used for the production of liquid biofuels (IEA-Bioenergy 2009). There are many technological routes available to convert biomass feedstock into final bioenergy products. Several conversion technologies have been developed to adapt to the unique physical nature and chemical composition of various biomass feedstocks. These include direct combustion (heat), co-firing/combustion (heat/power), gasification (heat/power), anaerobic digestion (heat/power/fuel:

methane), fermentation (fuel: bioethanol), trans-esterification (fuel: bioediesel), and photosynthesis (fuel: hydrogen) (IEA-Bioenergy 2009). These various conversion technologies will dictate overall environmental performances. For example, ethanol production through biochemical or thermochemical conversions is expected to result in different levels of decreasing GHG emissions. However, these conversion-related differences are likely to be small in relation to those associated with feedstock production (Williams et al. 2009). In addition, emissions of methane or nitrous oxide from agricultural field and indirect land-use change may contribute to a more complicated overall picture (Cherubini and Strømman 2011). Side and rebound effects, as well as market mechanisms, of large-scale production of biofuels also affect food markets, resource scarcity, and environmental quality, while these factors are often left out in a sustainability assessment (Guinée et al. 2011; van der Voet et al. 2010). Moreover, bioenergy systems may involve a unit process with input–output flows, which often make it difficult to differentiate between economic (products) and elementary (resource use or emissions) flows.

Recently, there have been tremendous numbers of LCA studies describing bioenergy in order to support policy making. The growing debate on bioenergy and other bio-based products contributed to the acceleration of the development of LCA methodology. However, it is difficult to draw general conclusions from the set of studies due to large variations in outcomes. Sources of these variations include real-world differences, data uncertainties, incompleteness of included impacts, and methodological choices (van der Voet et al. 2010). More specifically, the methodological choices are related to the selection of a functional unit, system boundary, land-use aspects, biogenic carbon, treatment of multi-functional processes, data variability, and regionalized impact assessment (Cherubini and Strømman 2011; van der Voet et al. 2010; Guinée et al. 2009; Finnveden et al. 2009). This indicates that bioenergy poses more methodological challenges than other renewable energy. Moreover, these issues are insufficiently comprehensively addressed by current LCA studies.

This chapter is aimed at providing a systematic overview on the above-mentioned key issues in conducting LCA of bioenergy. Detailed comparison of methodological choices among different LCAs of bioenergy systems can be found in recent surveys such as those of Cherubini and Strømman (2011), van der Voet et al. (2010), Wiloso et al. (2012), and Singh et al. (2010). The structure of this chapter will follow the first three phases of the LCA framework (ISO 2006), including goal and scope definition, inventory analysis, and impact assessment as follows:

- Goal and scope definition:
  - Attributional and consequential LCA
  - Functional unit
- Inventory analysis:
  - System boundary
  - Land use and land-use change

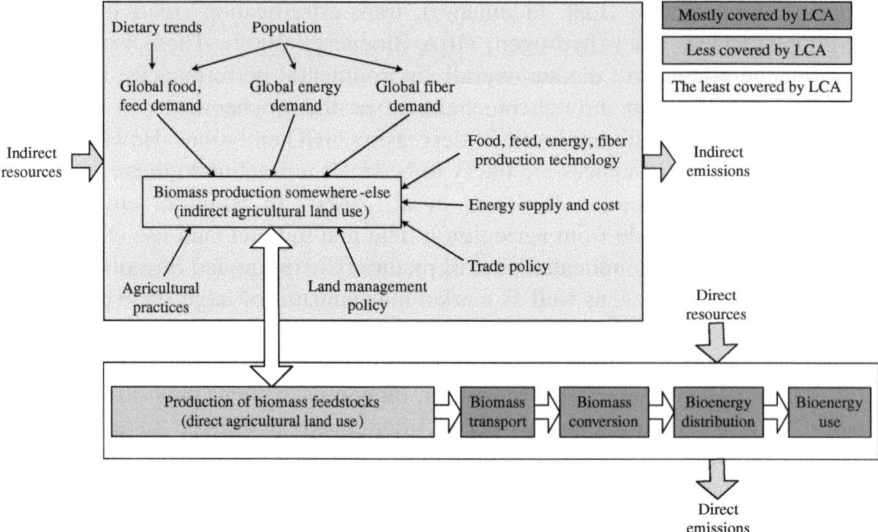

Fig. 1 Direct and indirect effects of a generic bioenergy system (modified from Sheehan 2009). Different shading intensity indicates present coverage in LCA studies

- Biogenic carbon
- Treatment of multi-functional processes
- Data variability

• Impact assessment:

- Impact categories
- Regionalized impact assessment

A generic bioenergy system that spanned from a cradle-to-grave boundary is presented in Fig. 1. The system covers biomass production, biomass transport, biomass conversion, and bioenergy distribution and use. In the upstream chain, the production of biomass feedstock is connected with agricultural land use, direct and indirect. The association of the biomass feedstock with land-use aspects is currently recognized as the central feature in conducting an LCA of bioenergy systems.

2 Goal and Scope Definition

Questions related to the overall objective of LCA studies should be formulated in the goal and scope definition. The goal is closely related to the context in which an LCA study is done, and the scope includes making choices concerning the methodology to use in the subsequent modeling (Baumann and Tillman 2004).

Goal and scope definition is an important initial step since the choice of methodology used depends on the purpose of the individual study. These methodological choices include system boundary, treatment of multi-functional processes, types of required inventory data, and functional unit. The first three topics are described in the following section (attributional and consequential LCA), while the last one is described separately.

2.1 Attributional and Consequential LCA

A clear definition of the goal and scope should specify the types of LCA needed. They can be attributional or consequential (ALCA and CLCA for short). In general, the goal of ALCA is to assess the environmental burden of a product, assuming a status quo situation, while the goal of CLCA is to assess environmental consequences of a change in demand (Thomassen et al. 2008). These different LCA principles require a systematic approach to reduce uncertainty due to freedom of choosing the methodology (Finnveden et al. 2009).

ALCA describes the environmentally relevant flows to and from a life cycle and its subsystems (Finnveden et al. 2009). The attributional method is less important for policy decisions as its purpose is not to support changes. ALCA, however, is useful in obtaining insight into the main environmental impacts related to existing bioenergy products. This is done to better describe the effect of changing feedstocks, changing production processes, or improving efficiency. Another type of application with a more direct relevance to a bioenergy system is the use of ALCA to identify main hot spots in the life cycle chain, the share of certain emissions, or flows to an impact category. This can be a first step in realizing process improvements from a sustainability point of view. An example of this is the LCA study of a generic life cycle of bioenergy with a boundary system as shown in Fig. 1 but without involving the indirect effects. This is in contrast with a CLCA concept, which also includes the indirect effects.

CLCA describes how relevant environmental flows (resource use and emissions) will change in response to possible decisions (Finnveden et al. 2009). Referring to this definition, Cherubini and Strømman (2011) concluded that the CLCA appears as the most broadly applied in bioenergy systems as compared to ALCA. They revealed that almost three-fourths of the reviewed studies compare the environmental impacts with those of a fossil reference system. This is done to address the needs of policy makers in order to decide on relevant bioenergy options. The assessment, however, needs further clarification since not all comparison studies necessarily qualify as consequential.

A distinction between foreground and background systems is especially useful in the CLCA approach. Background systems are often based on databases representing average data of aggregated industrial processes, such as electricity. When a chain of processes are being considered as a foreground system, the proposed technology needs to be specifically known and marginal data are required

(Finnveden et al. 2009). Other distinctive characteristics of CLCA are that unit processes within a system boundary are included to the extent of their expected change caused by a demand and that co-products are handled by system expansion (Weidema 2003). To summarize the main characteristics of these approaches, a comparison between ALCA and CLCA is given in Table 1.

CLCA is, in principle, only preferable within certain limits since the uncertainties in the modeling stage may outweigh the insight gained from it (Cherubini and Strømman 2011). This is related to the fact that the reference system should always refer to the scope and context of the study. For example, the bioenergy system is typically compared with a fossil reference system producing the same amount of products and services. In most cases, however, studies use conventional extraction of crude oil as a benchmark, thereby ignoring the increasing carbon footprint arising from the extraction of non-conventional oil such as oil sands, shale oil, and deep-ocean drilling (Harvey and Pilgrim 2011). Similarly, when the bioenergy pathway delivers some co-products able to replace existing products, the reference to the substituted products should also be defined in the fossil reference system. The same applies to the case when the production of feedstock for biofuels uses land that was previously storing carbon such as forests. In this case, the previous land use should be taken into consideration for the determination of carbon emissions due to land-use change (Singh et al. 2010). Also, when the same feedstock is used for another function, the reference system should include the alternative biomass use. In our view, this last example is the crucial aspect of CLCA in the case of a bioenergy system. This requires a CLCA approach to include the production of biomass feedstocks, resulting in a wider system boundary. This feedstock, consequently, is no longer available for other purposes (such as food, feed, or fiber), so new land to produce an extra feedstock may be needed. The above requirements may increase the uncertainty of the assessment; hence, the adoption of CLCA approach must be treated carefully.

A famous issue in CLCA is the coverage of indirect land use in a biofuel system. Based on the study of Searchinger et al. (2008), Zamagni et al. (2012)

Table 1 Main characteristics of ALCA and CLCA (based on Thomassen et al. 2008)

Characteristics	ALCA	CLCA
Synonym	Status quo, descriptive	Change-oriented
Type of questions	Accounting	Assessing consequences on changes
Type of required inventory data	Average, historical	Marginal, future
Knowledge on the cause–effect chains	Physical mechanisms	Physical and market mechanisms
Functional unit	Represents static situation	Represents change in volume
System boundaries	Static processes	Affected processes by change in demand
Treatment of multi-functional processes	Co-product allocation (partition)	System expansion
Assessment quality	Sensitive to uncertainty	Higher sensitivity to uncertainty

pointed out that most of the previous LCA studies provided only a limited analysis to the life cycle of biofuel system. They failed to account for the indirect effects (i.e., those taking place outside the biofuel value chain) by excluding emissions from land-use change. As shown in Fig. 1, indirect effects may result from the competition for land currently used for food, feed, or fiber to fuel production (Hedegaard et al. 2008). Interaction between various influencing factors and displacement mechanisms can occur in many forms. The main challenge now is how to quantitatively measure the indirect impact of biofuel development on other chains (food, feed, and fiber) that is modeled based on global economic interaction. A CLCA was also used to address problems like the environmental consequences of including the production of second-generation biofuels from biomass residues compared to a current palm oil biodiesel production system in Malaysia (Lim and Lee 2011) or to investigate the expected indirect effects of the development of a grass biomethane industry in Ireland (Smyth and Murphy 2011).

Currently, there is no clear distinction between ALCA and CLCA in most policy guidelines of a country or region, partly due to unresolved debate in framing direct/indirect effects and allocation of co-products (Brander et al. 2009; van Dam et al. 2010). This conclusion is based on at least three policy guidelines (UK's Renewable Transport Fuel Obligation (RTFO), EC's Renewable Energy Directive (RED), and US's Renewable Fuel Standard (RFS)) that tend not to distingusih ALCA and CLCA. For example, EC's RED and UK's RTFO include only direct land-use change, while US's RFS includes both direct land-use change and indirect land-use change; EC's RED is based on energy allocation, while UK's RTFO and US's RFS prefer system expansion (van Dam et al. 2010). These conditions may result in a combination of the two approaches within a single analysis and, consequently, an unfair comparison of results derived from different methods (Brander et al. 2009).

2.2 Functional Unit

A product system is defined based on a functional unit of a product, specified in relation to the nature of a system, geographical, and time boundary. The main role of a functional unit is to be used as a reference to quantitatively connect inputs and outputs of a life cycle inventory (LCI). In this way, LCA results of the same functional unit can be compared between one another provided that, among other things, the system boundaries are similar and the scales are normalized. A proper functional unit that positively reflects the reality is very important in LCA studies. This is important since different choices of functional units from the same system may result in different outcomes when compared to each other. A nice illustration on the effect of different functional units on the results of biofuel LCAs is given by van der Voet et al. (2010).

Theoretically, a functional unit in the form of one MJ would be more appropriate to compare the best use of biomass feedstock for bioenergy of different forms (heat, electricity, biofuel). However, in practice, the functional units vary among studies. Based on current reviews, typical functional units commonly used in LCAs of biofuel systems are volume or mass of input biomass feedstock, volume or mass of biofuel, caloric value of biofuel, driving distance of a car, and agricultural land area (van der Voet et al. 2010; Cherubini and Strømman 2011; Wiloso et al. 2012). These choices of functional units are driven by the main questions or goals of the LCA study. For example, to compare the benefit of gasoline and biofuel systems as transportation fuels will lead to a functional unit in terms of 1 km driving distance. Land area required to produce biomass feedstock is an extremely important parameter since bioenergy can compete against food, feed, or fiber under land availability constraints. However, there are only a few bioenergy LCAs based on this parameter. One of them is the study by Lim and Lee (2011) that used a one-year use of one-hectare palm oil plantation as a functional unit to produce both biodiesel and bioethanol.

3 Inventory Analysis

An LCI of a product or process quantifies economic and environmental inputs and outputs around the system boundary. It is constructed as a flow model of a technical system according to the system boundary decided in the goal and scope definition. The model is basically a mass and energy balance over a system, but only environmentally relevant flows are considered. Activities of the LCI also include data collection of all activities in the system and calculation of the environmental loads in relation to the functional unit (Baumann and Tillman 2004). There are five key aspects specific to bioenergy systems that need further elaboration, i.e., system boundary, land use, biogenic carbon, multi-functional processes, and data variability.

3.1 System Boundary

In LCA of bioenergy studies, the choice of system boundary is often arbitrary. With the basic cradle-to-grave principle of LCA, everything should be included; however, in practice, many processes are left out for different reasons. System boundaries define what are to be included and what are not. In general, capital goods and wastes as an input feedstock are cut off from the system. This implies that the emissions by the production of capital goods and wastes are not taken into account.

As previously indicated in Sect. 2.1, one of the main issues related to CLCA is the identification of the processes to be included in the analyzed system, which implies the way in which boundaries are defined. In the case of biofuels, for example, the system boundary is expanded to include emissions and resources used, directly and indirectly, as a result of the consequential effects of introducing biofuels to the global economy. In this regard, the rule is to include only relevant affected processes, defined as those that respond to changes in demand or supply driven by the decision at hand (Zamagni et al. 2012). In doing so, the resulting functional unit of the whole system may consist of multiple functions, including the main system and those processes added into the system boundary. However, when a comparative analysis must be conducted, it may be difficult to guarantee the functional equivalency between the systems compared since the processes included could serve different functions. Such a resulting multi-functional system raises some concerns about whether it can still be considered a functional unit (Zamagni et al. 2012). In this case, differences in system boundaries are rather crucial. Therefore, they must be specified, unambiguous, consistent, and in-line with the actual goal and scope of the study (van der Voet et al. 2010). This may be the most difficult problem to address.

The cradle-to-gate approach is sufficient for comparing various production technologies to make the same biofuel from different feedstocks, while the cradle-to-grave is the best approach for comparing, for example, the utilization of certain biofuels with fossil fuels (Singh et al. 2010). Cradle-to-gate studies are performed by excluding the use and waste treatment stages, but it is, of course, admissible only when there is no difference between these stages. To illustrate this, a comparison between a plastic cup and a paper cup for drinking tea can be used. In this case, the upstream stages (the growing of tea plants, the processing of tea leaves, and the boiling of water) are likely the same, but the waste treatment of plastic cups is obviously different from that of paper cups (Heijungs and Wiloso 2012).

The same system boundary with a difference in functions will have a different basis of comparison. For example, electricity generated from municipal solid waste is not very efficient and usually shows no improvement over a fossil fuel alternative. However, when a waste management aspect is included in the electricity generation, this extended new waste-to-energy system boundary will likely favor over the waste management alone (without electricity generation) or over a fossil fuel system (van der Voet et al. 2010).

3.2 Land Use and Land-Use Change

Although the majority of global GHG emissions have been blamed on the use of fossil fuels, there has recently been growing recognition that land use also significantly contributes to the emissions. The increased understanding of the effects of land-use change needs further consideration in bioenergy systems. In this regard, a UNEP-SETAC guideline on land-use impacts (soil quality, biodiversity, and ecosystem services) has been proposed (Koellner et al. 2012), but there is currently no widely acceptable way to incorporate land-use impacts in an LCA study. The main reason may be that this aspect is very difficult to quantify.

In agricultural land use, there are three time periods in examining the long-term consequences of agricultural activities, i.e., the period before (transformation), during (occupation), and after (restoration) agriculture (Mila i Canals et al. 2007). Based on these time frames, one may refer to land use as an activity during the occupation period and land-use change as land transformation or a change in the properties of the land surface area. This could be a new type of land use at a single point in time such as deforestation or agricultural expansion (Mila i Canals et al. 2007). Similarly, IPCC refers to land-use change as land conversion but also, interestingly, as changes in carbon pools without land conversion (IPCC 2001). In fact, the precise place of land use and land-use change in the LCA framework is not clear. For example, besides as an activity, land use can also be an inventory item, just like CO_2 (certain land area occupied for certain period of time). Additionally, land-use change can be an activity (a unit process, e.g., clearing of forest) (Heijungs et al. 1992). Even impacts of land use or land-use change are frequently indicated simply with the term land use.

Mitigating the competition for land can only be established if the complexity of the competition dynamics is fully addressed. Each of the contributing factors (energy, food, feed, and fiber demand) cannot be treated in isolation (Harvey and Pilgrim 2011). All these factors are intimately interconnected, particularly in large-scale development of bioenergy (McKone et al. 2011). Although the competition of land used for food, fiber, and energy was recognized a long time ago, quantification attempts involving competition aspects have been made only quite recently (Searchinger et al. 2008). Drivers for increased bioenergy use (e.g., policy targets for renewables) can lead to increased demand for biomass, leading to competition for land currently used for food production and, possibly, indirectly causing new and sensitive areas to be converted into arable land (IEA-Bioenergy 2009). These interconnected factors in the complexity of direct and indirect land use are previously illustrated in Fig. 1, while activities, resource use, and emissions typically involved in land use and land-use change are shown in Fig. 2.

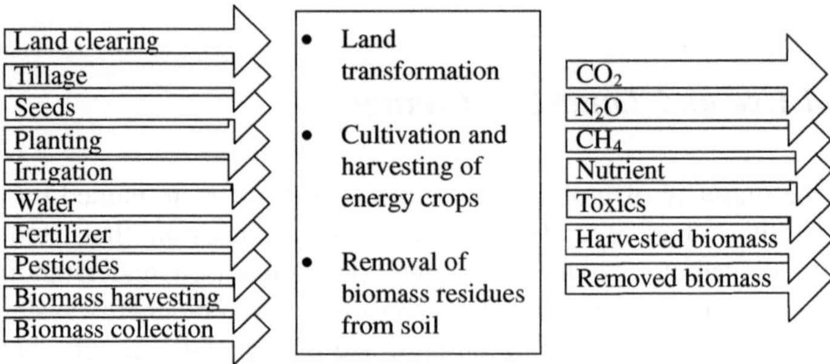

Fig. 2 Inventory of activities, resource use, and emissions in the agricultural chain of biomass feedstock

3.2.1 Direct Impacts

Land use and land-use change, in relation to biomass supply for bioenergy, are characterized as having various input–output inventories, resulting in different contributions to impact categories that affect different areas of protection. Relevant impact categories include global warming, eutrophication, acidification, toxicity, water use, and land use. These impacts are induced by input–output components and activities in the agricultural chain including land transformation, cultivation of energy crops, and removal of biomass residues from soil, as shown in Fig. 2. Typical inventories include, for example, the use of fossil fuels in tractors for land clearing, tillage, planting, and harvesting; the application of seeds, fertilizer, and pesticides; and the use of water for irrigation. Important GHG emission species related to agricultural activities are N_2O and CH_4 in addition to CO_2. Land-use-related activities may directly affect the quality of land (natural environment) as an area of protection. This quality in terms of ecosystem services include soil quality, biomass productivity, and biodiversity (Mila i Canals et al. 2007). The characterization of these land-use impact categories, however, is less developed compared to other categories.

3.2.2 Indirect Impacts

In principle, indirect land use will have the same inventory components and relevant impact categories as that of direct land use. Indirect land use refers to the changes in land use that take place elsewhere as a consequence of the development of bioenergy systems. In the LCA methodology, this indirect impact may have a broader meaning, including any relevant effects to different chains, for example, if large-scale bioenergy production affects food production chains. As an illustration, if fertile land previously used for food crops (such as corn, soybeans, or palm) is transformed to produce bioenergy, this could lead to farmers clearing wild lands elsewhere in the world to meet the displaced demand for food crops (Tilman et al. 2009).

The paper by Searchinger et al. (2008) has pointed out the significant contribution of indirect impacts on the LCA of bioenergy systems. The authors argued that, based on a sustainability criterion, fuel oil is better than most biofuels. There are two connected arguments put forth. First, biofuel development provoked a rise in the price of food, leading to the stimulation and expansion of food production. Second, the subsequent displacement of food production into new areas of cultivation (indirect land-use change) resulted in a release of CO_2 into the atmosphere. It holds biofuel production responsible for global climate change in ways not measured by previous LCA studies (Harvey and Pilgrim 2011). The above explanation on indirect impact changes the entire nature of LCA to one which must be able to model global economic interaction (Sheehan 2009). In addition to indirect land use, other types of indirect impacts may be needed to properly assess the total GHG emissions implications of substituting biofuels for gasoline. In this regard, Liska and Perrin (2009) illustrated that livestock and military security also had a significant impacts in the case of the US bioenergy system. The inclusion of these indirect effects in the bioenergy system understudy can change the direction of the final results. There is, however, much scientific uncertainty in measuring these indirect emissions related to both bioenergy and fossil oil systems, thus creating a problem on how to properly calculate them.

3.3 Biogenic Carbon

One of the important aspects in bioenergy systems is related to biogenic (short-cycle) carbon. Although under debate, it has been recognized that bioenergy is not carbon neutral since it requires a significant input of fossil fuels. In practice, many studies exclude biogenic carbon from biofuel LCAs, rather than including it initially as an extraction and later as an emission. This convention is so widespread that in the majority of biofuel LCA case studies, the aspect of biogenic carbon is not even mentioned (van der Voet et al. 2010).

The neutrality of biogenic carbon is part of the natural carbon cycle over a relatively short period of time, resulting in stable atmospheric carbon. As illustrated in scenario 1 of Fig. 3, this is the case when the emission of biogenic carbon in the form of naturally decayed or burned biomass is compensated by the same amount of photosynthetic carbon sequestered by naturally grown vegetation. However, this cycle can no longer be ‘neutral’ if the input–output inventory is out of balance. This occurs, for example, when large-scale bioenergy systems

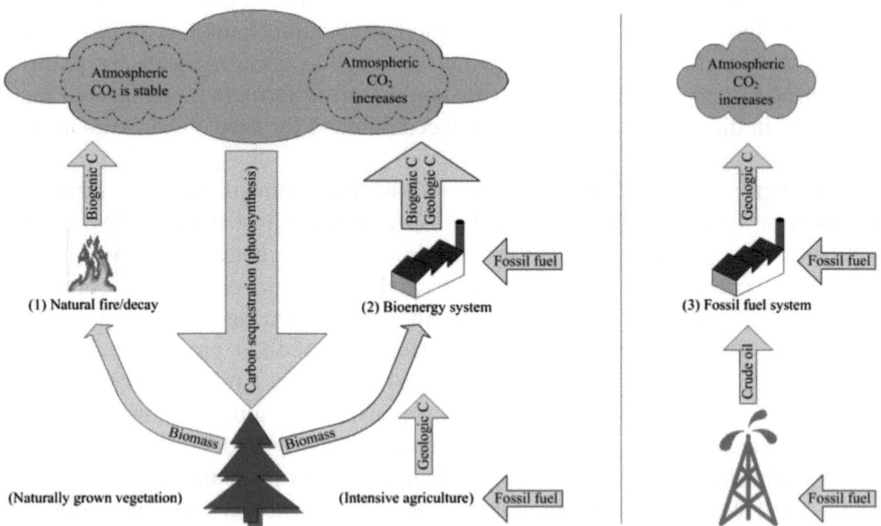

Fig. 3 Biogenic carbon cycle versus ‘irreversible’ geologic carbon emission

introduced are involving significant amounts of fossil fuel and agricultural input (scenario 2). In this case, the bioenergy system may emit more total carbon than the sequestration capacity of trees, resulting in net accumulation of CO_2 in the atmosphere. On the other hand, carbon emissions from fossil fuel combustion are considered as an irreversible one-way process (scenario 3). It transfers geological carbon, locked underground, over long-term geological time into the atmosphere. This process increases atmospheric carbon levels with time. Therefore, to properly assess the benefit of the bioenergy system over fossil fuel systems, it is necessary to account for all relevant input–output flows in the inventory phase of LCA studies, including carbon sequestration and carbon emissions of both biogenic and geologic sources. From various options provided in LCA studies, the bioenergy product with the larger GHG saving, among other criteria, would be the preferred energy system.

There are at least two points to make with respect to carbon neutrality of bioenergy. First, if there is anything neutral, it is LCA, as an analytical tool, that makes the conclusion. If biofuels are carbon neutral, this will result from the LCA study instead of being a starting point of the LCA study. Second, there are several situations where the carbon neutrality of bioenergy is challenged. One of these occurs in the situation when some of the CO_2 absorbed is not released as CO_2, instead, as CH_4, a greenhouse gas that is much stronger than CO_2. This may happen, for instance, when the biotic feedstock is subject to a process of incomplete burning or anaerobic decomposition with leakages occurring along the way. Another case is a plantation with two co-products (e.g., palm oil and palm kernel oil) where part of the absorbed CO_2 is allocated to each of the co-products. In chains with only one product, exclusion of biogenic carbon can result in the same outcome as long as the issue of CH_4 does not arise. However, in cases of chains with co-products, it makes a difference. Allocation may place the credits for extracted CO_2 in a different part of the multi-product chain, while ignoring biogenic CO_2 would not have this effect (van der Voet et al. 2010). A recent review indicated that carbon sequestration, if included at the biomass generation stage, can offset the GHG emissions from all parts of the life cycle chains at a high ethanol percentage ($\geq$85 %) (Wiloso et al. 2012). A final example challenging the bioenergy neutrality is the fact that there is a time difference between CO_2 fixation and release. A specific dynamic LCA method has been developed to account for such situations.

The most appropriate way to treat carbon cycles is to view them as genuine cycles. During tree growth, a certain amount of atmospheric CO_2 is fixed but is ultimately released as CO_2 or CH_4 when the wood is landfilled, is incinerated, or decays naturally. At the systems' level, the fixation of CO_2 during tree growth is subtracted from the CO_2 emitted during waste treatment of discarded wood (Guinée et al. 2009). For fossil fuels, carbon fixation has taken place as a natural process millions of years ago, but carbon emissions occur immediately when these fuels are burned.

The rationale behind different treatments between biogenic carbon and geologic carbon is because, for example, forestry (the process that fixates the CO_2) is considered as a unit process. It is an intentional activity, controlled by humans, requiring inputs and producing outputs. The creation of fossil fuels is a spontaneous process without human intervention. The forestry is, thus, an activity that should be included in the flow diagram of an LCA study, whereas the process of fossil fuel formation should not (Guinée et al. 2009). Wegener Sleeswijk et al. (1996), in their report on the application of LCA to agricultural products, propose to not include biogenic carbon dioxide in the analysis if the entire life cycle is being analyzed. If the study is based on cradle-to-gate analysis, carbon sequestration must either be included, or it must be explicitly stated that this fixation is being excluded from the study. If this is not done, there is a danger that if other researchers use the results of the study, they will include, say, the emission of CO_2 during combustion of biodiesel fuel, while fixation of CO_2 was omitted in the cradle-to-gate analysis.

There is currently no consensus regarding how to treat biogenic carbon at the policy level. The Intergovernmental Panel on Climate Change (IPCC) currently considers biomass to be carbon neutral, suggested by the adoption of a stock-change method rather than an input–output flow approach in carbon accounting (Levasseur et al. 2012a). In this case, if biogenic carbon is released later in the life cycle, CO_2 emissions are not accounted for to avoid double counting. As discussed in Johnson (2009), a life cycle-based method such as the British specification PAS 2050 (BSI 2011) suggests the same approach as IPCC, not considering biogenic carbon uptakes and emissions, while the International Reference Life Cycle Data System ILCD (EC-JRC-IES 2010) recommends the opposite. Similar to PAS 2050, EU Directive (2009) also excludes the capture of CO_2 in the cultivation of biomass and emissions from biofuel use from the calculation of GHG emissions by setting their values equal to zero. The rationale behind these differences is the argument that the combustion or decay of woody biomass is simply part of the global cycle of biogenic carbon, and over a long period of time, it does not increase the amount of carbon in circulation due to compensation by photosynthetic processes. Meanwhile, in the conventional LCA practices, all flows including carbon uptake and emissions should be accounted for in the inventory stage without considering the time scale. To deal with this time frame issue, Levasseur et al. (2012a) proposed to treat biogenic carbon as temporary storage with dynamic LCA. The argument behind this approach is that the concentration of CO_2 in the atmosphere is temporarily reduced and some radiative forcing is avoided. This is favorable in the short term as it also allows 'buying time,' while technology develops in the field of GHG emission reduction and mitigation (Levasseur et al. 2012b).

3.4 Treatment of Multi-Functional Processes

Various forms of bioenergy products are ideally derived from feedstocks produced with much lower life cycle GHG emissions than traditional fossil fuels and with little or no competition with food production. According to Tilman et al. (2009), feedstocks in this category may include perennial plants grown on degraded lands, crop residues, sustainably harvested wood and forest residues, double crops and mixed cropping systems, municipal and industrial wastes. These various feedstocks and bioenergy products in LCA should be treated with proper allocation and recycling procedures to attribute environmental burden of multi-functional processes to their input or output flows.

A multi-functional process is a unit process, yielding more than one functional flow including co-production, combined waste processing, and recycling. Co-production is a multi-functional process having more than one functional outflow and no functional inflow. Recycling is a multi-functional process having one or more functional outflows and one or more functional inflows. Combined waste processing is a multi-functional process having no functional outflow and more than one functional inflow. The most relevant multi-functional processes in bioenergy systems with reference to the types of input and output inventory are the first two cases as illustrated in Fig. 4. Guinée (2002) distinguishes two steps in solving the multi-functionality problem. The first concerns avoiding burden allocation in accordance with the ISO preference. This is done by specifying the system boundary to a unit operation level (e.g., individual machines) to reduce the number of multi-functional processes or by system expansion. It is accomplished by extending the analyzed product system to include additional functions related to the co-products or recycled wastes. The system then includes more than one functional unit. The term system expansion is sometimes used to refer to the substitution method. The second step concerns solving the remaining multi-functionality problems by allocation on the basis of mass, energy, or economic values. Further discussion on the procedure to deal with allocation procedures and system expansion can be found in Tillman et al. (1994) and Heijungs and Guinée (2007).

If some waste streams from agriculture are used to make bioenergy products, how the waste was produced is not included in the inventory. It is assumed that its production is free of environmental burden. This, however, requires a clear distinction between products and wastes. To distinguish products from wastes, the economic value of flows can be used as the determining factor. A product is a flow between two processes with a positive economic value, whereas a waste is a flow between two processes with a negative economic value (Guinée et al. 2009). However, there are quite a few cases where we do not know for certain if the price of an agricultural residue is positive or negative, especially when it remains within

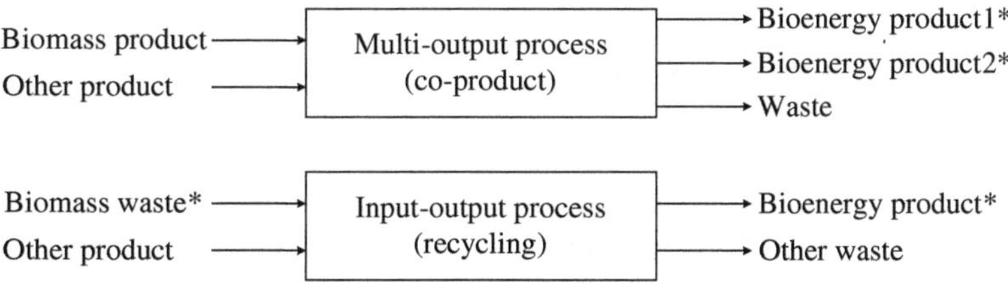

Fig. 4 Relevant multi-functional processes in bioenergy systems (*= functional flows)

one company or farm. An example of this is that someone may pay to have their residues picked up, while someone else must pay to receive it. Further, due to technological developments, fluctuations in markets, and governmental intervention, goods may rapidly turn into a waste or the other way around (Heijungs and Wiloso 2012).

3.5 Data Variability

An LCA depends on a large number of input elements, and these elements are often based on data of varying quality. The variability in input quality will, in turn, influence the robustness of outcome estimates. This is an important issue that deserves more attention in LCA. A strong challenge for LCA in addressing uncertainty is to provide and track metrics of data quality with respect to how data are acquired (measurements, assumptions, expert judgment), to what extent the data have been validated (checked with respect to mass and energy balance), and how well the data capture technological, spatial, and temporal variations. Some of these uncertainties and variabilities cannot be reduced with the current knowledge (through improvements in data collection or model formulation) because of their spatial and temporal scale and complexity (McKone et al. 2011).

When developing LCIs, one needs quantitative data on the inflows and outflows of the included processes such as resource use, emission data, energy use, and waste production. The limited accuracy and availability of LCI data are generic problems of LCAs. Uncertainty can be due to various reasons that may stem from geographical, temporal, and technological differences. In the case of bioenergy systems, common sources of uncertainty include variability in agriculture yield as it depends on soil conditions, weather, and agricultural practices; variability in biomass conversion technology at different development status; and regional variability as the data are known only for certain countries (Heijungs and Wiloso 2012). Despite the above difficulties, doing LCA is now much easier than ten years ago since there are now a number of online data repositories for different continents. Some of these databases are quite extensive, though mostly for the USA and EU. The Ecoinvent database, for example, contains thousands of processes from electricity production to transport by truck and from palm oil production to pesticide production (Ecoinvent 2010).

3.5.1 Agricultural Process Variability

Data variability in the agricultural chain of bioenergy systems is an issue in LCI. For example, there are a large number of potential biomass feedstocks with different characteristics. This presents substantial challenges for current LCA approaches because of the vast scope of information needed to address so many alternatives (McKone et al. 2011). The production of biomass feedstock is likely to involve hundreds to thousands of decision-makers, unlike oil companies that have a less hierarchical structure for decision-making (McKone et al. 2011).

Data gaps and uncertainties are typical to agricultural processes because field measurements are difficult to obtain. Different feedstocks, types of soil, agricultural practice, and climate conditions result in various emission levels so that it is difficult to generalize the environmental performance of biofuels. For example, in the debate around palm oil biodiesel, the emissions from soil related to the agricultural process depend heavily on local circumstances, while the GHG benefits over fossil fuels are global in nature. These emissions vary from very positive to very negative. Such differences are problematic in the sense that they would offer an uncertain basis for policy making (van der Voet et al. 2010).

As previously mentioned, there are three time periods examined to determine the long-term consequences of agricultural activities. The period before agriculture is highly uncertain since the history of when the transformation was taking place is usually unknown. Similarly, the restoration period after the cessation of agriculture activities is highly dynamic. In relation to restoration time, McLauchlan (2006) mentioned that some systems may reach the condition of pre-agricultural time after decades to millennia. From the above description, it is clear that periods before and after agriculture are not easy to adopt in the assessment of land-use impact, mainly due to lack of data availability to follow such a long-term soil quality dynamic. Furthermore, topography, soil, and climate variability within a region prevent direct scaling of LCA balances to geographical scales (Schmer et al. 2008).

3.5.2 Conversion Process Variability

Data gaps and uncertainties related to bioenergy technological routes, particularly on an industrial scale, are not fully resolved. Many advance bioenergy processes are still in a stage of development, and data will become more informative as technologies are deployed. This fact makes LCA methodology difficult to apply during the early phases of a major technology shift (McKone et al. 2011). This is especially true for immature technologies where validation is presently not possible. In the case of second-generation bioethanol, for example, most of the LCA studies use advanced process configurations that are still in developing stages and no existing commercial scale can be referred to for validation. In this regard, there is a risk of under- or over-estimating the real impacts of the current production technology; Therefore, sensitivity analysis is necessary (Wiloso et al. 2012).

There are many technological routes which can be used to convert raw biomass feedstock into bioenergy products. These different technologies all have a different development status as illustrated in Fig. 5. For example, the production of heat by direct combustion of biomass is historically practiced and still the leading bioenergy application throughout the world. For a more energy efficient use, modern and large-scale heat applications are often combined with electricity production (combined heat and power) systems. The use of biomass residues for second-generation biofuels production would significantly decrease the potential pressure

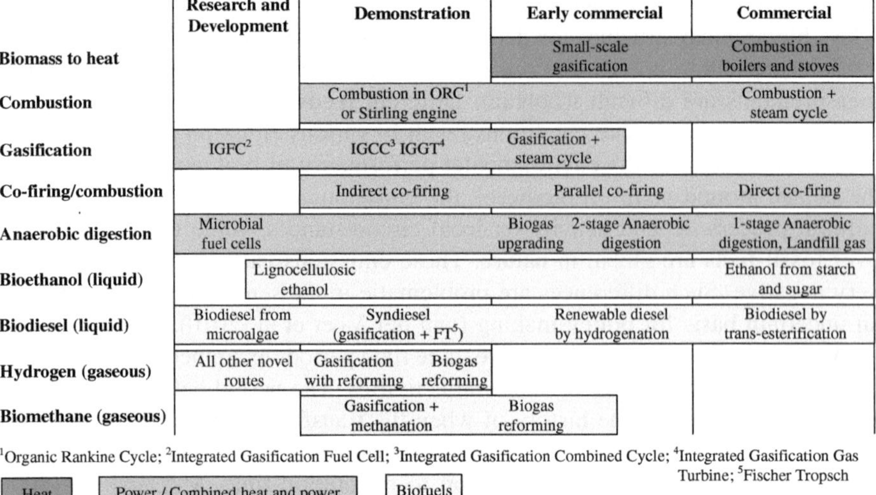

Fig. 5 State of the art of the conversion technologies for bioenergy (modified from IEA-Bioenergy 2009)

on land use and improve GHG emission reductions when compared to some first-generation biofuels, leading to lower environmental risk. These second-generation technologies mainly use lignocellulosic feedstocks for the production of ethanol, synthetic diesel, or aviation fuels. In this regard, they are still immature and require further development to demonstrate reliable operation on a commercial scale (IEA-Bioenergy 2009).

3.5.3 Regional Variability

Data gaps in bioenergy LCA are also present with respect to coverage of feedstock types and of geographical areas with an over-representation of Europe and North America (Cherubini and Strømman 2011). Economic and political interactions that influence land use can cause more variation as the system boundary expands across ecosystems and political borders (Singh et al. 2010). Many studies also show that water consumption varies significantly, depending on regional irrigation requirement and practices (Borrion et al. 2012).

4 Impact Assessment

In general, environmental impact assessment can be regarded as either potential impact or real impact. But in LCA, only potential impact or maximum possible impact is considered (Baumann and Tillman 2004). In addition, impact category should be mutually independent in order to avoid double counting of environmental burden. Life cycle impact assessment (LCIA) consists of seven activities, i.e., selection of impact category, classification, characterization, normalization, grouping, weighting, and data quality analysis. According to the ISO standard for LCA (ISO 2006), the first three are mandatory, while the rest are optional. Two aspects of LCIA that need further elaboration with regard to bioenergy systems are impact categories and regionalized impact assessment.

4.1 Impact Categories

It is important to properly select the set of relevant impact categories in the bioenergy systems under study. Areas of protection in environmental impact assessment include ecosystem health, human health, resource availability, and man-created environment. Assessment of bioenergy production from specific biomass is suggested to be based on a complete set of impact categories, including climate change, ozone depletion, human and ecotoxicity, photo-oxidant formation, acidification, eutrophication, land-use impacts, and depletion of abiotic resources. But McKone et al. (2011) suggested a balance between being comprehensive and being parsimonious. Failure to address a key impact can lead to incomplete or unreliable information, creating biased decisions. Clearly, the set of chosen impact categories need to be fixed accordingly in the formulation of goal and scopes of the study, but a default minimum would restrict the risk of biased decisions.

Early LCA studies were often limited to net energy output and global warming. The net energy output is an important parameter because, in many cases, the process of producing fuels from the feedstock is energy intensive and, therefore, limits the overall benefit. This parameter (net energy output), however, only determines the technical feasibility of the bioenergy systems rather than being an impact itself. For global warming, the result of the LCI is a list of GHG emissions of all processes in the chain, which are then added up and translated into CO_2 equivalents (so-called carbon footprints). According to the recent review by Cherubini and Strømman (2011), approximately 90 % of bioenergy LCAs include global warming in their evaluation while primary energy demand rates second (71 %). Other impact categories, mainly acidification and eutrophication, are estimated by 20–40 % of the studies. Only 9 % included the land-use category in their impact assessment. The reason for including global warming in most of the studies is because climate policy dominates the scene, while other impacts are not considered as important. In addition, some of them are site specific, which may limit the generalization of the result. Also, there is significantly less agreement in the quantification methods of some impact categories. Particularly notorious are the impacts related to land use, water use, biodiversity, and genetically modified organisms (Heijungs and Wiloso 2012). With the increasing pressure of a growing population, water use is now also considered as increasingly relevant. Water footprints specify water requirements on a cradle-to-gate basis, and their studies in bioenergy systems are now emerging. It is concluded that energy from biomass has, by far, the largest water footprint compared with other energy sources (van der Voet et al. 2010).

According to van Dam et al. (2010) in IPCC (2011), environmental impacts of bioenergy systems can be distinguished by two classifications based on the coverage of impacted areas. The first is global or regional in nature, including GHGs, acidification, eutrophication, water availability, and air quality. The second is local coverage, including soil quality, biodiversity, water availability, and air quality. Other important classifications related to bioenergy systems are genetically modified organisms and food security (replacement of staple crops and safeguarding local food security). Recent LCA studies typically include a wider scope of impacts supported by sufficient databases and characterization models. Standard life cycle impact assessment methods are available, namely ReCiPe, EDIP, TRACI, LIME, and CML-IA. These methods include selected set of impact categories.

4.2 Regionalized Impact Assessment

Regionalized impact assessment is important in bioenergy system as the boundary also includes agricultural systems. Therefore, assessment criteria should reflect the regional or local conditions of the specific bioenergy system under study. For truly global impact categories like climate change and stratospheric ozone depletion, this is not a problem since the impact is independent of where the emission occurs. For the other impacts, however, they are often regional or even local in nature. In this case, a global set of standard conditions can disregard large and unknown variations in the actual exposure of the sensitive parts of the environment. Sometimes, differences in sensitivities of the receiving environment can have a stronger influence on the resulting impact than differences in inherent properties of the substance (Potting and Hauschild 1997; Bare et al. 2003). In general, these spatial differentiations relate to the characteristics of both the emitting source and the receiving environment (Finnveden et al. 2009). LCA can address net changes across large geographical areas, but it must also address how the impacts will be experienced on local or regional scales. Accurate assessments must not only capture spatial variation in appropriate scales (from global to farm level) but also provide a process to aggregate spatial variability that can be applied on all geographical scales (McKone et al. 2011).

Several groups have worked on developing site-dependent characterization for LCIA. Recently, methods supporting site-dependent characterization of a range of non-global impact categories were published for processes in Europe, the USA, and within some countries (Finnveden et al. 2009). There are some differences between these data sets partly related to the different definitions of the characterization factors (Seppälä et al. 2006). For example, the variation in acidification impact can be as high as three orders of magnitude between different countries within Europe (Potting et al. 1998). For eutrophication, the uncertainty associated with field emissions contributes more than the uncertainty associated with emissions from the other system components (Basset-Mens et al. 2006).

Inherent differences associated with variability in soil types and complex interactions with local climates must be considered in order to obtain a more representative value in relation to location-dependent aspects. Other types of influencing variability are different soil management and vegetation types. Similarly, dryer climates will rely increasingly on irrigation placing pressure on groundwater supplies. In this regard, the impacts of biofuels on water are highly regional (Sheehan 2009). This issue is of concern for LCA methods in general as well as a challenge specific to biofuel development.

5 Future Trends

Most of the assumptions and data used in LCA studies of existing bioenergy systems are related to conditions and practices in Europe and North America, but more studies are now becoming available for other regions such as Brazil, China, and Southeast Asia (Cherubini and Strømman 2011). First-generation biofuel options based on sugar or starch feedstock are currently available commercially, but lignocellulosic biofuels are expected to be deployed over the year 2020 (IPCC 2011). In this regard, LCA studies of prospective bioenergy options are more uncertain than LCA studies on current bioenergy feedstocks. The way that uncertainties and parameter sensitivities are handled is an important aspect to be developed. Another important aspect to be resolved in the LCA of bio-based renewable energy systems is the proper way to define system boundaries, particularly in relation to direct and indirect effects of land use and land-use change. Further, consensus on the treatment of biogenic carbon should also be prioritized.

6 Conclusion

Bio-based renewable energy sources are presently the largest global contributor to renewable energy as alternative sources of heat, electricity, and biofuel. From the perspective of LCA, they pose more methodological challenges than other renewable energy systems. One of the main reasons is that biomass feedstocks are produced through agricultural systems that are a notorious case to LCA. Agricultural land use has been indicated as the major contributor of GHG emissions in the bioenergy life cycle chain. However, this is not conclusive since quantification methods in terms of functional unit, system boundary, the treatment of biogenic carbon and multi-functional processes, and regionalized impact assessment are not agreed upon. In addition, the inherent variability in the agricultural data and immature production technology increase the uncertainty of the result of LCA studies. There is homework to do for harmonizing LCA framework for bioenergy systems from the point of view of LCA methodology development and demand for policy consideration.

Acknowledgments Financial support from NFP-Nuffic, the Netherlands, for EIW is gratefully acknowledged. Appreciation is also given to Research Center for Chemistry, Indonesian Institute of Sciences (LIPI, Indonesia), for giving an opportunity for EIW to pursue a PhD study at Leiden University, the Netherlands. We would like to thank Patrik Henriksson of CML, Leiden University, for fruitful discussion. Also, comments from anonymous reviewers are greatly appreciated.

References

Bare JC, Norris GA, Pennington DW, McKone TE (2003) TRACI, the tool for the reduction and assessment of chemical and other environmental impacts. J Indust Ecol 6(3–4):49–78

Basset-Mens C, van der Werf HMG, Durand P, Leterme P (2006) Implications of uncertainty and variability in the life cycle assessment of pig production systems. Int J Life Cycle Assess 11(5):298–304

Baumann H, Tillman AM (2004) The Hitch Hiker's guide to LCA: an orientation in life cycle assessment methodology and application. Studentlitteratur, Sweden

Borrion AL, McManus MC, Hammond GP (2012) Environmental life cycle assessment of lignocellulosic conversion to ethanol: a review. Renew Sust Energy Rev 16:4638–4650

Brander M, Tipper R, Hutchison C, Davis G (2009) Consequential and attributional approaches to LCA: a guide to policy makers with specific reference to greenhouse gas LCA of biofuels. Technical paper TP-090403-A, Ecometrica April 2009. http://ecometrica.com/assets/approachesto_LCA3_technical.pdf. Accessed 12 Oct 2012

BSI (2011) PAS 2050: specification for the assessment of the life cycle greenhouse gas emissions of goods and services. British Standards Institution, London

Cherubini F, Strømman AH (2011) Life cycle assessment of bioenergy systems: state of the art and future challenges. Bioresour Technol 102:437–451

EU Directive (2009) Directive 2009/28/EC of the European Parliament and of the Council of 23 April 2009 on the promotion of the use of energy from renewable sources. Official J Eur Union 52(L140):16–62

EC-JRC-IES (2010) International reference life cycle data system (ILCD) handbook—general guide for life cycle assessment—detailed guidance, 1st edn. European commission—joint research centre—institute for environment and sustainability. Publications Office of the European Union. Luxembourg

Ecoinvent (2010) Ecoinvent data v2.2. Swiss center for life cycle inventories

Finnveden G, Hauschild MZ, Ekvall T, Guinée JB, Heijungs R, Hellweg S, Koehler A, Pennington D, Suh S (2009) Recent developments in life cycle assessment. J Environ Manag 91:1–21

Guinée JB (ed) (2002) Handbook on life cycle assessment: operational guide to the ISO standards. Kluwer Academic Publisher, Dordrecht

Guinée JB, Heijungs R, van der Voet E (2009) A greenhouse gas indicator for bioenergy: some theoretical issues with practical implications. Int J Life Cycle Assess 14:328–339

Guinée JB, Heijungs R, Huppes G, Zamagni A, Masoni P, Buonamici R, Ekvall T, Rydberg T (2011) Life cycle assessment: past, present, and future. Environ Sci Technol 45:90–96

Harvey M, Pilgrim S (2011) The new competition for land: food, energy, and climate change. Food Policy 36:S40–S51

Hedegaard K, Thyo KA, Wenzel H (2008) Life cycle assessment of an advanced bioethanol technology in the perspective of constrained biomass availability. Environ Sci Technol 42(21):7992–7999

Heijungs R, Guinée JB (2007) Allocation and "what-if" scenarios in life cycle assessment of waste management systems. Waste Manag 27(8):997–1005

Heijungs R, Wiloso EI (2012) LCA of second generation bioenergy and biomaterials: progress and problems. In: Proceedings of the the 2nd Korea-Indonesia workshop and international symposium on bioenergy from biomass, Serpong, Tangerang, Indonesia

Heijungs R, Guinée JB, Huppes G, Lankreijer RM, Udo de Haes HA, Wegener Sleeswijk A, Ansems AMM, Eggels PG, van Duin R, de Goede HP (1992) Environmental life cycle assessment of products: guide and backgrounds. CML Leiden University, Leiden

IEA-Bioenergy (2009) Bioenergy—a sustainable and reliable energy source. International Energy Agency. http://www.ieabioenergy.com/LibItem.aspx?id=6479. Accessed 11 April 2012

IPCC (2001) Special report on land use, land-use change and forestry. IPCC special reports on climate change. http://www.grida.no/publications/other/ipcc_sr/?src=/climate/ipcc/land_use/044.htm#s2-2-1. Accessed 4 Sept 2012

IPCC (2011) Bioenergy. In: IPCC special report on renewable energy sources and climate change mitigation. Cambridge University Press, Cambridge

ISO (2006) Environmental management—life cycle assessment—requirements and guidelines (ISO 14044). International Organization for Standardization, Geneva

Johnson E (2009) Goodbye to carbon neutral: getting biomass footprints right. Environ Impact Assess Rev 29:165–168

Koellner T, de Baan L, Beck T, Brandao M, Civit B, Goedkoop M, Margni M, Milài Canals L, Müller-Wenk R, Weidema B, Wittstock B (2012) Principles for life cycle inventories of land use on a global scale. Int J Life Cycle Assess. doi:10.1007/s11367-012-0392-0

Levasseur A, Lesage P, Margni M, Samson R (2012a) Biogenic carbon and temporary storage addressed with dynamic life cycle assessment. J Indust Ecol. doi:10.1111/j.1530-9290.2012.00503.x

Levasseur A, Lesage P, Margni M, Brandao M, Samson R (2012b) Assessing temporary carbon sequestration and storage projects through land use, land-use change and forestry: comparison of dynamic life cycle assessment with ton-year approaches. Climatic Change. doi:10.1007/s10584-012-0473-x

Lim S, Lee KT (2011) Parallel production of biodiesel and bioethanol in palm-oil- based biorefineries: life cycle assessment on the energy and greenhouse gases emissions. Biofuels Bioprod Bioref 5:132–150

Liska AJ, Perrin RK (2009) Indirect land use emissions in the life cycle of biofuels: regulations vs science. Biofuels Bioprod Bioref 3:318–328

McKone TE, Nazaroff WW, Berck P, Auffhammer M, Lipman T, Torn MS, Masanet E, Lobscheid A, Santero N, Mishra U, Barrett A, Bomberg M, Fingerman K, Scown C, Strogen B, Horvath A (2011) Grand challenges for life-cycle assessment of biofuels. Environ Sci Technol 45:1751–1756

McLauchlan K (2006) The nature and longevity of agricultural impacts on soil carbon and nutrients: a review. Ecosystems 9:1364–1382

Mila i Canals L, Bauer C, Depestele J, Dubreuil A, Knuchel RF, Gaillard G, Michelsen O, Wenk RM, Rydgren B (2007) Key elements in a framework for land use impact assessment within LCA. Int J Life Cycle Assess 12(1):5–15

Potting J, Hauschild MZ (1997) Predicted environmental impact and expected occurrence of actual environmental impact. Part 2: spatial differentiation in life-cycle assessment via the site-dependent characterisation of environmental impact from emissions. Int J Life Cycle Assess 2:209–216

Potting J, Schöpp W, Blok K, Hauschild MZ (1998) Site-dependent life-cycle assessment of acidification. J Indust Ecol 2:63–87

Schmer MR, Vogel KP, Mitchell RB, Perrin RK (2008) Net energy of cellulosic ethanol from switchgrass. PNAS 105(2):464–469

Searchinger T, Heimlich R, Houghton RA, Dong F, Elobeid A, Fabiosa J, Tokgoz S, Hayes D, Yu TH (2008) Use of U.S. croplands for biofuels increases greenhouse gases through emissions from land-use change. Science 319:1238–1240

Seppälä J, Posch M, Johansson M, Hettelingh J (2006) Country-dependent characterization factors for acidification and terrestrial eutrophication based on accumulated exceedence as an impact category indicator. Int J Life Cycle Assess 11:403–416

Sheehan JJ (2009) Biofuels and the conundrum of sustainability. Curr Opin Biotechnol 20:318–324

Singh A, Pant D, Korres NE, Nizami AS, Prasad S, Murphy JD (2010) Key issues in life cycle assessment of ethanol production from lignocellulosic biomass: challenges and perspectives. Bioresour Technol 101:5003–5012

Smyth BM, Murphy JD (2011) The indirect effects of biofuels and what to do about them: the case of grass biomethane and its impact on livestock. Biofuels Bioprod Bioref 5(2):165–184

Sørensen B (2002) Renewable energy, 2nd edn. Academic, San Diego

Thomassen MA, Dalgaard R, Heijungs R, de Boer I (2008) Attributional and consequential LCA of milk production. Int J Life Cycle Assess 13(4):339–349

Tillman AM, Ekvall T, Baumann H, Rydberg T (1994) Choice of system boundaries in life cycle assessment. J Clean Prod 2(1):21–29

Tilman D, Socolow R, Foley JA, Hill J, Larson E, Lynd L, Pacala S, Reilly J, Searchinger T, Somerville C, Williams R (2009) Beneficial biofuels—the food, energy, and environment trilemma. Science 325:270–271

van Dam J, Junginger M, Faaij APC (2010) From the global efforts on certification of bioenergy towards an integrated approach based on sustainable land use planning. Renew Sust Energy Rev 14(9):2445–2472

van der Voet E, Lifset RJ, Luo L (2010) Life-cycle assessment of biofuels, convergence and divergence. Biofuels 1:435–449

Wegener-Sleeswijk A, Kleijn R, Meeusen-van Onna MJG, Leneman H, Sengers HHWJM, van Zeijts H, Reus JAWA (1996) Application of LCA to agricultural products; 1. core methodological issues; 2. supplement to the LCA guide; 3. methodological background. Centre of Environ Sci. Leiden

Weidema BP (2003) Market information in life cycle assessment. Environmental project no. 863. Danish Environmental Protection Agency, Copenhagen

Williams PRD, Inman D, Aden A, Heath GA (2009) Environmental and sustainability factors associated with next-generation biofuels in the US: what do we really know. Environ Sci Technol 43:4763–4775

Wiloso EI, Heijungs R, de Snoo G (2012) LCA of second generation bioethanol: a review and some issues to be resolved for good LCA practice. Renew Sust Energy Rev 16:5295–5308

WRI (2005) Navigating the numbers: greenhouse gas data and international climate change policy. World Resources Institute, Washington DC

Zamagni A, Guinée J, Heijungs R, Masoni P, Raggi A (2012) Lights and shadows in consequential LCA. Int J Life Cycle Assess 17:904–918

The Application of Life Cycle Assessment on Agricultural Production Systems with Reference to Lignocellulosic Biogas and Bioethanol Production as Transport Fuels

Nicholas E. Korres

Abstract The need for new approaches in agricultural production such as these of integrated agricultural systems for food and energy production necessitates the rapprochement of these systems in terms of their environmental burden. This in combination with the importance of lignocellulosic materials for biofuel production makes the system under examination extremely complex. The feedstock production, transport, processing, and conversion of cellulosic materials have not been attempted to any real degree anywhere in the world; hence, a number of sustainability issues related to energy inputs and environmental quality need to be examined. This highlights the importance of LCA as an important optimization tool. Nevertheless, the interactions and intra-, interrelationships necessitate a thorough study of the system under examination and a good knowledge of life cycle thinking.

1 Introduction

Research on potential climatic changes under an atmospheric composition modified by human activity through greenhouse gas (GHG) emissions, a function of CO_2, CH_4, and N_2O emissions, has indicated that the rise in average global temperature is likely (Owen 2001). Agriculture releases significant amounts of carbon dioxide (CO_2), methane (CH_4), and nitrous oxide (N_2O) to the atmosphere (Korres et al. 2011) while eutrophication, acidification, and natural resource depletion usually due to non-judged agricultural practices impose significant environmental threats to soil, water bodies, and biodiversity.

N. E. Korres (✉)
Freelance consultant in Sustainable Agriculture and Bioenergy, 26 Grigoroviou Str, Patisia
GR-11141 Athens, Greece
e-mail: nkorres@yahoo.co.uk

A. Singh et al. (eds.), *Life Cycle Assessment of Renewable Energy Sources*,
Green Energy and Technology, DOI: 10.1007/978-1-4471-5364-1_3,

Recent changes in the agricultural sector and related market niche along with consumer's preferences for eco-friendly production methods necessitate changes in the traditional model of "productivism agricultural era" which is focused on production of food and fiber (Wilson 2007) in favor of a "post-productivism agricultural era" which focuses on environmental management and "production of nature" (Marsden 1999). In support to this, the 2003 mid-review of the common agricultural policy (CAP) marked agricultural support payments conditional upon compliance with certain environmental standards (EC 2003). Holmes (2006) reported that the transition of agricultural structure from "productivism" to "post-productivism" model will be achieved through its multifunctional role. Marsden and Sonnino (2008) considered multifunctional agriculture as a part of a sustainable rural development paradigm within an agro-industrial model. This could result in generation of advantages toward increase non-farm income from the emerging opportunities such as bioenergy production. Earlier, at the beginning of 1990s, Sachs and Silk (1991) characterized farming systems as type 1 integrated food and energy systems (IFES) and type 2 IFES. This categorization was based on the way they were designed in terms of integration and intensification toward the simultaneous production of food and energy. More particularly, type 1 IFES is characterized through the production of feedstock for food and for energy on the same land and multiple-cropping patterns or agroforestry systems (Bogdanski et al. 2010). Type 2 IFES seek to maximize synergies between food crops, livestock, fish production, and sources of renewable energy through the adoption of agro-industrial technology (such as anaerobic digestion) that allows maximum utilization of all by-products and encourages recycling and economic utilization of residues (Bogdanski et al. 2010).

It is therefore imperative to examine and reconsider practices that alleviate the environmental burden of agricultural and bioenergy production systems. One way to achieve this is through the application of life cycle assessment (LCA) which allows for a detailed analysis of material and energy fluxes under various production schemes. This includes indirect inputs to the production process and associated wastes and emissions and the downstream fate of products (Korres et al. 2010).

The purpose of this chapter is to describe how the applications of LCA in agricultural production systems in relation to bioenergy use in transportation sector, particularly biogas and bioethanol production using lignocellulosic materials as feedstock, can assist in decision-making processes toward sustainable production practices. The development of generic guidelines and corresponding commentary on important issues for each LCA phase can assist greatly toward proper applicability of LCA in both sectors.

The need for this kind of approach is justified by discussions on bioenergy production sustainability in terms of carbon dioxide emissions reduction but also by consumer needs for environmental friendly production practices and products. It should be noted, however, that bioenergy is considered renewable and sustainable form of energy under certain conditions (Perley 2008). For example, to maintain the carbon dioxide balance, biomass harvest must not exceed growth increment while carbon dioxide emitted during production, transportation, and processing must be taken into account. The conversion efficiency of the product should be considered together with its end use to limit the risk of policy failure. The appropriateness of different bioenergy production systems in economic, environmental, and social terms will depend to a large extent on national and local circumstances. In planning a bioenergy strategy, analysis of different options and their broad impacts should be carried out to ensure that policy objectives will be met (Anonymous 2008). It is understood that a well-integrated plan of food and energy production may be one of the best ways to improve food and energy security and simultaneously reduce poverty in a climate-smart way (Bogdanski et al. 2010).

2 Agriculture and Energy: A Strong Interchangeable Relationship

Agriculture and energy have always been tied by close links although the nature and strength of the relationship keep changing over time (FAO 2008). In modern agricultural production, energy consumption is one of the major factors that establishes security and abundance in food supply chain. This is very true as agriculture became increasingly reliant on chemical fertilizers, the use of pesticides, the introduction of new hybrid varieties, the application of irrigation in arid regions, and the introduction of powered farm machinery. Fossil fuels, especially oil and natural gas, have enabled the intensification of farm productivity. Natural gas provides the hydrogen and energy used to produce most nitrogen fertilizers and both gas and oil are the sources for other agricultural chemicals, including pesticides and herbicides (Heinberg and Bomford 2009). In addition, food storage, processing, and distribution are often energy intensive activities. Consequently, higher energy costs have a direct and strong impact on agricultural production costs and food prices (Bata and Bhonot 2011). Nevertheless, environmental, economical, and social needs require a rapprochement of agricultural and farming systems toward sustainable production (Korres et al. 2011). The recent emergence of gaseous and liquid biofuels based on agricultural crops as transport fuels has reasserted the linkages between energy and agricultural output markets. Demand for agricultural feedstocks for bioenergy production will be a significant factor for agricultural markets and for world agriculture over the next decade and perhaps beyond (FAO FAO 2008). Particularly, the demand for biofuel feedstocks may help reverse the long-term decline in real agricultural commodity prices, creating both opportunities and risks (FAO 2008). This, although fossil fuels are expected to remain the bulk of the primary energy mix, can be seen as renewable energy is on the rise and will continue to be so in the future. The world's total primary energy demand amounts to about 12,274.6 million tonnes of oil equivalent (Mtoe) per year whereas biomass, including agricultural and forest products and organic wastes and residues, accounts for 10 % of this total (BP 2012).

2.1 Ethanol Production from Starch and Sugar Crops

Efforts, worldwide, to replace conventional fuels with biofuels can be seen by the high growth in ethanol and biogas during the past decade. According to the US Energy Information Administration (EIA 2013) total world liquid biofuels production increased almost sixfold over the period 2000–2010, more specific from 315,000 to 1,856,000 barrels per day. Ethanol (from *Zea mays* or maize) has been the leading biofuel in the Unites States and in Brazil (from *Saccharum officinarum* or sugarcane), (Moschini et al. 2012) whereas biodiesel (from *Brassica napus* or rapeseed) is by far the leading biofuel in EU (Korres et al. 2011).

The superiority of maize as main ethanol feedstock stems mainly from its advantage over other feedstocks in economic efficiency of conversion into ethanol (i.e., fuel yield of maize for e.g., is higher than that of barley and sorghum) (Board 2009). Dry milling and wet milling are the two processes for (first-generation) ethanol production from maize with the former being the most common. The coproducts from the conversion of maize to ethanol are known as distillers dried grains (DDGs) and maize oil from dry and wet milling, respectively, that can serve as a portion of livestock feed rations (Aines et al. 1986—cited in Board 2009). Grain sorghum (*Sorghum bicolor*) is also used as feedstock for the production of bioethanol (NSP 2012) and produces roughly the same amount of ethanol as maize although the crop's yield per unit area is lower than that of maize. Sorghum also produces DDGs and is completely interchangeable with maize in the ethanol production process (NSP 2008). Barley (*Hordeum vulgare*) is also being used in three US ethanol plants (RFA 2008). Research on hulless barley varieties as a potential feedstock to increase ethanol output in comparison to conventional barley varieties hence making this feedstock more attractive is under process (Board 2009).

Crops with high sugar content (i.e., sugarcane and *Beta vulgaris* or sugar beet) are easier to process into ethanol than starch crops since the sugar required by fermentation is already present. The fermenting and distilling technology for ethanol production from these crops is not much different than that used in breweries (Board 2009). One ton of sugarcane produces about 19.3 gallons of ethanol, a greater ethanol output per acre compared to maize. Sugar beet shows a great potential for ethanol production which with current conversion technologies yields an ethanol output per unit area close to that of sugarcane. Nevertheless, sugar beet at present is a high-cost input for biofuel production and is not used for that purpose (Salassi and Fairbanks 2006). Sweet sorghum (*Sorghum* spp.), which contains carbohydrates in fractions of both sugar and starch, is another feedstock candidate (Lau et al. 2006) as well as energy cane, a breed of sugarcane that produces high amounts of sugar and stalk for ethanol conversion. Nevertheless, despite the high ethanol yields from first-generation bioethanol (Fig. 4), sustainability criteria, environmental and economic concerns diverse bioenergy market toward second-generation bioethanol, i.e., bioethanol; production from lignocellulosic materials.

3 Why Lignocellulosic Materials Should be Used as Feedstock for Biofuel Production? A Glimpse

The use of biomethane as a transport fuel has recently started to gain attention in many European countries (Mathiasson 2008) whereas biogas production from biomass has been promoted in many developing regions including Asia, Latin America, and some regions of West Africa (Eisentraut 2010).

Lignocellulosic biomass (i.e., agricultural, industrial and forest residuals) is the most investigated type of feedstock as one of the most abundant resources with wide availability in most of the countries worldwide (Kim et al. 2002; Jorgensen et al. 2007). Pitkanen et al. (2003) reported that lignocellulosic materials could support the sustainable production of liquid transportation fuels. In support, Kim and Dale (2004) estimated that 49.1×10^6 L $year^{-1}$ of bioethanol can be produced by the utilization of the crop's dry waste material worldwide (approx. 73.9×10^5 t), an amount 16 times higher than the current-world ethanol production.

It has been projected that a major part of the European renewable energy will originate from farming and forestry (Korres et al. 2013) while at least 25% of all EU bioenergy in the future can originate from biogas, produced from wet organic materials such as animal manure, whole-crop silages, wet food, and feed wastes (Holm-Nielsen and Oleskowicz-Popiel 2008).

Growing demands for CO_2-neutral transportation fuels and the desire to achieve a reduced dependence on fossil resources have been the major driving forces for the substantial increase in the amounts of bioethanol produced by fermentation of biomass (Rass-Hansen et al. 2007) and the amount of biogas produced by the anaerobic digestion of various lignocellulosic materials (Korres et al 2013). Furthermore, the utilization of fermentable sugars from lignocellulosic materials for "green" ethanol and/or biogas production (Farrell et al. 2006; Demirbas 2008; Ni and Sun 2009) given the need for sustainable energy production and use (Prasad et al. 2007) deserves a closer examination.

In addition, policy incentives can turn the interest of the bioenergy/renewable energy market in favor of lignocellulosic materials. As reported in the newsletter of the European Biomass Industry Association (EUBIA 2012), the European Commission presented recently its policy for biofuels through a proposal which aims to limit global land conversion for biofuel production and raise the climate benefits for biofuel used in the EU. Increases in the minimum GHG savings threshold for new installations to 60%, inclusion of indirect land use change factors in the reporting of GHG savings, and limitation in the amount of food crop-based biofuels and bioliquids are suggested. Finally, the importance of market incentives for biofuels "with no or low indirect land use change emissions, and in particular the second- and third-generation biofuels produced from feedstock that do not create an additional demand for land, including algae, straw, and various types of waste" is highlighted. In the USA, GHG reductions and the establishment of a sustainable bioenergy industry are aimed to be achieved through the Energy

Independence and Security Act (EISA) in which GHG reduction thresholds from a 2005 baseline are imposed. The EISA dictates 20% reductions for renewable fuels, 50% for advance fuels, 50% for biomass-based fuels, and 60% for cellulosic biofuels (EPA 2010). It is obvious from the above that the role of lignocellulosic materials in bioenergy arena will be proved in the near future very significant. This is the reason why this chapter will be focused on the lignocellulosic biomass and second-generation biofuels (i.e., biogas and bioethanol) rather that the typical first-generation liquid biofuels. Finally, as reported by FAO (2008), the production of liquid biofuels in many countries is not currently economically viable without subsidies, given existing agricultural production and biofuel processing technologies and recent relative prices of commodity feedstocks and crude oil.

3.1 Biogas Production from Lignocellulosic Materials

Lignocellulosic materials and non-food/organic waste used for biogas production can be crops such as grass or maize silage (Korres et al. 2010; McEniry et al. 2013; Neureiter 2013), agricultural residues, and by-products (Nuri et al. 2008; Parawira et al. 2008; Holm-Nielsen et al. 2009; Rao et al. 2010; Eze and Ojike 2012) algae (Bruton et al. 2009; US DOE 2010; Brennan and Owende 2010; Benzie and Hynes 2013;), industrial and organic wastes (Salminen and Rintala 2002; Dearman and Bentham 2007; Zhang et al. 2007; Fountoulakis and Manios 2009; Beno et al. 2009; Ortner et al. 2013; Singh 2013) or other lignocellulosic or organic materials which are suitable for bacterial biodegradation (Fig. 1).

A discussion on the variability (Korres and Nizami 2013) and best management practices to increase bioenergy production can be found in Demirbas (2009) and Cherubini and Ulgiati (2010). In general, the overall biogas production process can be divided into three distinguished phases, namely the input phase, the biogas plant/processing phase, and the output phase (Fig. 2).

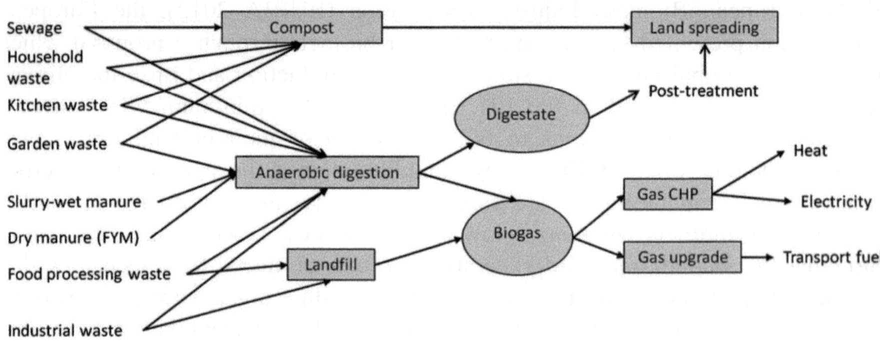

Fig. 1 Biogas route map

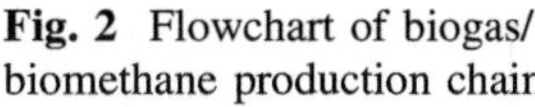

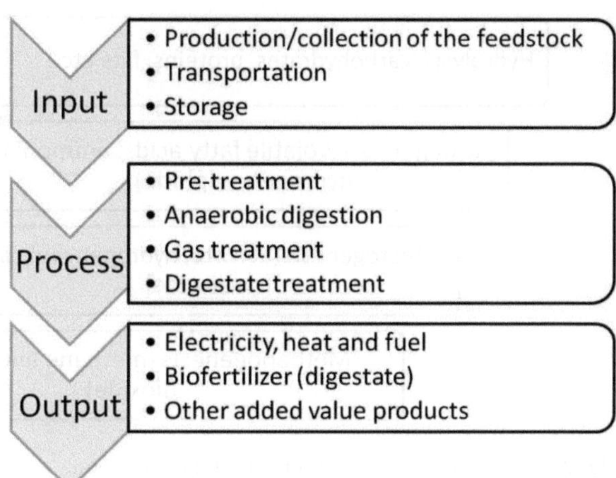

Fig. 2 Flowchart of biogas/biomethane production chain

Anaerobic digestion (the second activity within "process" phase as shown in Fig. 3) is a versatile biochemical process by which organic matter is converted to biogas under anaerobic conditions (Korres et al. 2011). This is achieved as a result of the consecutive biochemical breakdown of polymers to methane and carbon dioxide in an environment in which various microorganisms harmoniously grow and produce/reduced end products (McCarty 1982).

Four successive biological processes are involved in the anaerobic degradation of organic matter, namely hydrolysis, acidogenesis, acetogenesis, and methanogenesis (Fig. 3). Complex polymers, as stated above, are converted into monomers by extra-cellular enzymes during hydrolysis while these monomers are transformed mainly into volatile fatty acids (acetic, propionic, and butyric acids) during acidogenesis. Acetate, carbon dioxide (CO_2), and hydrogen (H_2) are produced from volatile fatty acids during acetogenesis which is finally converted into methane (CH_4) during methanogenesis (Bernet and Beline 2009).

The biogas produced during anaerobic digestion is composed of CH_4 (55–75%), CO_2 (25–45%), and trace elements such as hydrogen sulfide (H_2S which range from 0 to 2,000 ppm) and ammonia (NH_3 within the ranger of 0–590 ppm) (Rasi et al. 2007). In addition, trace amounts of hydrogen (H_2), nitrogen (N_2), carbon monoxide (CO), saturated or halogenated hydrocarbons, and oxygen (O_2) are occasionally present. The biogas is usually saturated with water vapor (H_2O) (Rasi et al. 2007). If the biogas is to be used as a transport fuel or to be injected in the natural gas grid for other use, it has to be upgraded or scrubbed (i.e., removal of corrosive components, particles, water and increase heating value to approximately 50 MJ/kg or methane content of 97%) to gain natural gas standards. The upgraded and pressurized biogas is then ready to be used (Nilsson 2001).

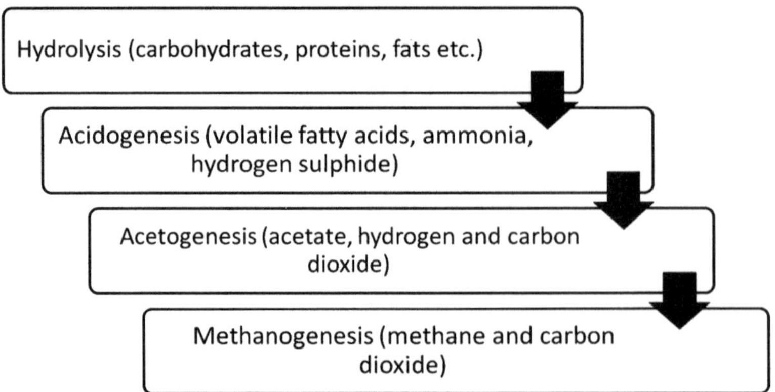

Fig. 3 Biological processes during anaerobic digestion

3.2 Bioethanol Production from Lignocellulosic Materials

Cellulosic ethanol is chemically identical to first-generation bioethanol (i.e., CH_3CH_2OH), but it is produced from different feedstock via a more complex process (cellulose hydrolysis).

In contrast to first-generation bioethanol, which is derived from sugar or starch produced by food crops (e.g., wheat, maize, sugar beet, sugarcane, sweet sorghum), (Fig. 4), cellulosic ethanol may be produced from agricultural residues, other lignocellulosic materials, or energy crops (EUBIA, European Biofuels Technology Platform, undated). Agricultural crop residues are lignocellulosic biomass (non-grain, non-root portion of agricultural crops) that remains in the field after harvest (Nelson 2007). The most common residues include the stalks, ears, and cobs from corn (stover) and straw from wheat crop, sugarcane bagasse, barley hull, wheat barn, rice husks, and rice washing drainage (Singh et al. 2010). Oilseed crops, e.g., sunflower or soybeans, produce fewer residues than grain crops and in most cases are not considered for soil sustainability reasons. Residues from other crops, e.g., cotton and pruning from orchard and vineyards, may be available but their use, due to their limited amount in most temperate climates, as lignocellulosic feedstock for bioethanol production is not feasible (Singh et al. 2010).

As mentioned earlier, lignocellulosic materials are abundant in most countries and they are generally considered to be more sustainable although they need to be hydrolyzed into simple sugars prior to distillation. This may be achieved using either acid or enzyme hydrolysis. More specifically, the production process of bioethanol from lignocellulosic materials consists of the feedstock pre-treatment, hydrolysis, fermentation, product separation and distillation and post-treatment of the liquid fraction (Fig. 5) (Balat et al. 2008; Hendriks and Zeeman 2009). According to Di Nicola et al. (2011), the classic method used in the fermentation of the hydrolyzed biomass is the separate hydrolysis and fermentation, in which

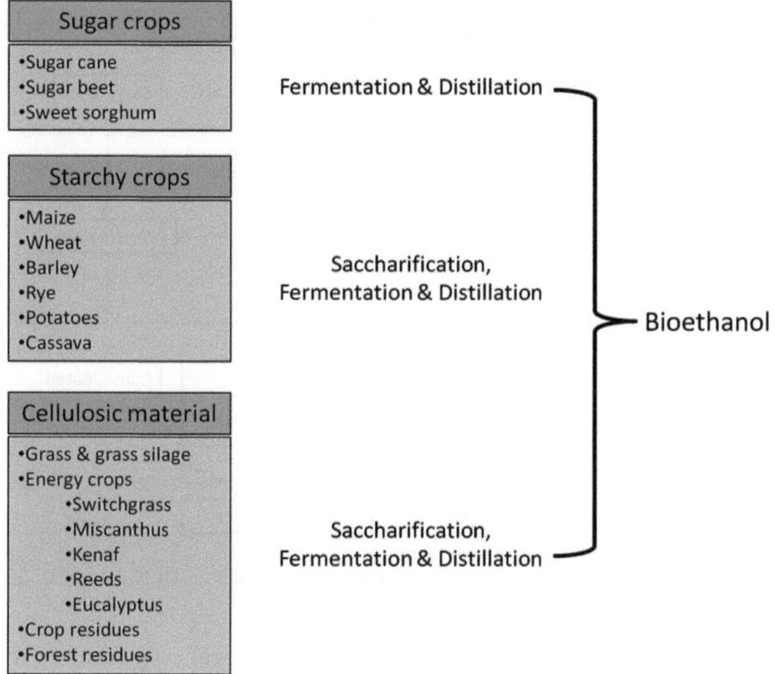

Fig. 4 Production of bioethanol by various feedstocks and processes (based on FAO 2008)

the two processes are completed in different units. A commonly used alternative is simultaneous saccharification and fermentation, in which hydrolysis and fermentation are completed in the same unit whereas a last option is represented by consolidated bioprocessing (Di Nicola et al. 2011).

Pre-treatment is an important phase toward improvements of the production rate and higher yield of monomeric sugars during hydrolysis in which unconverted fractions (from the pre-treatment) of cellulose and hemicellulose are converted into monomeric sugars. Hydrolysis is completed either chemically by acids or enzymatically by addition of cellulases (Fig. 5). The monomeric sugars produced at this phase include both pentoses and hexoses which can be fermented to ethanol. The latter can be fermented quite easily, but the fermentation of pentoses is a selective process which can only be done by a few organisms. Ethanol itself is an inhibitor for the fermenting yeasts and bacteria along with the furans, phenolic, carboxylic acids, and other soluble lignin compounds that are formed during fermentation (Hendriks and Zeeman 2009). Ethanol is recovered from the fermentation broth by distillation (Hendriks and Zeeman 2009) whereas the process residuals (e.g., cellulose, hemicellulose, lignin, and other solid materials) can be used to produce heat or to be converted into octane boosters or for the production of chemicals (Wyman 1994).

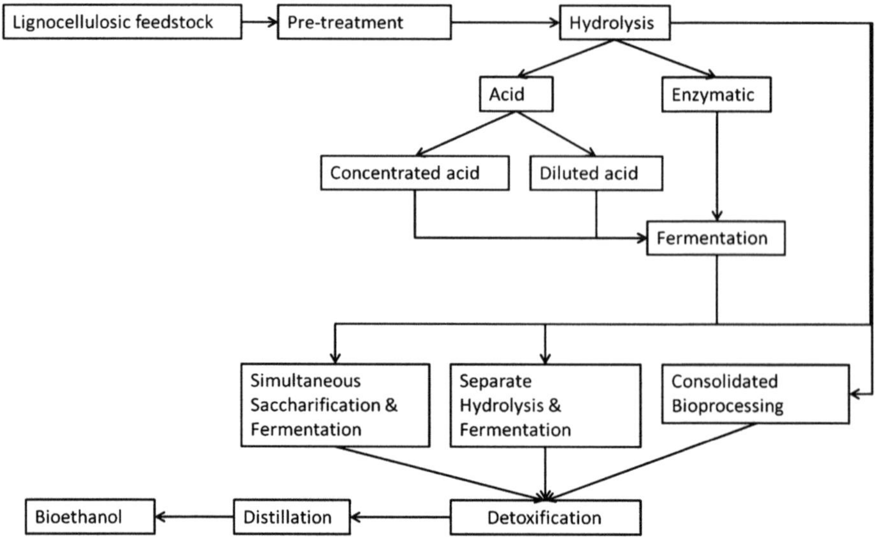

Fig. 5 Flowchart of lignocellulosic bioethanol production (based on Di Nicola et al. 2011)

Sustainability issues concern the effects of agricultural crop residues removal on soil erosion, loss of soil fertility, texture, and moisture (Nelson 2007). The actual potential to produce cellulosic ethanol is multifaceted. Large-scale production, transport, processing, and conversion of cellulosic materials have not been attempted to any real degree anywhere in the world although demonstration plants for commercial-scale production of cellulosic ethanol are under development in EU and USA. In addition, a number of pilot plants are developing thermochemical/biochemical routes to create bioethanol from commercial waste and MSW.

4 Life Cycle Assessment

4.1 General Principles

LCA is a process used for the evaluation of the environmental burdens associated with products and processes. It seeks to identify and quantify energy and materials consumed and waste released to the environment, thereby enabling the evaluation and comparison of environmental improvement options. The assessment includes the entire life cycle of the product, process, or activity (SETAC 1993). A product's life cycle is generally broken down into stages. The number of stages can vary; six stages are often distinguished, namely (1) product design; (2) raw material extraction and processing; (3) manufacturing of the product; (4) packaging and distribution to the consumer; (5) product use and maintenance; and (6) end-of-life management, i.e., reuse, recycling, and disposal (Udo de Haes and van Rooijen 2005). Although the methodology is by no means finalized and a number of important issues still must be resolved, LCA is currently standardized by ISO 14040 series (ISO 1997).

The technical framework for the LCA methodology as it is defined in ISO 14040 consists of four phases, namely (1) goal and scope definition; (2) inventory analysis; (3) impact assessment; and (4) interpretation (Fig. 6). (SETAC 1993; ISO 1997; Wenzel et al. 1997; Frankl and Rubik 2000). These phases are not followed just one after the other, but they form an iterative process, which can be followed in different rounds achieving increasing level of detail (from screening LCA to full LCA) or which may lead to changes in the first phase because of the results of the last phase (Korres 2013). More particularly, the first phase (i.e., goal and scope definition) sets the boundaries for the analysis and defines the level of detail and the functional basis for comparison. This is of crucial importance for avoidance of confusing results and misleading interpretations (Frankl and Rubik 2000). Thus, during this phase, questions should be considered about the purpose of the LCA; the spatial and temporal scope of the LCA; the functional units to be assessed; the target group; the decisions and the extent of these decisions supported by LCA; and finally, the product/solution to be assessed along with alternatives for comparison. The second phase (inventory analysis) quantifies input and output flows of materials, energy, water, and emissions or other pollutants used in each process through the entire production chain and present these in a process flow chart (Figs. 2 and 5). Since this phase can affect the complete LCA, it is necessary to follow the precise standards for data collecting, calculation procedures, allocation rules (SETAC 1993; ISO 1997). In this phase, the system to be assessed and the system's boundaries should be clearly defined. The third phase (impact assessment) quantifies and clusters effects of the resource use and emissions into a number of environmental impact categories (i.e., selection of impact categories, classification and characterization of environmental impacts based on the inventory analysis) which may be weighted according to their importance and goal and

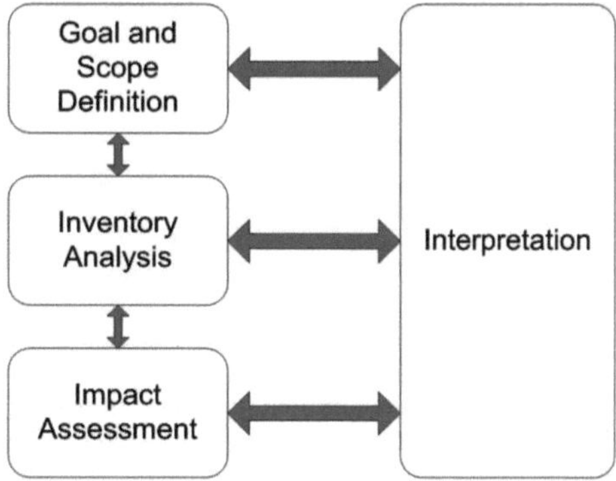

Fig. 6 Life cycle assessment framework (based on ISO 1997)

scope of the LCA. General impact categories are human health, biotic natural environment, natural resource depletion, abiotic environment, and man-made biotic and abiotic environment (Korres 2013). These impacts are operationalized by specific impacts such as global warming, ozone depletion, acidification or eutrophication, ecotoxicity, land use, and habitat loss (Korres 2013). In the characterization phase, the impacts are analyzed, quantified, and calculated, requiring scientific knowledge about load–response relationships. For that purpose, the inventory data need to be analyzed by modeling approaches, like the use of equivalency factors (e.g., ozone depletion potential) or toxicological data. The last phase (interpretation) reports the results in a comprehensible way and evaluates the opportunities to reduce the environmental impact of the product or service under examination according to the goal and scope of the study. As a basis for a decision-making process, the results of the LCA can be used for improvements and support evidences for other environmental concepts, tools, and systems such as ecolabeling, environmental management system (Fawer 2001). In contrast to other environmental management tools, which tend to focus on specific life stages of a product or process, LCA analyzes the entire life cycle, looking up and down the supply chain, from raw material extraction to final disposal. A simplified biofuel pathway is shown in Fig. 7 where system boundaries, inventory analysis (data for inputs), and impact assessment (environmental effects) are clearly shown.

More particularly, the feedstock production phase (left part) which includes crop production and husbandry management along with processing of the feedstock (middle part) and some of the important markets (right part) into which biofuels and their coproducts are traded. Examples of bioethanol production coproducts include animal feed from corn ethanol or bagasse, from sugarcane ethanol, which can be used for the production of heat or electricity though its

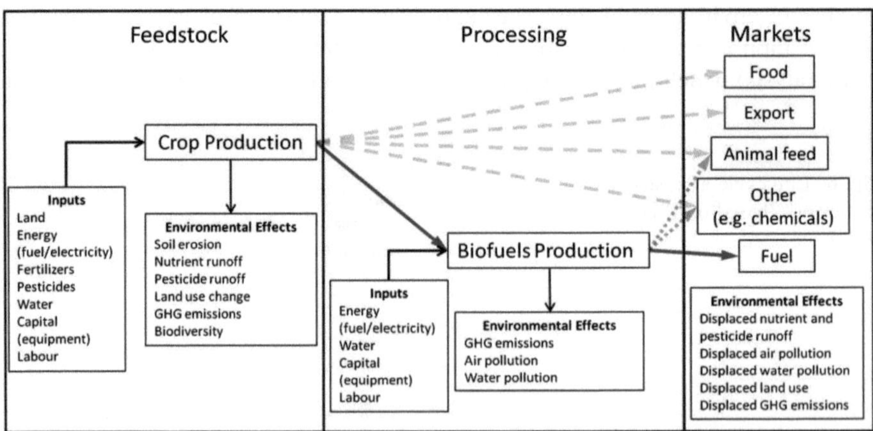

Fig. 7 A simplified biofuel pathway in which inputs and related environmental effects are depicted (*dotted lines* represent pathways irrelevant to this chapter) (based on Kammen et al. 2007)

combustion (Quintero et al. 2008). Another well-known example is the digestate, a residue of anaerobic digestion, which can be used as a substitute of mineral fertilizers (Lukehurst et al. 2010; Korres et al. 2011; Taylor et al. 2012; Chambers and Taylor 2013).

It is worth mentioning that some of the inputs and related environmental effects throughout the biofuel production chain may be indirect such as the energy and related emissions from the manufacture of feedstock production inputs, e.g., fertilizers, herbicides, lime or from the pre-treatment of the feedstock (e.g., macerating) or other activities, e.g., mixing and water-pumping activities in anaerobic digestion plant (Korres et al. 2011; Korres 2013). It is also vital to note—and to reflect in biofuel analyses—that the indirect impacts of biofuel production and in particular the destruction of natural habitats (e.g., rainforests, savannah, or in some cases, the exploitation of "marginal" lands which are in active use, even at reduced productivity, by a range of communities, often poorer households and individuals) to expand agricultural land, may have larger environmental impacts than the direct effects. The indirect GHG emissions of biofuels produced from productive land that could otherwise support food production (reference system for comparisons) may be larger than the emissions from an equal amount of fossil fuels (Delucchi 2006; Farrell et al. 2006). Attention to these issues is vital if biofuels are to become a significant component of sustainable energy and socio-economic systems (Kammen et al. 2007).

In addition, biofuel production and usage also displaces some environmental effects because they substitute in fuel and other markets for products that have their own environmental effects. The extent to which the coproducts of biofuel production displace other products and their environmental impacts (rather than stimulate additional consumption) depends on the elasticity of demand in the relevant markets (the more inelastic the demand, the greater the displacement), the way in which the coproducts affect supply curves, and other market and non-market (i.e., political and regulatory) factors (Kammen et al. 2007).

4.2 LCA and Agricultural Production

Concerns about the environmental impacts of agricultural production systems and energy sectors have led to considerable publicity about the importance of applying LCA technique for minimizing these burdens. LCA has been promoted as one of the best ways of determining the real impacts of agricultural products (Loerincik et al. 2008; Harris and Narayanaswamy 2009) and consequently has been proved an important tool for possible mitigation options and eco-friendly production. The application of mineral and organic fertilizers, soil management practices, animal production systems, and manure management (Mummey et al. 1998; Steinfeld et al. 2006; Smith et al. 2007) are some factors which enhance the environmental burden of agricultural production including GHG emission, eutrophication, acidification, among others.

4.2.1 LCA in Agriculture: A Challenging Complexity

Agriculture does not consume resources in a linear sense, as for example, many industrial processes, and is not therefore a pure "cradle-to-grave" process (Haas et al. 2000). The same authors argued that the term "LCA" in agriculture could be misleading since the main agricultural processes are made within a farm and are based on renewable resources. As they have suggested, "the term eco-balance used for LCA in French or German is regarded to fit more accurately."

Agriculture LCA has several differences and greater complexity from LCA of industrial processes, the most important is that agriculture utilizes land and soil. The balances of soil nutrients such as nitrogen (N), phosphorus (P), and potassium (K), through fertilizer application and plant uptake, need careful consideration. Modeling nitrogen dynamics in the soil, for example, requires (1) the quantification all N losses and (2) understanding of the interactions between these losses. A conceptual model which is known as "hole-in-the-pipe" (Fig. 8) (Firestone and Davidson 1989) depicts the flows of inorganic nitrogen through the microbial processes of nitrification and denitrification. Nitrogen oxides escape through "leaks" in the pipe which symbolizes the actual nitrification and denitrification processes occurring in the soil (the size of the pipe is variable mainly due to varying availability of C and N). The size of the "holes" through which N gases can "leak out" is determined by soil moisture content, water-filled pore space as well as by other soil conditions such as pH and temperature.

It becomes obvious that estimating long-term balances requires the use of simulation modeling, which most probably must be adapted to the local context considering variations in soil texture, rainfall, altitude, etc.

Many agricultural systems are interlinked and therefore changes to one system, for example, arable crops used for animal feed or grass silage for animal feed will affect other systems e.g., animal production systems or bioenergy production systems. Further complications occur with systems which are included or interact with other, e.g., as in the case of beef production which is partly derived from the dairy sector. Hence, there can be difficulties assigning environmental impacts between various product components particularly when the animals which may be reared in geographically diverse areas including lowlands and/or highlands incorporated into the LCA.

In addition, agriculture contributes to GHG emissions by the consumption of fuel or electrical energy, both directly (i.e., in the operation and maintenance of

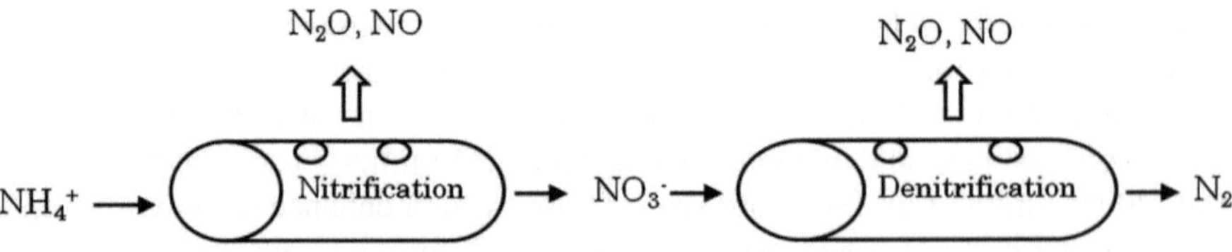

Fig. 8 Soil *N* dynamics (based on Firestone and Davidson 1989)

machinery used during crop establishment and husbandry but also for the maintenance of livestock housing) and indirectly (i.e., in the form of manufactured mineral fertilizers and pesticides but also as embodied energy in machinery and human labor) (Korres 2013). The level of energy consumption and GHG emissions depends mainly on the production system (e.g., organic or conventional) but also on the product mix (e.g., the mix of crops and livestock and/or bioenergy production). It has been shown, for example, that abandonment of fossil fuel-derived nitrogen and synthetic pesticides in organic farming consumes less energy and consequently contributes less to GHG gas emissions than conventional agriculture (Carlsson-Kanyama 1998; Pimentel and Pimentel 2003; Wallen et al. 2004; Weber and Matthews 2008).

Besides the approach to input use, soil management practices, such as tillage, irrigation, use of cover crops (Mummey et al. 1998) in cropping systems and storage of slurries and manures in livestock systems, influence GHG gas emissions. In the context of choice of the cropping system, crop rotation has a strong influence on emissions. For example, adapting crop rotations to include more perennial crops, thereby avoiding use of bare and fallow land, reduces GHG gas emissions from agriculture by accumulating soil carbon stocks (Smith et al. 2007).

When multiple-cropping systems are practiced either as sequential cropping or intercropping (Korres 2005), the issue of land use, which is one of the most fundamental factors that influences directly (e.g., tillage) or indirectly (e.g., by the collection of crop residues to be used for biogas or bioethanol production) carbon stocks arises. Cultivation, generally leads to reduction in soil organic carbon (Reay and Grace 2007) which, without counteracting husbandry practices such as winter cover crops (Rajagopal and Zilberman 2007), is exacerbated by crop residues removal, e.g., corn stover (Wilhelm et al. 2004) (Fig. 9).

A number of management practices are available to increase soil carbon inputs in croplands through the use of crop rotations with high residue yields, or reducing the gap between successive crops in annual crop rotations (i.e., the fallow period) or increasing fertilizer and manure use efficiency through the justification of their use. In addition, soil carbon losses, on annual croplands, can be reduced by decreasing the frequency and intensity of soil tillage, in particular through conversion to no-till practices (Paustian et al. 1997; Huggins et al. 1998).

In addition, Lal (2004) reported integrated pest management and drip irrigation along with conservation tillage management as low carbon intensity practices. It has been shown that carbon sequestration can dramatically influence the sustainability of biofuels and particularly biogas (Korres et al. 2010); hence, cropping system should be considered if precision, completeness, representativeness, and comparability in LCA are to be secured (Korres 2013).

The statements above highlight the complexity of LCA technique in agricultural systems but also in bioenergy production particularly when land use and land use change are considered. It has been reported that indirect GHG emissions of biofuels produced from productive land that could otherwise support food production may be larger than the emissions from an equal amount of fossil fuels (Delucchi 2006; Farrell et al. 2006). Kammen (2007) stated that attention to these

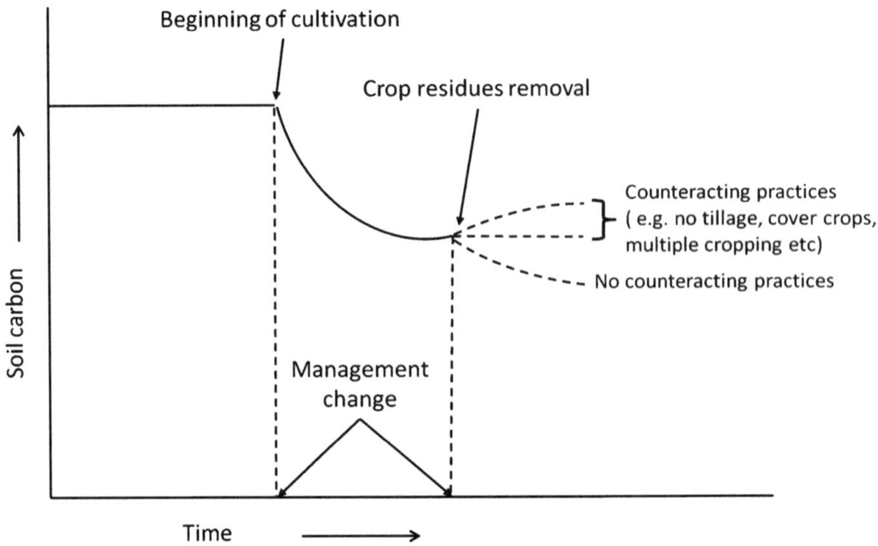

Fig. 9 Soil carbon alterations with management practices (based on US DOE 2006)

issues is vital if biofuels are to become a significant component of sustainable energy and socioeconomic systems.

Nevertheless, this chapter, as mentioned earlier, focuses on agricultural LCA that deal specifically with biogas and bioethanol production. In the following sections, issues in relation to four distinctive phases of the LCA as they have been defined by ISO 14040 (ISO 1997) series will be discussed. Grant et al. (2008) stated that when LCA is applied to GHG emissions in agriculture all pre-farm, on-farm, and post-farm emissions of carbon dioxide, methane, and nitrous oxide should be included. As such, a full LCA of emissions takes into account downstream emissions but also those associated with the fuel and other inputs during pre and on-farm activities known as upstream emissions (Grant et al. 2008). In other words, all inputs are traced back to primary resources, for example, electricity is generated from primary fuels like coal or oil (Williams et al. 2006). Fertilizers that are based on ammonium use methane as a feedstock and source of energy. Other fertilizers, such as phosphate and potassium, also require energy for extraction from the ground, processing, packing, and delivery. Machinery, including tractors and processing equipment, requires steel, plastic, and other materials for their manufacture. This involves energy costs in addition to the direct diesel use (Harris and Narayanaswamy 2009).

4.2.2 LCA Framework

Key components and critical phases of agricultural LCA will be dealt in this section are the following: (1) goal and purpose of agricultural LCAs; (2) LCA system boundary; (3) functional unit(s); (4) life cycle inventory (LCI) and allocation methods for coproducts and foreground and background data sources-data quality and assessment; and (5) environmental impact assessment (EIA) and impact categories.

Goal and Scope Definition

The goal and scope definition phase of an LCA in agriculture includes several decisions that are of relevance for all subsequent steps, i.e., LCI, Life cycle impact assessment (LCIA), and interpretation (Frischknecht and Jungbluth 2007). As Svoboda (1995) stated that the goal of LCA is not to arrive at the answer but, rather, to provide important inputs to a broader strategic planning process. Use of LCA assists to focus attention on "hot spots" for optimizing the environmental performance of systems and broadens the debate to include the wider environmental impacts of alternatives (Cowell 1999). A wide variety and goals were found to exist in the literature of agricultural LCAs although as Harris and Narayanaswamy (2009) mentioned that agricultural LCAs generally compare the environmental impact of farming practices or types of animal feed. As such, some LCA practitioners in fiber production and textile industry used the technique to examine the energy difference between various types of textiles (Woolridge et al. 2006) or to examine methodological problems and solutions for textile products (Dahllof 2005) or to determine the energy required to produce one metric tonne (1,000 kg) of raw cotton (including both seed and lint, in the field) across a range of global production practices (Matlock et al. 2008). Comparison of production practices is also illustrated by a study on bread-making wheat production where the relative environmental impacts of conventional versus less intensive agricultural production systems are compared (Cowell 1999). In LCA of grassland-based production systems, mainly in dairy production systems, the goal and scope of the study concerns mainly the eco-friendliness of the system under examination. Casey and Holden (2006) examined Irish suckler beef units, comparing GHG emissions of conventional Irish agri-environmental scheme versus organic farms. The same authors (Casey and Holden 2005), in another study, focussed on GHG emissions from an Irish dairy unit and assessed various scenarios to be considered toward GHG emissions reduction. LCA studies in Australia and New Zealand on dairy industry concerned the environmental impacts of the dairy supply chain and the implications of intensification on their eco-efficiency, respectively, so that dairy companies could improve environmental performance of their business (Nicol and Sage 2003; Basset-Mens et al. 2009).

Functional Unit

The functional unit (FU) is dependent on the goal of the study and the system boundary and is generally chosen to reflect the way each commodity is traded. The reference unit, that denotes the useful output, is known as the FU and has a defined quantity and quality, for example, 1 tonne of bread-making wheat. In many studies, the FU is typically one unit of weight of product (e.g., kg or tonne), or one hectare of land used (Haas et al. 2000; Haas et al.2001; Nicol and Sage 2003; Casey and Holden 2006). Consensus is needed on the FU for livestock, e.g., 1 kg of bone free leaving the gate, 1 kg of live weight leaving the farm gate, or 1 t of carcass dead weight. The choice can help avoid allocation (Harris and Narayanaswamy 2009).

System Boundaries

System boundaries are a set of criteria specifying which unit processes/tasks are part of a product system (ISO 14040: 2006). The system boundary should include as far as possible all relevant life cycle stages and processes (EC 2010). Hayashi et al. (2005) reported that one of the methodological characteristics of agriculture LCA studies, since they are analyzing production processes, is that their system boundaries are defined as the cradle-to-gate type (Baumann and Tillman 2004—cited in Hayashi et al. (2005)). Nevertheless, the system boundary will largely depend on the goal of the study if, for example, the scope is for environmental improvement of the farm or the whole supply chain to consumer (Harris and Narayanaswamy 2009).

Life Cycle Inventory

A LCI analysis is the process of quantifying the energy and raw material requirements, atmospheric emissions, waterborne emissions, solid wastes, and other releases for the entire life cycle of a product, process, or activity. In this stage, all relevant data are collected and organized. The evaluation of comparative environmental impacts or potential improvements without LCA is not possible whereas the level of accuracy and detail of the collected data should be reflected throughout the remainder of the LCA process (WEC 2004).

The key steps of a LCI include the (1) development of a flow diagram of the processes under evaluation (Figs. 2, 5 and 10); (2) development of data collection plan (Korres, 2013); (3) collection of the data based on certain rules and protocols regarding their quality (Korres 2013) or calculation when possible/necessary (see below); and (4) evaluation and reporting of the results (Korres 2013).

A generic flow process model for agricultural production within each operational unit or task consists of four main tasks, namely seedbed preparation, sowing/planting, field operations, and harvesting. The field operations are divided into fertilization, irrigation and weed and pest control (Fig. 10). The task and subtasks during crop production stage can be characterized as mechanical and/or non-mechanized (e.g., human labor). Fertilization can be characterized either as conventional (inorganic), organic (e.g., manure or digestate or mulching), or mixed (combination of inorganic/organic).

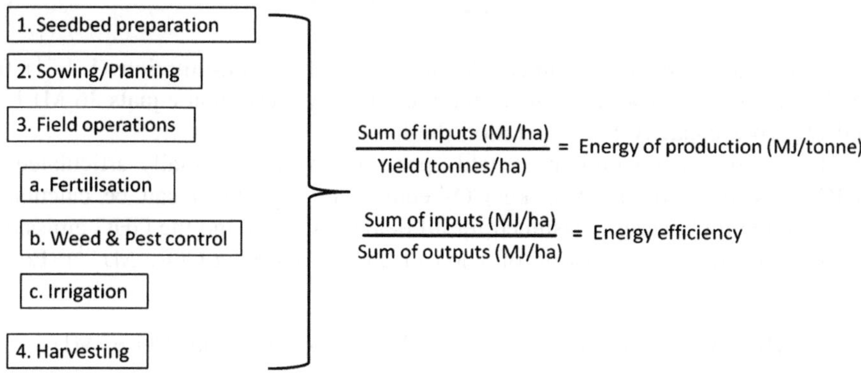

Fig. 10 A Generic flow process in agriculture and related energy inputs (based on Matlock et al. 2008)

The types of input energy (IE) analyzed by LCA, according to Eq. (1), include

$$IE = E_F + E_I + E_{IR} + E_{ED} + E_L \tag{1}$$

E_F energy in the fuel consumed by machinery (MJ ha^{-1}); E_I indirect energy consumed to produce applied inputs such as seeds, pesticides, fertilizers expressed in MJ ha^{-1}; E_{IR} direct energy used, where necessary, in irrigation expressed in MJ ha^{-1}; E_{ED} indirect embodied energy expended in machinery manufacture or farm buildings, depreciated over their useful life expressed in MJ ha^{-1}; E_L indirect energy consumed in human labor expressed in MJ ha^{-1}.

In this chapter, only the first three forms of energy will be discussed as the use of indirect embodied energy and indirect energy consumed by human labor, in most cases, excluded from the LCI in biofuels (Korres 2013). In several studies, emissions associated with the production of machines, buildings, and roads are excluded because the lack of relevant data (Cederberg and Mattsson 2000; Cederberg and Stadig 2003; Casey and Holden 2005).

The energy associated within each task (e.g., seedbed preparation, sowing) can be either documented from a review of contemporary literature as accurately as possible to ensure that subsequent decisions based on arguments embodied in the findings are valid (Sapsford and Jupp 2006; Korres 2013) or calculated. Direct mechanical energy (i.e., energy from fuel consumption) for each task can be calculated, for example, by multiplying the estimated fuel requirements (for tractor or harvester) to complete a task (volume of fuel per unit area of production) by the energy per unit volume of fuel (i.e., 37.6 MJ L^{-1} for diesel fuel) (Larson and Fangmeier 1978; Griffith and Parsons 1983). Alternatively, an algorithm for the calculation of the direct energy consumed during the field operations as suggested by Korres et al. (2010) and it is expressed by Eq. (2) can be used.

$$E_F = \sum_{1}^{i}((F_{ci} \times f_c)/O_{ci}) \tag{2}$$

where E_F fuel energy consumed (MJ ha^{-1}), F_{ci} fuel consumption (L h^{-1}) for i field operation, f_c heating value of the fuel, usually diesel that equals 36 MJ L^{-1} and O_{ci} work capacity for i operation (ha h^{-1}).

The environmental burden of field operations which is usually articulated as GHG emissions expressed as kg CO_2 equivalent (kg CO_2e) can be calculated based on the fuel energy consumed for all field operations and the GHG emissions produced from the combustion of 1 MJ of diesel (0.888 kg CO_2e MJ^{-1}) (Eq. 3) (Korres et al. 2010):

$$\text{GHG emissions}(\text{kg}\,CO_2/\text{ha}) = FE(\text{MJ}/\text{ha}) \times 0.0888(\text{kg}\ CO_2\text{e}\ /\text{MJ}) \tag{3}$$

As it can be noticed, the energy per unit volume of the fuel used in agricultural operations (diesel in this case) varies between 36 and 37.6 MJ L^{-1} and according to Saunders et al. (2006) up to 41.2 MJ L^{-1}. This adds up to uncertainty of calculated parameters in LCA and signals the importance in data collection in terms of accuracy, representativeness, and consistency (Korres 2013).

According to (Romanelli and Millan 2005), indirect energy in farm inputs can be calculated as a fraction of the solid (e.g., seeds, lime, and fertilizers) and liquid (e.g., pesticides in liquid form).

The energy enclosed in the solid fraction of farm inputs depends on the application rate and the enclosed energy within each and can be obtained using Eq. (4).

$$E_{si} = Q_t \times E_{ci} \tag{4}$$

where Q_t quantity of input applied per hectare (kg ha^{-1}); E_{ci} = energy content (energy index) of a solid input (MJ kg^{-1}). Some representative energy indices of solid and liquid inputs are mentioned below (Table 1).

Indirect energy consumed by the liquid fraction of crop production inputs can be calculated based on Eq. (5):

$$E_{Li} = \frac{E_{li} \times a.i \times V_p}{V_a} Q \tag{5}$$

where E_{li} energy content of the liquid input (MJ L^{-1}); *a.i.* concentration of the active ingredient in the commercial product (%); V_p used volume of the commercial product (L); V_a volume to be applied (L); Q application rate (L ha^{-1}). The calculations of emissions from the consumption of indirect (solid and liquid) inputted energy in crop production (e.g., fertilizers, herbicides, lime) but also emissions due to the application of fertilizer (e.g., direct and indirect nitrous oxide emissions) or these from lime and pesticides (e.g., volatilization) are described in detailed in Korres et al. (2010) and Korres (2013).

The energy consumed by irrigation can be estimated either as energy values directly from the literature (Tsatsarelis 1991; Wanjura et al. 2002; Yilmaz et al. 2004; Oren and Ozturk 2006) or as energy calculated from volume of diesel fuel

Table 1 Energy indices of various farm's solid and liquid inputs in crop production

Input	Energy content (MJ kg^{-1})	Source
Seeds		
Grass	12.0, 12.2	1, 2, 13
Maize	15.4	3, 12
Cereals	10.5	4, 12
Fertilizers		
Nitrogen	50.0, 74.0	1, 2, 4, 5, 6, 7, 8, 9, 13
Urea	8.6	11, 12
Ammonium Nitrate	77.0	11, 12
Ammonium Sulfate	22.0	11, 12
Anhydrous ammonia	68.0	11, 12
Potassium Nitrate	14.6	11, 12
Calcium Nitrate	16.7	11, 12
Phosphorous(P_2O_5)	8.6, 12.6	1, 5, 6, 7, 8, 9, 10, 11, 12, 13
Simple Phosphate	9.8	11, 12
Triple Phosphate	6.7	11, 12
Diammonium Phosphate	44.1	14
Potassium (K_2O)	6.7	1, 5, 6, 7, 8, 9, 10, 11, 12, 13
KCl	7.2	11, 12
Potassium Sulfate	3	11, 12
Pesticides		
Herbicides	254.6, 264.0	1, 2,3, 6, 7, 10, 13
Insecticide	184.7	3
Fungicide	97.1	3
Lime		
Lime, Limestone	1.2, 1.7	1, 6, 8, 9, 11, 12, 13

1 Kelm et al. 2004; 2 Rosenberger et al. 2001; 3 Pimentel 1980; 4 Pellizzi 1992; 5 Gerin et al. 2008; 6 Wells 2001; 7 Dalgaard et al. 2001; 8 Cropgen 2004a, b; 9 Elsayed and Mortimer 2001; 10 Cropgen 2004a, b; 11 Ferraro 1999; 12 Romanelli and Milan 2005; 13 Smyth et al. 2009; 14 Hetz 1992

used (Rogers and Alum 2007) or as energy calculated based on the irrigated area and the amount of pumped water needed from a groundwater reservoir (Eq. 6) (Romanelli and Millan 2005), a case which is applicable in many arid and semi-arid climates.

$$E_{IR} = \frac{f_{ee} \times P_e \times U_d \times ND}{A_i} \tag{6}$$

where f_{ee} enclosed energy in electrical energy (i.e., according to Pimentel 1984, this equals 12 MJ kW^{-1} h^{-1}); P_e power of the motor driving the pumping system (kW); U_d average daily use (h); ND period of irrigation (days); A_i total irrigated area by the system (ha). In this calculation, consideration should be taken in regard to the specifications of the device along with physical/topographical characteristics of the water source (e.g., depth of well).

GHG emissions due to irrigation can be calculated based on the following equation (Eq. 7):

$$\text{GHG emissions}(\text{kg CO}_2\text{e}) = \frac{Q \times EF}{1000} \tag{7}$$

where Q (activity) is the electricity used expressed in kWh and EF is the relevant emission factor expressed in kg CO_2 e/kWh. With respect to lignite power plants, significant variations in cumulative GHG emissions have been quoted in the literature, ranging from approx. 800–1,700 g CO_2 e q/kWh$_e$ (Weisser 2007). The great variation in the emissions of current lignite power plants indicates the importance of thermal plant efficiency and operating mode, since most GHG emissions occur at the combustion stage (Weisser 2007).

Allocation

Agricultural LCA is often complex because in addition to the main product there are usually coproducts, so that appropriate environmental impacts need to be assigned to each product, a process known as allocation. There may also be by-products or waste and emissions to the environment, for example, nitrate (NO_3) to water and nitrous oxide (N_2O) to the air (Harris and Narayanaswamy 2009). Allocation may be performed on a mass, volume, economic, or energy basis although it is best to avoid allocation through system expansion or division of processes into product-specific subprocesses (Vikman et al 2004; Labutong et al. 2012) Economic allocation has, in the past, been utilized but studies of beef and dairy products have shown this to increase uncertainty. The need for allocation can be dependent on choice of FU and system boundary. Allocation was not needed in one beef study because the FU choice (live weight) and system boundary (cradle-to-farm gate) meant that the by-products occurred outside the farm gate (by-products occur post-processing). Karlsson (2003) estimated GHG emissions and mitigation costs for a range of biomass-based cogeneration systems under different methodological assumptions. The choice of a FU was given strong consideration, since the proportion between the products may differ between the studied and the reference system (i.e., an alternative, typically a "job as usual", system with which the system under examination is compared). This can be dealt with by considering one product as the FU and the other as a by-product and then assuming that the difference in generation of the by-product is balanced by another energy system in the reference scenario (allocation by subtraction) (Vikman et al 2004).

4.3 Environmental Impact Assessment

The life cycle impact assessment (LCIA) phase of an LCA is the evaluation of potential human health and environmental impacts of the environmental resources and releases identified during the LCI. Impact assessment should address ecological effects and human health effects; it may also address resource depletion.

A LCIA attempts to establish a linkage between the product or process and its potential environmental impacts. Critical questions for example:

- What are the impacts of that much quantity of CO_2 or that much quantity of methane emissions being released into the atmosphere by a typical beef farm annually?
- Which is more damaging to air pollution?

Typical midpoint environmental impact categories considered mostly in LCA are as follows: (1) the greenhouse effect (global warming potential), (2) eutrophication potential, (3) acidification potential, (4) formation of photochemical oxidants, (5) particles, and (6) energy balance (Borjesson et al. 2011). More particularly, the global warming potential refers to the increase in the average temperature of the Earth's surface, due to an increase in the global warming potential, caused by anthropogenic emissions of global warming gases such as carbon dioxide, methane, nitrous oxide, fluorocarbons, e.g., CFCs and HCFCs, and others. Acidification refers to the accumulation of acidifying substances, e.g., sulfuric acid, hydrochloric acid in the water particles in suspension in the atmosphere which are deposited onto the ground by rains; acidifying pollutants have a wide variety of impacts on soil, groundwater, surface waters, biological organisms, ecosystems, and materials, e.g., buildings. Eutrophication which is a process whereby water bodies, such as lakes or rivers, receive excess chemical nutrients, typically compounds containing nitrogen or phosphorus that stimulate excessive plant growth, e.g., algae. Nutrients can come from many sources, such as fertilizers applied to agricultural fields, deposition of nitrogen from the atmosphere, erosion of soil containing nutrients, and sewage treatment plant discharges (Wenisch and Monier 2007). An LCIA provides a systematic procedure for classifying and characterizing these types of environmental effects. GHGs emissions, for example, from different sources are indexed according to their global warming potential. According to the intergovernmental panel on climate change (IPCC 2001), over a 100-year time span, carbon dioxide (CO_2) assumes the value of 1 whereas the two other GHGs of importance in agriculture LCA are methane (CH_4) and nitrous oxide (N_2O) which, according to a re-evaluation of the IPCC in 2001, take a value of 23 and 296 respectively. Hence, the volume of GHG emissions in terms of CO_2e can be calculated using Eq. (8) (IPCC 2001; EC 2009):

$$\text{GHG}(\text{kg of } CO_2e) = CO_2(\text{kg}) + 23 \times CH_4(\text{kg}) + 296 \times N_2O(\text{kg}) \tag{8}$$

Midpoint impact assessment approaches reflect the relative potency of the stressors at a common midpoint within the cause–effect chain. Analysis at a midpoint minimizes the amount of forecasting and effect modeling incorporated into the LCIA, thereby reducing the complexity of the modeling and often simplifying communication. Midpoint modeling can minimize assumptions and value choices, reflect a higher level of societal consensus, and be more comprehensive than model coverage for endpoint estimation. (Bare et al. 2003). Endpoints depicted in Fig. 11 belong to a larger, more generic impact category, e.g., "skin

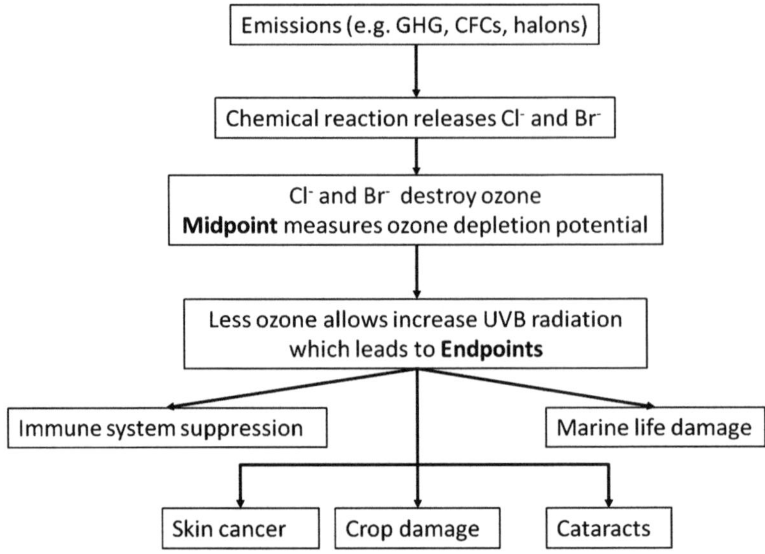

Fig. 11 Midpoint versus endpoint LCIA (based on Bare et al. 2003)

cancer" and "cataract" into human health, "marine life damage" into natural environment, "crop damage" into man-made biotic environment or natural resources.

4.4 Interpretation

LCA interpretation is a systematic process to evaluate the results of the inventory analysis and impact assessment, to select the preferred product, process, or service with a clear understanding of the uncertainty and the assumptions used to generate the results (SAIC 2006). In practice, the process is iterative as the results from subsequent stages will often require previous assumptions about system boundaries, required data, data quality, etc. to be modified. LCA offers various opportunities to reduce or mitigate the environmental impact throughout the whole life cycle of a product, process, or activity.

5 LCA and Biofuels Production

To identify savings in energy and emissions from biofuel production and utilization, a thorough evaluation of the corresponding life cycle is to be carried out carefully; LCA is an effective tool for this, which accounts for the relative environmental impacts of biofuel life cycle with respect to "base case" such as fossil fuel-based life cycle (Sreejith et al. 2013). Additionally, nearly all LCA studies on the role of biofuels in mitigating global warming and boosting energy security have concluded that "second-generation" (or "advanced") biofuels which rely on non-food feedstocks and offer improved energy and GHG profiles are necessary to make wider use of biofuels feasible worldwide (Earley and McKeown 2009).

5.1 LCA of Biogas Production from Lignocellulosic and Non-Food Feedstocks

Production of biogas is an integrated process in which many stages and combinations are involved. The overall biogas production can be divided into three distinguished phases, namely the input phase (i.e., production/collection of the feedstock, transportation, and storage), the biogas plant/processing phase (i.e., pretreatment, anaerobic digestion per se, gas treatment, and digestate treatment), and the output phase (i.e., production of various goods and value-added products as in biorefineries).

5.1.1 System Boundaries

The employment of LCA in biogas production necessitates the expansion of the typical agriculture LCA boundaries (Fig. 12a) to include transport and process energy flows and related environmental burdens (Fig. 12b) for biomethane production and transportation of digestate, an anaerobic residue, back to the field.

5.1.2 Goal and Scope

The goal of an LCA study shall unambiguously state the intended application to the intended audience of the study whereas the scope should be adequately defined so as to ensure compatibility with the goal (Singh et al 2010).

5.1.3 Functional unit

In all bioenergy assessment systems, the choice of an appropriate FU, as the basis for comparisons, is of major importance (Ekvall and Finnveden 2001). In practice, the FU consists of a qualitatively defined function or property (e.g., environmental impact) and quantified unit (e.g., 1 m^3 or 1 MJ of fuel). There is significant diversity in relation to the FU used in LCA, particularly in the case of biofuels. Korres et al. (2010) defined the FU as m^3 biomethane $year^{-1}$. In addition, according

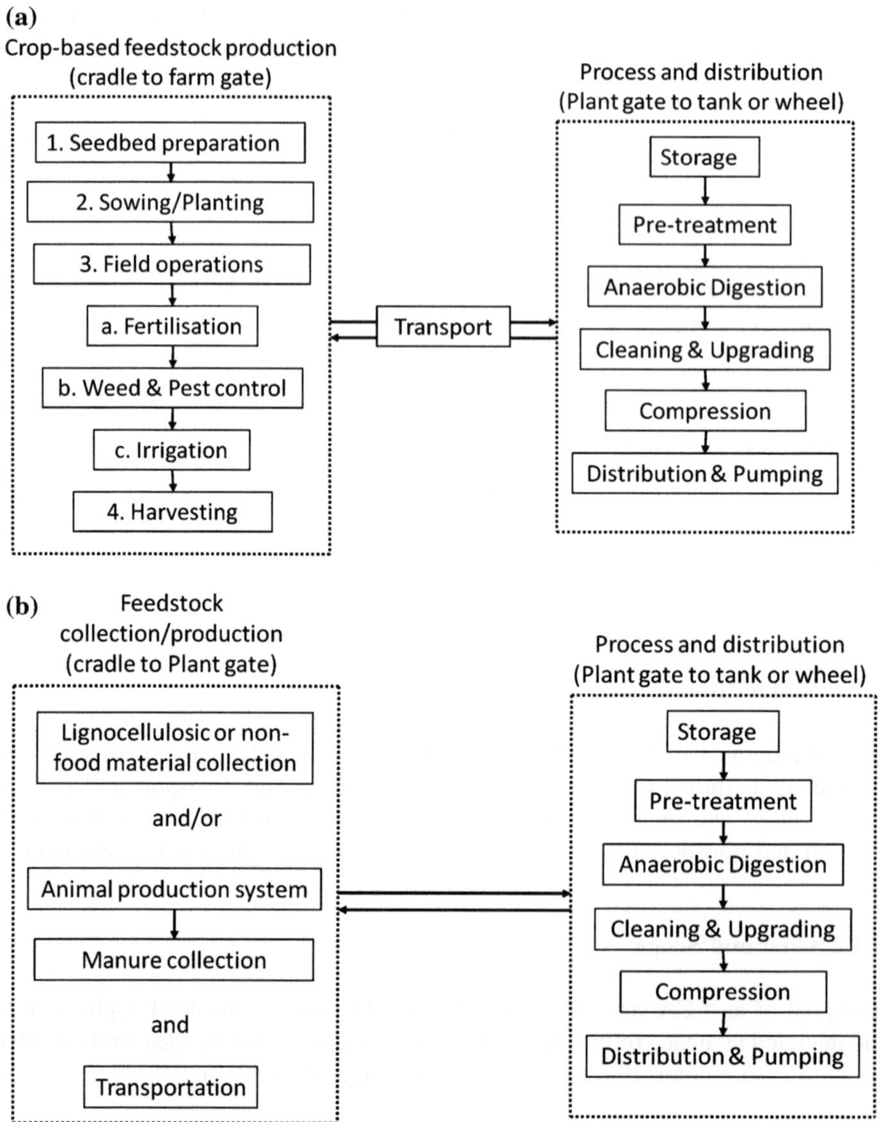

Fig. 12 Flowchart of biogas/biomethane production from **a** a crop-based, and **b** agriculture or non-food and manure feedstocks (animal-based production system is included)

to the 2009 EU Renewable Directive (EC 2009) for the evaluation of grass biomethane sustainability as a transport fuel, the environmental impact in terms of GHG emissions was expressed as g CO_2 equivalent (CO_2e) MJ^{-1} energy replaced. Borjesson et al. (2011) in their LCA of biofuels in Sweden including biogas from organic waste, manure, and crops used "environmental impact per MJ fuel" as FU.

5.1.4 Reference System

A typical objective of LCA is to discover essential differences in potential environmental impacts between two alternative systems fulfilling the same functions (Lindfors et al. 1999). Therefore, the choice of the reference system to which the bioenergy system is compared is critical because the estimated benefits of bioenergy based on the replacement of the assumed energy system can differ significantly depending on the chosen reference system. In the case of biofuels, diesel production is used as reference system.

5.1.5 Life Cycle Inventory

The same process as it has been described in Korres (2013) should be followed. Very briefly, the preparation of data collection and data collection spreadsheet followed by data collection and data validation by which data are related to FU are the major first steps of the LCI. According to Fava et al. (1994), the major LCA inventory stages to be considered while collecting data for LCI development and which are generally applicable to biogas production include activities such as raw materials (feedstock) acquisition and energy consumption; manufacturing, processing and formulation; distribution and transportation; use/re-use/maintenance; recycling; and waste management.

5.1.6 Energy Consumed in AD Plant

The energy consumed in the AD plant for the production of biogas can be calculated based on the Eq. (9).

$$E_{AD} = E_{\text{Direct}} + E_{\text{Indirect}} = E_H + E_{Mc} + E_{MX} + E_P + E_L \quad (9)$$

where E_H energy (direct) required to heat the digester up to an operational temperature; E_{Mc} energy for feedstock pre-treatment (e.g., macerating); E_{MX} electrical energy for feedstock mixing during digestion process; E_P electrical energy necessary for water pumping or, as in the case of continuously stirred tank reactor, for recirculation of liquid from one digester to the other; E_L energy loss. A detailed analysis and calculations of the net energy (and related emissions) in bioreactors which involve the thermal and the electrical energy necessary to run the bioreactor are reported in anonymous (2007); Smyth et al. (2009); Ruggeri et al. (2010), and Korres et al. (2010). Very briefly, the outcome of a GHG LCA on grass biomethane (produced by anaerobic digestion and used as a transport fuel in place of diesel) conducted by Korres et al. (2010) is presented in Table 2. The largest contributors were emissions from the anaerobic digestion process followed by crop production. Indirect emissions from the production of nitrogen and potassium fertilizers were the major contributors to agricultural emissions, and, in the

Table 2 Summary of GHG emissions (base-case scenario) from grass biomethane

	Emissions (g CO_2e MJ^{-1} energy replaced)	Emissions (kg CO_2e $ha^{-1}year^{-1}$)
Feedstock production		
Crop production	9.01	893
Herbicide volatilization	0.05	5.44
Lime dissolution	5.55	550
N2O emissions	5.29	525
Total agricultural emissions	19.90	1,973
Transportation	0.89	88
Biomethane production process		
Anaerobic digestion plant	25.49	2,524
Upgrading	12.64	1,251
Total processing emissions	38.13	3,775
Biogas losses	10.82	1,071
Total	69.74	6,904

Note Based on Korres et al. (2010)

biomethane production process, the largest source of emissions was from digester heating. When compared with emissions from fossil diesel grass biomethane production under the base-case scenario, which includes the production of grass silage and transportation of feedstock to anaerobic digestion plant and digestate back to field, GHG emissions savings were estimated to 21.5%.

Nevertheless, cumulative GHG emissions savings under various sensitivity analysis scenarios resulted in GHG emissions savings of up to 89.4% (Korres et al. 2011).

5.2 LCA Bioethanol Production from Lignocellulosic and Non-Food Feedstocks

Much of the analysis for bioethanol production has focused on the outcome of net energy during its production (Shapouri et al. 2003; Murphy and Power 2008) (Fig. 13).

Figure 13 summarizes the results of several studies on fossil energy balances for different types of bioethanol fuel (and conventional gasoline and diesel) in which as wide variation is revealed concerning the estimated energy balances across different feedstocks mainly depending on factors such as feedstock productivity, agricultural practices, and conversion technologies (FAO 2008). Conventional petrol and diesel have fossil energy balances of around 0.8–0.9, because some energy is consumed in refining crude oil into usable fuel and transporting it to markets. If a biofuel has a fossil energy balance exceeding these numbers, it

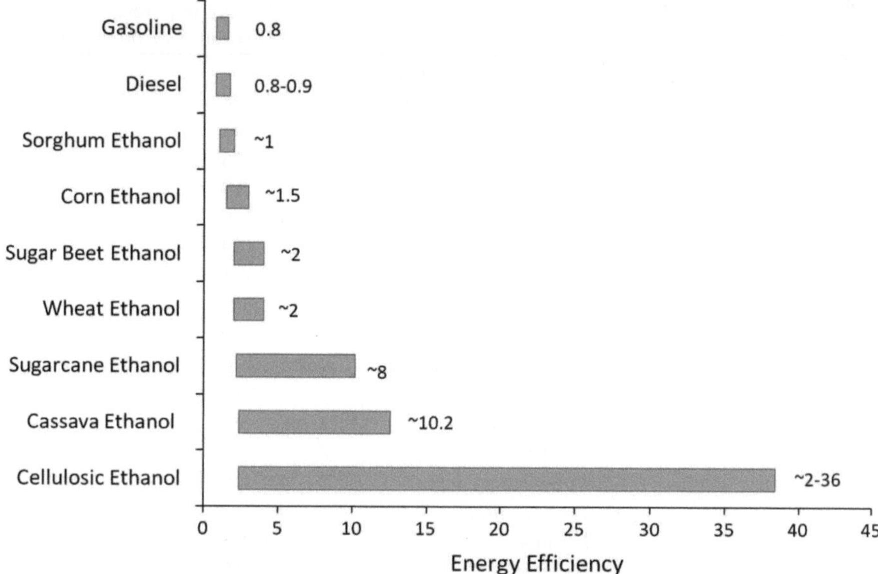

Fig. 13 Estimated ranges of fossil (*gasoline* and *diesel*) energy balances and selected bioethanol types (based on Worldwatch Institute 2006; Rajagopal and Zilberman 2007; FAO 2008; Earley and McKeown 2009)

contributes to reducing dependence on fossil fuels. Albeit to widely varying degrees, all bioethanol types are making a positive contribution in this regard. It is worth to mention that the favorable fossil energy balance of sugarcane-based ethanol depends not only on feedstock productivity but also on the fact that it is processed using biomass residues from the sugarcane (bagasse) as energy input (FAO 2008). Table 3 displays in detail the energy balance breaking down to its secondary components for sugarcane bioethanol.

The range of estimated fossil fuel balances for cellulosic feedstocks is even wider, reflecting the uncertainty regarding this technology and the diversity of potential feedstocks and production systems. Similarly, the net effect of biofuels on GHG emissions may differ widely. Nevertheless, as stated by Farrell et al. (2006), energy ratios are sensitive to specification and assumptions and thus can produce uninterpretable values.

5.2.1 Feedstocks

The actual potential to produce cellulosic ethanol as mentioned above is multi-faceted. Because large-scale production, transport, processing, and conversion of cellulosic materials have not been attempted to any real degree anywhere in the world a number of sustainability issues related to energy inputs and environmental quality need to be examined in conjunction with production, harvest, and

Table 3 Energy balance based on average values of sugarcane to ethanol

	Energy requirement ($MJ\ ton^{-1}$ of processed cane)
Feedstock production	
Agricultural operations	38
Transportation	43
Fertilizers	66
Lime, herbicides, etc.	19
Seeds	6
Equipment	29
Total for production	201
Process (ethanol production)	
Electricity	0
Chemicals and lubricants	6
Building	12
Equipment	31
Total for process	49
Total energy input	250
Energy output	
Ethanol	1,921
Bagasse	169
Total energy output	2,090
Net energy balance (out/in)	8.4

Based on Macedo et al. (2003)

collection. For this reason, as Singh et al. (2010) stated, the main goal for LCA of lignocellulosic ethanol should be to evaluate the environmental impacts of the system under examination and to quantify the ecological benefits from the replacement of the conventional or reference system. It may also provide a tool for policy makers and consumers to determine the optimum eco-friendly fuel. The FU, depending on the goal of the study, must be expressed in terms of per unit output (kWh or km) basis. For transport services, the FU ought to be expressed in "per km distance travelled" and should not be expressed in "unit energy at fuel tank" due to variations of mechanical efficiency between different fuels and type of engine (Gnansounou et al. 2009; Murphy and Power 2009).

5.2.2 System Boundaries

Inconsistency of system boundaries in LCA analysis of lignocellulosic ethanol system through omission of the production of various inputs (e.g., thermochemical or biochemical approach for degradation of cellulosic feedstock, fertilizer, pesticides, and lime) along with bioethamol utilization (Luo et al. 2009; Gnansounou et al. 2009) could cause a significant variation on the outcome of the analysis. Table 4 represents clearly the similarities and differences between second (lignocellulosic) bioethanol and bioethanol produced by grain crops.

Table 4 Characterization of bioethanol production system boundaries based on several characteristics through production chain (IEA 2007)

Feedstock	Harvest technique	Feedstock conversion	Process heat	Sugar conversion to alcohol	Coproducts
Sugar crops (e.g., cane)	Cane stalk cut, mostly taken from field	Sugars extracted through baggase-crashing, soaking chemical treatment	Primarily from crashed cane (baggase)	Fermentation and distillation of alcohol	Heat, electricity and molasses
Grain crops (wheat, corn)	Starchy parts of plants harvested; stalks mostly left in the field	Starch separation, milling, conversion to sugars via enzyme application	Typically from fossil fuel	Fermentation and distillation of alcohol	Animal feed (e.g., DDG sweetener (if corn feedstock)
Cellulosic crops	Full plant harvested, grasses cut with regrowth	Cellulose conversion to sugar via saccharification (enzymatic hydrolysis); lignin use for process energy	Lignin and excess cellulose	Fermentation and distillation of alcohol	Heat, electricity, animal feed, bioplastics, etc.
Waste biomass (crop residue, forestry waste, municipal waste, mill waste, etc.)	Waste materials gathered, separated, cleaned to extract materials high in cellulose)	Cellulose conversion to sugar via saccharification (enzymatic hydrolysis); lignin use for process energy	Lignin and excess cellulose	Fermentation and distillation of alcohol	Heat, electricity, animal feed, bioplastics, etc.

If the LCA aims to compare biofuels with their fossil substitutes (i.e., gasoline), the utilization stage is crucial (Singh et al. 2010); the final energy produced from tank for a given end use (transport/heat/electricity) depends on the combustion performances of that engine using that fuel (Gnansounou et al. 2009; Murphy and Power 2009). Many researchers use the "well-to-tank" system boundary to compare environmental impact of biofuels with fossil fuels while many others use "well-to-wheel" or "cradle-to-grave" system (Singh et al. 2010).

5.2.3 Processing and Conversion

Current technologies for ethanol production from lignocellulosic material are based on chemical or enzymatic conversion of the substrate to fermentable sugars followed by fermentation process using a microorganism (Xiros and Christakopoulos 2009). However, enzymatic hydrolysis of lignocellulosic biomass without pretreatment is usually not so efficient due to the high resistance of the materials to microbial degradation (Taherzadeh and Karimi 2008).

In addition to biochemical hydrolysis (i.e., enzymatic or chemical/acidic hydrolysis) (Fig. 5 and Table 2), there is also the thermochemical approach to second-generation bioethanol production. Both of these approaches can be used to produce a wide variety of fuels. In the biochemical approach, enzymes (biological catalysts, usually obtained from microorganisms) or acid is used to break down cellulosic materials to sugars that are then fermented into alcohols (such as ethanol) by microorganisms. These are separated through distillation. In the thermochemical approach, heat, pressure, chemical catalysts, and water are used to break down biomass in much the same way that petroleum is refined. Thermochemical technologies include gasification, fast pyrolysis, and hydrothermic processing. These technologies can be used to convert almost any kind of biomass into fuel, from grass to turkey feathers, giving them a potential advantage over biochemical technologies that rely on developing specific enzymes to break down specific plant matter (Bransby 2007; Lange 2007).

5.2.4 GHG Emissions

GHG emissions and savings are the center of attention in most LCA studies in comparison to a reference system (Gnansounou et al. 2009; Liska et al. 2009) along with other midpoint impacts such as eutrophication, acidification, ozone depletion. Nevertheless, very few studies have considered these impacts since they are site specific, thus limiting generalization of the results and pollution shifting phenomena (Cherubini et al. 2009).

6 Conclusions

Certain restrictions in space and time did not permit a thorough approach to this important topic, the application of LCA onto a mixed production system. The author of this chapter choose to use the general framework as it has been described in ISO 14040 series and to explain the most important issues a scholar in this type of analysis but also other stakeholders can use without any difficulty. Nevertheless, LCA is a tool and a piece of art rather than a pure scientific working protocol. Of course, it should be based on solid scientific evidence, but the way a practitioner can bring all the evidence collected during LCI stage together requires deep knowledge of the system under examination, imagination, and a system dynamics thinking. I hope I have passed this message through.

The form of energy is considered within a LCA in agricultural systems depends on the goal of the study. The inherited complexity of agricultural systems necessitates the thorough study of the system under examination. Nevertheless, representativeness, consistency and accuracy of the analysis should not be put under question, particularly in the case of biofuels. This is because environmental and socioeconomic decisions can be made based on the results of LCA. For a useful discussion, it is necessary to declare explicitly the energy form considered but also, all energy sources should be documented separately. Especially, the consideration of whether and how to incorporate biomass energy depends on the subject and the goal of the study. The possible variation of total energy input and its impact on energy intensity and energy efficiency should be illustrated by scenarios in compliance with the specific questions to be answered. Relevant assumptions and system boundaries always have to be documented for interpretation of the results. Renouf et al. (2008) stated that many of the dominant environmental impacts from cropping systems result from dynamic systems within agricultural soils. The use of agricultural modeling techniques to model these processes can be proved beneficial to LCA's cause. Water, one of the most important natural resources, has received little attention in the LCA literature. It should be extremely useful if water footprint could be incorporated with LCA or vice versa.

References

Aines G, Klopfenstein t, Stock R (1986) "Distillers Grains" Nebraska cooperative extension MP51. University of Nebraska, Lincoln

Anonymous (2007) Energy balance optimisation for an integrated arable/livestock farm unit. Cropgen, Renewable energy from crops and agrowastes, SES6- CT-2004-502824, University of Vienna (BOKU-IFA)

Anonymous (2008) Forests and energy-key issues. FAO Forestry Paper 154, Food and Agriculture Organization of the United Nations, Rome 2008

Balat M, Balat H, Oz C (2008) Progress in bioethanol processing. Prog Energy Combust Sci 34(5):551–573

Bare JC, Norris GA, Pennington DW et al (2003) TRACI-The tool for the reduction and assessment of chemical and other environmental impacts. J Ind Ecol 6(3):49–78

Basset-Mens C, Ledgard S, Boyes M (2009) Eco-efficiency of intensification scenarios for milk production in New Zealand. Ecol Econ 68(6):1615–1625

Bata RK, Bhonot J (eds) (2011) Energy demand-agriculture. Teri energy data directory and yearbook 2010. The Energy and Resources Institute, 2011, Teri Press, New Delhi

Baumann H, Tillman AM (2004) The hitch hiker's guide to LCA. Studentlitteratur, Lund, Sweden

Beno Z, Boran J, Houdkova L, Dlabaja T, Sponar J (2009) Cofermentation of kitchen waste with sewage sludge. Chem Eng Trans 18:677–682

Benzie JAH, Hynes S (2013) Suitability of microalgae and seaweeds for biomethane production. In: Korres NE, O'Kiely P, Benzie JAH, West JS (2013) (eds) Bioenergy production by anaerobic digestion. Using agricultural biomass and organic waste, Pubs Earthscan from Routledge, London

Bernet N, Beline F (2009) Challenges and innovations on biological treatment of livestock effluents. Bioresour Technol 100:5431–5436

Board (Biomass Research and Development Board) (2009). Increasing feedstock production for biofuels: economic drivers, environmental implications, and the role of research. Biomass Research and Development Initiative, http://www.esd.ornl.gov/eess/IncreasingBiofuelsFeedstockProduction.pdf. Accessed Feb 2013

Bogdanski A, Dubois O, Jamieson C, Krell R (2010) Making integrated food- energy systems work for people and climate: an overview. Environment and natural resources management paper 45, environment climate change [bioenergy] monitoring and assessment, food and agriculture organization of the United Nations, Rome 2010

Borjesson P, Tufvesson L, Lantz M (2011) Life cycle assessment of biofuels in Sweden. Lund University, Department of Technology and Society, Environmental and energy systems studies, report no. 70 May 2010

BP (2012) BP statistical review of world energy June 2012. www.bp.com/statisticalreview. Accessed Dec 2012

Bransby D (2007) Cellulosic biofuel technologies, a report sponsored by the Alabama department of economic and community affairs through the southeast biomass state and regional partnership. Auburn University, Auburn, February 2007

Brennan L, Owende P (2010) Biofuels from microalgae: a review of technologies for production, processing, and extractions of biofuels and co-products. Renew Sustain Energy Rev 14:557–577

Bruton T, Lyons H, Lerat Y, Stanley M (2009) A review of the potential of marine algae as a source of biofuel in Ireland. Sustainable Energy Ireland

Carlsson-Kanyama A (1998) Climate change and dietary choices-how can emissions of greenhouse gases from food consumption be reduced? Food Policy 23:277–293

Casey JW, Holden NM (2005) Analysis of greenhouse gas emissions from the average Irish milk production system. Agric Syst 86:97–114

Casey JW, Holden NM (2006) Greenhouse gas emissions from conventional, agri-environmental scheme and organic Irish suckler-beef units. J Environ Qual 35:231–239

Cederberg C, Mattsson B (2000) Life cycle assessment of milk production-a comparison of conventional and organic farming. J Cleaner Prod 8:49–60

Cederberg C, Stadig M (2003) System expansion and allocation in life cycle assessment of milk and beef production. Int J LCA 8:350–356

Chambers B, Taylor M (2013) The use of digestate as a substitute for manufactured fertiliser. In: Korres NE, O'Kiely P, Benzie JAH, West JS (eds) Bioenergy production by anaerobic digestion. Using agricultural biomass and organic waste, Pubs Earthscan from Routledge

Cherubini F, Ulgiati S (2010) Crop residues as raw materials for biorefinery systems: a LCA case study. Appl Energy 87:47–57

Cherubini F, Bird ND, Cowie A, Jungmeier G, Schlamadinger B, Woess-Gallasch S (2009) Energy- and greenhouse gas-based LCA of biofuel and bioenergy systems: key issues, ranges and recommendations. Resour Conserv Recycl 53(8):434–447

Cowell SJ (1999) Use of environmental life cycle assessment to evaluate alternative agricultural production systems. In: Proceedings of the 52nd N. Z. plant protection conference 1999, pp 40–44

Cropgen (2004a) D22: energy balance optimisation for an integrated arable/livestock farm unit. University of Jyvaskyla. http://www.cropgen.soton.ac.uk/deliverables.htm. Accessed Jan 2013

Cropgen (2004b) D25: life cycle energy balances on a number of crop species. University of Vienna (BOKU-IFA). http://www.cropgen.soton.ac.uk/deliverables.htm. Accessed Jan 2013

Dahllof L (2005) LCA methodology issues for textile products. Masters Thesis, Technical report no. 2004:8. Chalmers University of Technology, Goteborg, Sweden

Dalgaard T, Halberg N, Porter JR (2001) A model for fossil energy use in Danish agriculture used to compare organic and conventional farming. Agric Ecosyst Environ 87(1):51–65

Dearman B, Bentham RH (2007) Anaerobic digestion of food waste: comparing leachate exchange rates in sequential batch systems digesting food waste and biosolids. Waste Manage (Oxford) 27:1792–1799

Delucchi M A (2006) Lifecycle analysis of biofuels. Report UCD-ITS-RR-06-08. Institute of Transportation Studies, University of California, Davis. www.its.ucdavis.edu/people/faculty/delucchi. Accessed Jan 2013

Demirbas A (2008) Products from lignocellulosic materials via degradation processes. Energy Sources: Part A 30:27–37

Demirbas MF (2009) Biorefineries for biofuel upgrading: a critical review. Appl Energy 86:S151–S161

Di Nicola G, Santecchia E, Santori G, Polonara F (2011) Advances in the development of bioethanol: a review. In: Aurelio M, Bernardes DS (eds) Biofuel's engineering process technology. Intecheweb.org Published by InTech, Rijeka, Croatia

U.S. DOE (2006) Breaking the biological barriers to cellulosic ethanol: a joint research agenda, DOE/SC-0095, U.S. Department of Energy Office of Science and Office of Energy Efficiency and Renewable Energy. www.doegenomestolife.org/biofuels. Accessed Ja 2013

U.S. DOE (2010) National algal biofuels technology roadmap. U.S., Department of Energy, Office of Energy Efficiency and Renewable Energy, Biomass Program. http://biomass.energy.gov. Accessed Sep 2012

Earley J, McKeown A (2009) Smart choices for biofuels. Worldwatch Institute and the Sierra Club, Washington

EC (2003) Council regulation EC/1782/2003 for establishing common rules for direct support schemes under the common agricultural policy and establishing certain support schemes for farmers. Official J Eur Union L270/1-L270/69

EC (2009) Directive 2009/28/EC of the European parliament and of the council of 23 April 2009 on the promotion of the use of energy from renewable sources and amending and subsequently repealing, directives 2001/77/EC and 2003/30/EC. Official J Eur Union

EC (2010) The international reference life cycle data system (ILCD) handbook: general guide for life cycle assessment—detailed guidance. European Commission, Joint Research Centre, Publications office of the European Union, Luxembourg

EIA (2013) US energy information administration, independent statistics and analysis, countries, international energy statistics. www.eia.gov. Accessed Jan 2013

Eisentraut A (2010) Sustainable production of second generation biofuels. Information paper, International Energy Agency

Ekvall T, Finnveden G (2001) Allocation in ISO 14041: a critical review. J Cleaner Prod 9:197–208

Elsayed M, Mortimer N (2001) Carbon and energy modelling of biomass systems: conversion plant and data updates. UK Department of Trade and Industry

EPA (2010) EPA finalizes regulations for the national renewable fuel standard program for 2010 and beyond. Technical report EPA420-F-10-007, United States Environmental Protection Agency, Office of Transportation and Air Quality. http://www.epa.gov/otaq/renewablefuels/420f10007.pdf. Accessed Nov 2012

EUBIA (2012) EC Policy-Overview of new directives. Eur Biomass Ind Assoc Newslett 12/12. www.eubia.org. Accessed Dec 2012

European Biofuels Technology Platform (undated). Cellulosic ethanol. http://www.biofuelstp.eu/cell_ethanol.html#ce1. Accessed Jan 2013

Eze JI, Ojike O (2012) Anaerobic production of biogas from maize wastes. Int J Phys Sci 7(6):982–987

FAO (Food and Agriculture Organization) (2008) The state of food and agriculture. Food and Agriculture Organization of the United Nations, Rome 2008

Farrell AE, Pelvin RJ, Turner BT, Jones AD, O'Hare M, Kammen DM (2006) Ethanol can contribute to energy and environmental goals. Science 311:506–508

Fava JA, Denison R, Jones B, Curran MA, Vigon B, Selke S, Barnum J (eds) (1994) A technical framework for life cycle assessment. Pubs Society of Environmental Toxicology and Chemistry (SETAC) and SETAC Foundation for Environmental Education, FL, USA

Fawer M (2001) Concepts of life cycle assessments (LCA). Seminar on environmental tools-a competitive option. EMPA, St. Gallen

Ferraro LA (1999) Proposição de método de avaliação de sistemas de produção e de sustentabilidade. MSc Thesis, USP/ESALQ, Piracicaba

Firestone MK, Davidson EA (1989) Microbiological basis of NO and N_2O production and consumption in soil. In: Andreae MO, Schimel DS (eds) Exchange of trace gases between terrestrial ecosystems and the atmosphere. Wiley, New York, pp 7–21

Fountoulakis MS, Manios T (2009) Enhanced methane and hydrogen production from municipal solid waste and agro-industrial by-products co-digested with crude glycerol. Bioresour Technol 100(12):3043–3047

Frankl P, Rubik F (eds) (2000) Life cycle assessments in industry and business, adoption patterns, applications and implications. Springer, Berlin, Heidelberg, New York

Frischknecht R, Jungbluth N (eds) (2007) Overview and methodology. Data v2.0 (2007). Ecoinvent report no. 1, Dubendorf, December 2007

Gerin PA, Vliegen F, Jossart JM (2008) Energy and CO_2 balance of maize and grass as energy crops for anaerobic digestion. Bioresour Technol 99(7):2620–2627

Gnansounou E, Dauriat A, Villegas J, Panichelli L (2009) Life cycle assessment of biofuels: energy and greenhouse gas balances. Bioresour Technol 100(21):4919–4930

Grant T, Beer T, Campbell PK, Batten D (2008) Life cycle assessment of environmental outcomes and greenhouse gas emissions from biofuels production in Western Australia. Report KN29A/WA/F2.9 to The Department of Agriculture and Food Government of Western Australia, September 2008

Griffith D, Parsons S (1983) Energy requirements for various tillage-planting systems. Purdue University Cooperative Extension Service, NRC-202. Purdue University, West Lafayette. http://www.ces.purdue.edu/extmedia/NCR/NCR-202-W.html. Accessed Jan 2013

Haas G, Wetterich F, Geier U (2000) Life cycle assessment framework in agriculture on the farm level. Int J LCA 5(6):345–348

Haas G, Wetterich F, Kopke U (2001) Comparing intensive, extensified and organic grassland farming in southern Germany by process life cycle assessment. Agr Ecosyst Environ 83:43–53

Harris S, Narayanaswamy V (2009) A literature review of life cycle assessment in agriculture. RIRDC publication no. 09/029, RIRDC project no. PRJ-002940, March 2009, Rural Industries Research and Development Corporation

Hayashi K, Gaillard G, Nemecek T (2005) Life cycle assessment of agricultural production systems: current issues and future perspectives. In: International seminar on technology development for good agriculture practice in Asia and Oceania, Epochal Tsukuba, 25–26 Oct 2005

Heinberg R, Bomford M (2009) The food and farming transition: toward a post-carbon food system. Post Carbon Institute and The Soil Association. www.postcarbon.org/food. Accessed Jan 2013

Hendriks ATWM, Zeeman G (2009) Pretreatments to enhance the digestibility of lignocellulosic biomass. Bioresour Technol 100(1):10–18

Hetz JH (1992) Energy utilization in Chilean agriculture. Agric Mech Asia Afr Lat Am 23:52–56

Holmes J (2006) Impulses towards a multifunctional transition in rural Australia: gaps in the research agenda. J Rural Stud 22:142–160

Holm-Nielsen JB, Oleskowicz-Popiel P (2008) The future of biogas in Europe: visions and targets until 2020. In: Biogas: a promising renewable energy source for Europe. AEBIOM Workshop: European Parliament Brussels, 11 Dec 2008

Holm-Nielsen JB, Al Seadi T, Oleskowicz-Popiel P (2009) The future of anaerobic digestion and biogas utilization. Bioresour Technol 100(22):5478–5484

Huggins DR, Buyanovsky GA, Wagner GH, Brown JR, Darmody RG, Peck TR, Lesoing GW, Vanotti MB, Bundy LG (1998) Soil organic C in the tallgrass prairie-derived region of the Corn Belt: effects of long-term crop management. Soil Tillage Res 47:219–234

IEA (International Energy Agency) (2007) Biofuels for transport: an international perspective

IPCC (2001) Climate change 2001: synthesis report. In: Watson RT et al. (eds) A contribution of working groups I, II, and III to the third assessment report of theintegovernmental panel on climate change. Cambridge University Press, Cambridge, United Kingdom, New York

ISO (1997) International organization of standardization. 1997 Environmental management—life cycle assessment—principles and framework (ISO/FDIS 14040), ISO TC 207

ISO (2006) ISO Norm 14044:2006 international standard. In: Environmental management—life CYCLE assessment—requirements and guidelines. International Organisation for Standardisation, Geneva

Jorgensen H, Kristensen JB, Felby C (2007) Enzymatic conversion of lignocellulose into fermentable sugars: challenges and opportunities. Biofuels Bioprod Biorefin 1:119–134

Kammen DM (2007) Transportation's next big thing is already here. May, GreenBiz.com, Climate Wise. URL: http://www.greenbiz.com/news/columns_third.cfm?NewsID=35189. Accessed Jan 2013

Kammen DM, Farrell AE, Plevin RJ, Jones AD, Nemet GF, Delucchi MA (2007) Energy and greenhouse impacts of biofuels: a framework for analysis. OECD Research Round Table: Biofuels: Linking support to performance. Version 7 Sep 2007

Karlsson A (2003) Comparative assessment of fuel-based systems for space heating. Division of Environmental and Energy Studies, Lund University, Lund

Kelm M, Wachendorf M, Trott H, Volkers K, Taube F (2004) Performance and environmental effects of forage production on sandy soils. III. Energy efficiency in forage production from grassland and maize for silage. Grass Forage Sci 59(1):69–79

Kim S, Dale BE (2004) Global potential bioethanol production from wasted crops and crop residues. Biomass Bioenergy 26:361–375

Kim KH, Tucker MP, Nguyen QA (2002) Effects of pressing lignocellulosic biomass on sugar yield in two-stage dilute-acid hydrolysis process. Biotechnol Prog 18(3):489–494

Korres NE (2005) Encyclopaedic Dictionary of Weed Science: theory and Digest. Pub Lavoisier SAS, Intercept Ltd. ISBN: 1-898298-99-8

Korres NE (2013) Life cycle assessment as a tool for assessing biomethane production sustainability. In: Korres NE, O'Kiely P, Benzie JAH, West JS (eds) Bioenergy production by anaerobic digestion. Using agricultural biomass and organic waste, Pubs Earthscan from Routledge

Korres NE, Nizami AS (2013) Variation in anaerobic digestion. Need for process monitoring. In: Korres NE, O'Kiely P, Benzie JAH, West JS (eds) Bioenergy production by anaerobic digestion. Using agricultural biomass and organic waste, Pubs Earthscan from Routledge

Korres NE, Singh A, Nizami AS, Murphy JD (2010) Is grass biomethane a sustainable transport biofuel? Biofuels, Bioprod Biorefin 4(3):310–325

Korres NE, Thamsiriroj T, Smyth B, Nizami AS, Singh A, Murphy JD (2011) Grass biomethane for agriculture and energy. In: Lichtfouse E (ed) Genetics, biofuels and local farming systems, sustainable agriculture reviews, vol 7. Springer Science & Business Media, pp 5–50

Korres NE, O'Kiely P, Benzie JAH, West JS (2013) Bioenergy Production by Anaerobic Digestion. Using Agricultural Biomass and Organic Waste. Pub Earthscan from Routledge/ Taylor and Francis Publishing Group. ISBN 978-0-415-69840-5

Labutong N, Mosley J, Smith R, Willard J (2012) Life-cycle modelling and environmental impact assessment of commercial scale biogas production. MSc Thesis, University of Michigan, School of Natural Resources and Environment

Lal R (2004) Carbon emission from farm operations. Environ Int 30:981–990

Lange JP (2007) Lignocellulose conversion: an Introduction to chemistry, process and economics. Biofuels, Bioprod Biorefin 1:39–48

Larson D, Fangmeier D (1978) Energy in irrigated crop production. Trans Am Soc Agric Eng 21:1075–1080

Lau MH, Richardson JW, Outlaw JL, Holtzapple MT, Ochoa RF (2006) The economics of ethanol from sweet sorghum using the MixAlco process. AFPC research report 06-2, 11 Aug, Texas A&M University. http://www.afpc.tamu.edu/pubs/2/446/RR%2006-2.pdf. Accessed Dec 2012

Lindfors LG, Christiansen K, Hoffman L, Kruger I, Virtanen Y, Juntilla V, Hanssen OJ, Ronning A (1999) Nordic guidelines on life-cycle assessment: nord 1995:20, 3rd edn. Nordic Council of Ministers, Copenhagen

Liska AJ, Yang HS, Bremer VR, Klopfenstein TJ, Walters DT, Erickson GE, Cassman KG (2009) Improvements in life cycle energy efficiency and greenhouse gas emissions of corn-ethanol. J Ind Ecol 13(1):58–74

Loerincik Y, Tatti E, Belloto S, Jolliet O, Mousset J, Lepochat S (2008) Results of a literature review: life cycle assessment of agriculture production. In: 6th International conference on LCA in the Agri-Food Sector, Zurich, 12–14 Nov 2008

Lukehurst CT, Frost P, Seadi T Al (2010) Utilisation of digestate from biogas plants as biofertiliser. IEA Bioenergy, Task 37

Luo L, van der Voet E, Huppes G, de Haes Udo HA (2009) Allocation issues in LCA methodology: a case study of corn stover-based fuel ethanol. Int J Life Cycle Assess 14:529–539

Macedo IC, Leal MRLV, Silva JEAR (2003). Greenhouse gas emissions in the production and use of ethanol in Brazil: Present situation (2002). Prepared for the Secretariat of the Environment, Sao Paulo

Marsden T (1999) Rural futures: the consumption countryside and its regulation. Sociologia Ruralis 39:501–520

Marsden T, Sonnino R (2008) Rural development and the regional state: denying multifunctional agriculture in the UK. J Rural Stud 24(4):422–431

Mathiasson A (2008) Vehicle gas utilization in Sweden: today and tomorrow. In: 2nd Nordic biogas conference. The Swedish Gas Association, Malmö

Matlock M, Thoma G, Nutter D, Costello T (2008) Energy use life cycle assessment for global cotton production practices. Final Report, 15 March 2008, Center for Agricultural and Rural Sustainability University of Arkansas Division of Agriculture

McCarty PL (1982) One hundred years of anaerobic treatment. In: Hughes DE, Stafford DA, Wheatley BI, Baader W, Lettinga G, Nyns EJ, Verstraete W, Wentworth RL (eds) Anaerobic Digestion, 1981. Elsevier Biomedical Press B.V., Amsterdam, pp 3–21

McEniry J, Korres NE, O'Kiely P (2013) Grass and grass silage: agronomical characteristics and biogas production. In: Korres NE, O'Kiely P, Benzie JAH, West JS (eds) Bioenergy production by anaerobic digestion. Using agricultural biomass and organic waste, Pubs Earthscan from Routledge

Moschini G, Cui J, Lapan H (2012) Economics of biofuels: an overview of policies, impacts and prospects. Paper prepared for presentation at the 1st AIEAA conference "Towards a Sustainable Bio-economy: Economic Issues and Policy Challenges", Trento, 4–5 June 2012

Mummey D, Smith J, Bluhm G (1998) Assessment of alternative soil management practices on N_2O emissions from US agriculture. Agric Ecosyst Environ 70:79–87
Murphy JD, Power N (2008) How can we improve the energy balance of ethanol production from wheat? Fuel 87:1799–1806
Murphy JD, Power N (2009) An argument for using biomethane generated from grass as a biofuel in Ireland. Biomass Bioenergy 33:504–512
National Sorghum Producers (NSP) (2008) www.sorghumgrowers.com. Accessed February2013
National Sorghum Producers (NSP) (2012) www.sorghumgrowers.com. Accessed Feb 2013
Nelson R (2007) Cellulosic ethanol/bioethanol in Kansas. Background Report Prepared for the Kansas Energy Council Biomass Committee, 15 May 2007
Neureiter M (2013) Maize and maize silage for biomethane production. In: Korres NE, O'Kiely P, Benzie JAH, West JS (eds) Bioenergy production by anaerobic digestion. Using agricultural biomass and organic waste, Pubs Earthscan from Routledge
Ni Y, Sun Z (2009) Recent progress on industrial fermentative production of acetone-butanol-ethanol by *Clostridium acetobutylicum* in China. Appl Microbiol Biotechnol 83:415–423
Nicol R, Sage C (eds) (2003) Evaluation of the environmental performance of the Australian dairy processing industry using life cycle assessment. Final report, Dairy Australia, June 2003
Nilsson M (2001) LCA och MKB av produktion av biogas. KTH
Nuri A, Keskin T, Yuruyen A (2008) Enhancement of biogas production from olive mill effluent (OME) by co-digestion. Biomass Bioenergy 32(12):1195–1201
Oren M, Ozturk H (2006) Cotton production in south-eastern Anatolia region of Turkey. J Sustain Agric 21(9):119–130
Ortner M, Drosg B, Stoyanova E, Bochmann G (2013) Industrial residues for biomethane production. In: Korres NE, O'Kiely P, Benzie JAH, West JS (eds) Bioenergy production by anaerobic digestion. Using agricultural biomass and organic waste, Pubs Earthscan from Routledge
Owen JW (2001) Water resources in life-cycle impact assessment. Considerations in choosing category indicators. J Ind Ecol 5(2):37–54
Parawira W, Read JS, Mattiasson B, Bjornsson L (2008) Energy production from agricultural residues: high methane yields in pilot-scale two-stage anaerobic digestion. Biomass Bioenergy 32(1):44–50
Paustian K, Collins HP, Paul EA (1997) Management controls on soil carbon. In: Paul EA, Paustian K, Elliott ET, Cole CV (eds) Soil Organic matter in temperate agroecosystems: long-term experiments in North America. CRC Press, Boca Raton, pp 15–49
Pellizzi G (1992) Use of energy and labour in Italian agriculture. J Agric Eng Res 52:111–119
Perley C (2008) The status and prospects for forestry as a source of bioenergy in Asia and the Pacific. FAO Regional Office for Asia and the Pacific, Bangkok, Thailand
Pimentel D (1980) Handbook of energy utilization in agriculture. CRC Press, Boca Raton
Pimentel D (1984) Energy flow in the food system. In: Pimentel D, Hall CW (eds) Food and energy resources. Academic Press, Orlando
Pimentel D, Pimentel M (2003) Sustainability of meat-based and plant-based diets and the environment. Am J Clin Nutr 78:660S–663S
Pitkanen J, Aristidou A, Salusjarvi L, Ruohonen L, Penttila M (2003) Metabolic flux analysis of xylose metabolism in recombinant *Saccharomyces cerevisiae* using continuous culture. Metab Eng 5:16–31
Prasad S, Singh A, Joshi HC (2007) Ethanol as an alternative fuel from agricultural, industrial and urban residues. Resour Conserv Recycl 50:1–39
Quintero JA, Montoya MI, Sánchez OJ, Giraldo OH, Cardona CA (2008) Fuel ethanol production from sugarcane and corn: comparative analysis for a Colombian case. Energy 33(3):385–399
Rajagopal D, Zilberman D (2007) Review of environmental, economic and policy aspects of biofuels. World Bank Policy Research Working Paper No. 4341. World Bank, Washington
Rao VP, Baral SS, Dey R, Mutnuri S (2010) Biogas generation potential by anaerobic digestion for sustainable energy development in India. Renew Sustain Energy Rev 14(7):2086–2094

Rasi S, Veijanen A, Rintala J (2007) Trace compounds of biogas from different biogas production plants. Energy 32:1375–1380

Rass-Hansen J, Falsig H, Jørgensen B, Christensen CH (2007) Bioethanol: fuel or feedstock? J Chem Technol Biotechnol 82:329–333

Reay DS, Grace J (2007) Carbon dioxide: importance, sources and sinks. In: Reay DS, Hewitt CN, Smith KA, Grace J (eds) Greenhouse Gas Sinks. Pub. CAB International, pp 1–10

Renewable Fuels Association (RFA) (2008) www.ethanolrfa.org. Accessed April 2008

Renouf MA, Wegner MK, Nielsen LK (2008) An environmental life cycle assessment comparing Australian sugarcane with US corn and UK sugar beet producers of sugars for fermentation. Biomass Bioenergy 32(12):1144–1155

Rogers D, Alum M (2007) Comparing irrigation energy cost. Kansas State Research and Extension Service MF-2360, Kansas State University, Manhattan, Kansas

Romanelli TL, Milan M (2005) Energy balance methodology and modelling of supplementary forage production for cattle in Brazil. Sci Agric 62(1):1–7

Rosenberger A, Kaul HP, Senn T, Aufhammer W (2001) Improving the energy balance of bioethanol production from winter cereals: the effect of crop production intensity. Appl Energy 68(1):51–67

Ruggeri B, Tommasi T, Sassi G (2010) Energy balance of dark anaerobic fermentation as a tool for sustainability analysis. Int J Hydrogen Energy 35:10202–10211

Sachs I, Silk D (1991) Final report of the food energy nexus programme of the United Nations University, 1983–1987. UNU-FEN

SAIC (2006) Life cycle assessment: principles and practice. Scientific Applications International Corporation (SAIC), report no. EPA/600/R-06/060. National Risk Management Research Laboratory, Office of Research and Development, US Environmental Protection Agency, Cincinnati, Ohio

Salassi M, Fairbanks JN (2006) The economic feasibility of ethanol production from sugar in the United States. USDA/OCE (Department of Agriculture, Office of the Chief Economist). http://www.usda.gov/oce/EthanolSugarFeasibilityReport3.pdf. Accessed Jan 2012

Salminen E, Rintala J (2002) Anaerobic digestion of organic solid poultry slaughterhouse waste: a review. Bioresour Technol 83:13–26

Sapsford R, Jupp V (2006) Data collection and analysis. The Open University, SAGE, London, California, New Delhi

Saunders C, Barber A, Taylor G (2006) Food miles-comparative energy/emissions performance of New Zealand's agriculture industry. Research report 285, Lincoln University New Zealand. http://www.lincoln.ac.nz/documents/2328_rr285_s13389.pdf. Accessed Mar 2012

Shapouri H, Duffield JA, Wang M (2003) The energy balance of corn ethanol revisited. Trans ASAE Am Soc Agric Eng 46:4, 959–968

Singh A (2013) Organic wastes for biomethane production. In: Korres NE, O'Kiely P, Benzie JAH, West JS (eds) Bioenergy production by anaerobic digestion. Using agricultural biomass and organic waste, Pubs Earthscan from Routledge

Singh A, Pant D, Korres NE, Nizami AS, Prasad S, Murphy JD (2010) Key issues in life cycle assessment of ethanol production from lignocellulosic biomass: a review. Bioresour Technol 101:5003–5012

Smith P, Martino D, Cai Z, Gwary D, Janzen H, Kumar P, McCarl B, Ogle S, ÓMara F, Rice C, Scholes B, Sirotenko O, Howden M, MacAllister T, Pan G, Romanenkov V, Schneider U, Towprayoon S (2007) Policy and technological constraints to implementation of greenhouse gas mitigation options in agriculture. Agric, Ecosyst Environ 118:6–28

Smyth BM, Murphy JD, O'Brien CM (2009) What is the energy balance of grass biomethane in Ireland and other temperate northern European climates? Renew Sustain Energy Rev 13(9):2349–2360

Society of Environmental Toxicology and Chemistry (SETAC) (1993) Guidelines for life-cycle assessment: a 'Code of Practice'. SETAC publications, Brussels

Sreejith CC, Muraleedharan C, Arun P (2013) Life cycle assessment of producer gas derived from coconut shell and its comparison with coal gas: an Indian perspective. Int J Energy Environ Eng 4:8. doi:10.1186/2251-6832-4-8
Steinfeld H, Gerber P, Wassenaar T, Castel V, Rosales M, de Haan C (2006) Livestock's long shadow: environmental issues and options. FAO, Rome 2006
Svoboda S (1995) Note on life cycle analysis. Pollution prevention in corporate strategy. National Pollution Prevention Center for Higher Education, University of Michigan, pp 1–9. http://www.umich.edu/~nppcpub/
Taherzadeh MJ, Karimi K (2008) Pretreatment of lignocellulosic wastes to improve ethanol and biogas production: a review. Int J Mol Sci 9:1621–1651
Taylor MJ, Rollett AJ, Williams DJ, Chambers BJ (2012) Digestate-a low carbon nitrogen fertiliser. In: 8th world potato congress, Edinburgh, 27–30 May 2012
Tsatsarelis C (1991) Energy requirements for cotton production in central Greece. J Agric Res 50:239–246
Udo de Haes HA, van Rooijen M (2005) Life cycle approaches. The road from analysis to practice. UNEP/ SETAC Life Cycle Initiative
Vikman P, Gustavsson L, Klang A (2004) Evaluating greenhouse gas balances and mitigation costs of bioenergy systems: a review of methodologies. Biomass-based climate change mitigation through renewable energy (BIOMITRE)—work-package 1 June 2004
Wallen A, Brandt N, Wennersten R (2004) Does Swedish consumer's choice of food influence greenhouse gas emissions? Environ Sci Policy 7:525–535
Wanjura D, Upchruch D, Mahan J, Burke J (2002) Cotton yield and applied water relationships under drip irrigation. Agric Water Manag 55(2):217–236
Weber C, Matthews HS (2008) Food-miles and the relative climate impacts of food choices in the United States. Environ Sci Technol 42:3508–3513
Weisser D (2007) A guide to life-cycle greenhouse gas (GHG) emissions from electric supply technologies. Energy 32:1543–1559
Wells C (2001) Total energy indicators of agricultural sustainability: dairy farming case study. New Zealand Ministry of Agriculture and Forestry, New Zealand
Wenisch S, Monier E (eds) (2007) Life cycle assessment of different uses of biogas from anaerobic digestion of separately collected biodegradable waste in France. Synthesis, ADEME, Gaz de France, RDC-Environment and Bio-Intelligence Service, September 2007
Wenzel H, Hauschild M, Alting L (1997) Environmental assessment of products. Methodology, tools and case studies in product development, vol 1. Chapman & Hall, London, Weinheim, New York
Wilhelm W, Johnson JMF, Hatfield JL, Voorhees WB, Linden DR (2004) Crop and soil productivity response to corn residue removal: a literature review. Agron J 96(1):1–17
Williams AG, Audsley E, Sandars DL (2006) Determining the environmental burdens and resource use in the production of agricultural and horticultural commodities. Main Report, Defra Research Project IS0205, Cranfield University and Defra, Bedford
Wilson GA (2007) Multifunctional agriculture. A transition theory perspective. CAB International, Wallingford
Woolridge AC, Ward GD, Phillips PS, Collins M, Gandy S (2006) Life cycle assessment for reuse/recycling of donated waste textiles compared to use of virgin material: an UK energy saving perspective. Resour Conserv Recycl 46(2006):94–103
World Energy Council (WEC) (2004) Comparison of energy systems using life cycle assessment. A Special Report of the World Energy Council, July 2004
Worldwatch Institute (2006) Biofuels for transportation: global potential and implications for sustainable agriculture and energy in the 21st century. Washington, DC
Wyman CE (1994) Ethanol from lignocellulosic biomass: technology, economics, and opportunities. Bioresour Technol 50:3–16
Xiros C, Christakopoulos P (2009) Enhanced ethanol production from brewer's spent grain by a *Fusarium oxysporum* consolidated system. Biotechnol Biofuels 2009(2):4. doi:10.1186/1754-6834-2-4

Yilmaz I, Akcoaz H, Burhan O (2004) An analysis of energy use and input costs of cotton production in Turkey. New Mediterr 3(2):58–64

Zhang R, El-Mashad HM, Hartman K, Wang F, Liu G, Choate C, Gamble P (2007) Characterization of food waste as feedstock for anaerobic digestion. Bioresour Technol 98:929–935

Life-Cycle Assessment of Biomethane from Lignocellulosic Biomass

Abdul-Sattar Nizami and Iqbal Mohammed Ismail

Abstract This chapter evaluates the life-cycle assessment (LCA) studies of biomethane produced from lignocellulosic biomass as a biofuel and it is released into the environment in comparison with other bioenergy systems. A case study of grass biomethane that is produced by anaerobic digestion (AD) of grass silage and used as a transport fuel is described. The production of biomethane from AD is a well-known technological procedure that fulfills the requirements imposed by the environment, agronomy, and legislation in developing rural economies and sustainable biofuel production. All across Europe, the biomethane yield from various lignocellulosic biomass ranges from 10 to 1,150 $m^3\ h^{-1}$. The LCA studies have been gaining importance over the past few years to analyze biofuel sources from cradle to grave in determining optimal biofuel strategies. Included in these, LCA studies is the indirect input of biofuel production processes, related emissions and waste as well as the fate of downstream products. Eighty-nine percent of greenhouse gas (GHG) emission savings are achieved by AD of grass silage to produce biomethane as a transport fuel.

A.-S. Nizami (✉) · I. M. Ismail
Center of Excellent in Enviromental Studies (CEES), King Abdulaziz Universty (KAU), PO Box 80216, Jeddah 21589, Saudi Arabia
e-mail: nizami_pk@yahoo.com

I. M. Ismail
e-mail: iqbal30@hotmail.com

A. Singh et al. (eds.), *Life Cycle Assessment of Renewable Energy Sources*, Green Energy and Technology, DOI: 10.1007/978-1-4471-5364-1_4,

1 Introduction

1.1 Lignocellulosic Biofuels, Renewable Directive, and Sustainability

Large-scale replacement of fossil fuels with renewable energy sources is necessary due to energy security and climate change in the form of greenhouse gas (GHG) emissions (Farrell et al. 2006). Thus, there is an emerging utilization of lignocellulosic biomass, which is the largest source of renewable carbohydrates for bioenergy production (Jørgensen et al. 2007). The lignocellulosic biomass is an attractive feedstock for anaerobic digestion (AD) that produces biomethane to be used as a biofuel. However, according to the EU renewable directive of 2009, "… the GHG emission saving from the use of biofuels and bioliquids taken into account… shall be at least 35 %, whereas from 2017, GHG emission savings shall be at least 50 %" (EC 2009). Thus, the renewable directive (EC 2009) promotes nonfood feedstocks including perennial grasses, forest, and agricultural residues, energy crops, organic fraction of municipal solid waste (OFMSW), and other like substrates for biofuel production. Grasses are one of the lignocellulosic biomass for producing enriched biomethane as a transport fuel (Peeters 2009; Eisentraut 2010; Singh et al. 2010a). Biomethane from lignocellulosic biomass has a better energy balance when compared to first-generation liquid transport biofuels (Korres et al. 2011). Many European countries are seeking biofuels to meet sustainability criteria and to achieve GHG emission savings targets (Korres et al. 2010).

1.2 Significance of LCA Studies for Biofuels

The generation of biofuels is facing the challenges of becoming full commercialization (Singh et al. 2010b), which is expected in near future due to improved process technologies and value-added products (IEA 2009). Thus, to ascertain optimal biofuel strategies, it is necessary to take into account environmental impacts of biofuel and bioproducts (by-products) from cradle to grave. The indirect input in the biofuel production process, related emissions and wastes as well as the fate of downstream products are all included in life-cycle assessment (LCA) studies. Thus, the overall assessment and impact evaluation of biofuels is carried out in a systematic manner. Nevertheless, LCA can also bring inaccurate and unsuitable actions for the industry, policy-making sectors, and people's perception if not exercised correctly (Korres et al. 2011).

1.3 Anaerobic Digestion: A Source of Biofuel

AD is a process where organic waste and lignocellulosic biomass are converted into biogas and digestate for value-added products. The organic wastes include slaughterhouse waste, agricultural slurries and residues, and OFMSW. According to Prasad et al. (2007), among the entire biomass available in the world lignocellulosic biomass consisting of industrial, agricultural, and forest residues is the mainstream feedstock for biogas production. Different potential feedstocks for biogas production are listed in Table 1. The biogas produced can be used for the production of electricity or heating purposes at combined heat and power (CHP) units. Biogas can be further purified and upgraded to enriched biomethane (~97 % CH_4, ~3 % CO_2 and some minor constituents) and can be injected into the gas grid or used as gaseous biofuel for transport and heating purposes. The enriched biomethane used as a transport fuel has recently started to gain consideration in many European countries, such as in Sweden, Austria, France, and Switzerland (Korres et al. 2011). All across Europe, the biomethane yield from various lignocellulosic biomass ranges from 10 to 1,150 $m^3 h^{-1}$ (Dena et al. 2009). The methane (CH_4) yield of various feedstocks is exemplified in Table 2. AD brings a promising perspective for stakeholders in the discussion of carbon trading and carbon neutral production chains, when doing an LCA study.

The aim of this chapter is to evaluate the LCA studies of biomethane produced from lignocellulosic biomass as a biofuel and its release into the environment in comparison with other bioenergy systems. A case study of grass biomethane, produced by AD of grass silage and used as a transport fuel, is described.

Table 1 Different feedstocks for biogas production

Agricultural residues	*Municipal waste and residues*
• Livestock manure	• Sewage sludge
• Animal mortalities	• Municipal solid waste
• Citrus waste	• Food residuals
• Green waste	• Organic fraction of municipal solid waste
• Agricultural slurries	
• Sugarcane bagasse	
Energy crops	*Industrial origin*
• Energy maize	• Wastewater
• Grass	• Industrial sludges
• Miscanthus	• Industrial byproducts
• Oilseed rape	• Slaughterhouse waste
• Sugar beet	• Animal fats
• Sweet sorghum	• Biosolids
• Switchgrass	• Spent beverages
• Willow	

Table 2 Methane yield of different feedstocks

Feedstocks	Methane yield (m^3 CH_4 kg^{-1} volatile solid added)	Feedstocks	Methane yield (m^3 CH_4 kg^{-1} volatile solid added)
Barley	353–658	Sorghum	295–372
Triticale	337–555	Peas	390
Alfalfa	340–500	Reed canary grass	340–430
Sudan grass	213–303	Flax	212
Jerusalem artichoke	300–370	Straw	242–324
Oats grain	250–295	Rice straw	278
Maize, whole crop	205–450	MSW	278–320
Grass	298–467	Food waste	373
Hemp	355–409	Wheat grain	384–426
Sunflower	154–400	Clover	300–350
Wheat straw	290	Potatoes	276–400
Oilseed rape	240–340	Chaff	270–316
Leaves	417–453	Kale	240–334
Sugar beet	236–381	Turnip	314
Rye grain	283–492	Rhubarb	320–490
Fodder beet	420–500	Miscanthus	179–218
Nettle	120–420	Sludge	260
Chicken litter	290	Pig manure	310
Cattle manure	160	Source separated food waste	300–529
OFMSW	158–400	Timothy	345–375
Cocksfoot	315		

Chandra et al. 2012; Jha et al. 2011; Li et al. 2010; Cho and Park 1995; Juanga 2005; Murphy et al. 2011; Browne and Murphy 2012; János and Elza 2008

2 Methodology

2.1 Life-Cycle Assessment

According to International Organization for Standardization 14000 (ISO 2006), there are four phases of an LCA procedure, including (1) the goal, scope definition, and functional unit, (2) inventory analysis, (3) impact assessment, and (4) interpretation. An LCA provides systematic view and complete assessment of a product throughout its life cycle (Payraudeau et al. 2007). It is important to consider the whole life cycle due to efficient energy management of renewable sources and their GHG emissions. The scientific community considers LCA as one of the best method for calculating the environmental burden associated with bioenergy production (Consoli et al. 1993). The renewable directive (EC 2009) has provided guidelines for the LCA of biofuels. An LCA of biofuels must evaluate GHG

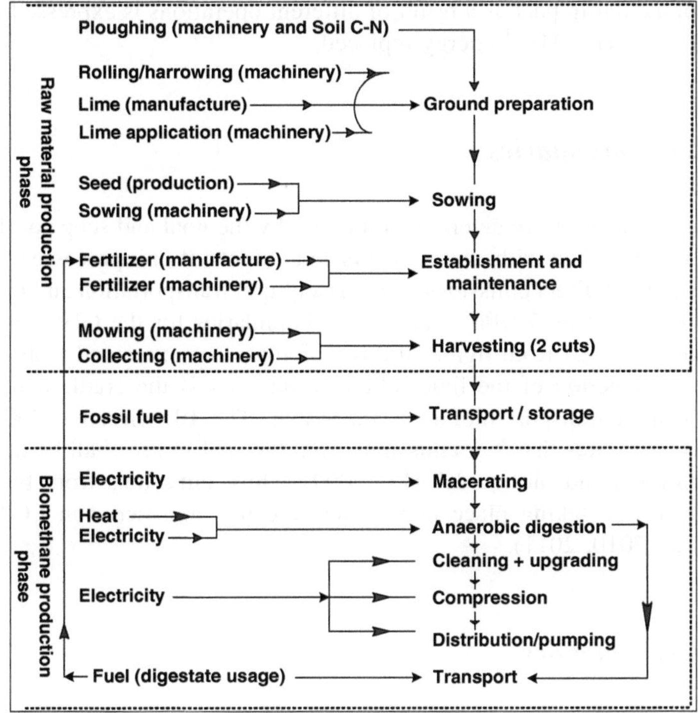

Fig. 1 A flow chart of lignocellulosic biomethane production system

emission reductions of carbon dioxide (CO_2), nitrous oxide (N_2O), and CH_4 in the global-warming potential (GWP) with relation to fossil fuel replacement (Korres et al. 2011). According to Gerin et al. (2008), there should be a net reduction and gain in GHG emissions and bioenergy, respectively, in LCA studies of biomethane produced from lignocellulosic biomass. In Figure 1, a comprehensive presentation of the whole cycle of lignocellulosic biomethane is shown, where GHG emissions are calculated based on energy inputs and outputs.

2.2 Goal, Scope, and Functional Unit

As a first step in conducting an LCA, goal, scope, and functional unit are defined. The goal addresses the intended applications to the intended audience, while scope has to be compatible with the goal of study and well defined (Singh et al. 2010a). The functional unit is an element of the product or system, which must be measurable and definable. It is used as a quantitative tool for the comparative analysis of bioenergy systems (Casey and Holden 2005). In AD, biomethane is the main product, and thus, the functional unit is described in m^3 biomethane per year.

The environmental impact as a result of different operations is expressed in g CO_2 equivalent (CO_2 eq.) MJ^{-1} energy replaced.

2.3 System Boundaries

The system boundaries are determined initially by the goal and scope of the study. They are further linked with energy inputs and outputs of unit processes, where all of the direct and indirect emissions from agriculture, transportation and process are calibrated (Singh et al. 2010a). The system boundaries for the GHG emission of biomethane produced from lignocellulosic biomass are examined from cradle to grave. The production of the lignocellulosic biomass is the cradle and enriched biomethane as a transport biofuel is the grave. The EU directive 2009/28/EC, Annex V, C-13 states that "… emissions from the fuel in use shall be taken to be zero for biofuels and bioliquids" (EC 2009). Thus, emissions from biomethane combustion (often taking place in vehicles) are not considered in LCA studies (Korres et al. 2010, 2011).

2.4 Impact Category

To determine the potential impact of GHG emissions of CO_2, N_2O, CH_4, the term GWP is used. GWP is defined as the collective outcome between the present instant and a certain time in the future resulted in a unit mass of gas released in the present (Casey and Holden 2005; Korres et al. 2010). A GWP of one (1) refers to the release of 1 kg CO_2 (Korres et al. 2011). According to EC (2009), the GWP of NO_2 and CH_4 on one (1) kg basis is 296 and 23, respectively. The following formula is used to calculate the volume of GHG emission in terms of CO_2 (EC 2009).

$$\text{GHG (t of } CO_2 \text{ eq.)} = CO_2\text{ (t)} + 23 \times CH_4\text{ (t)} + 296 \times N_2O\text{ (t)}.$$

3 LCA of Biomethane from Lignocellulosic Biomass

3.1 Sustainability Criteria and Energy Efficiency

The energy efficiency of a biofuel source is determined by considering all energy inputs and outputs over the entire product production cycle (Salter and Banks 2009). For example, biodiesel production in Europe is accomplished using rape seed oil that covers about 80 % of the land set aside for nonfood energy crops (Bauen 2005). Similarly, the rape seed biodiesel and wheat bioethanol both yield

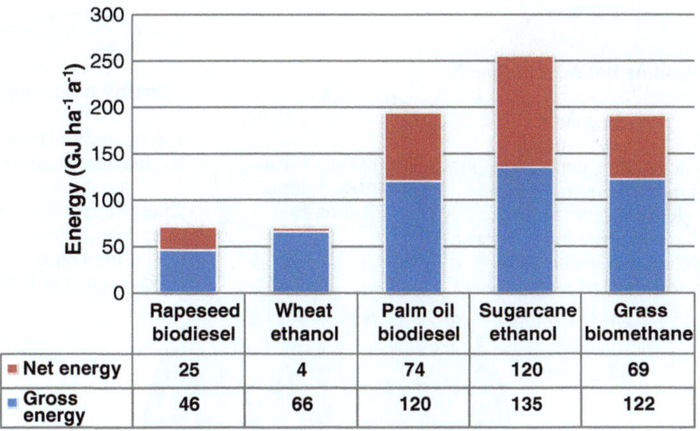

Fig. 2 The net and gross energy of different biofuels systems (Korres et al. 2010, 2011; Smyth et al. 2009)

less gross and net energy in comparison with palm oil biodiesel, grass biomethane, and sugarcane bioethanol (Fig. 2). The option to import substrates for biodiesel production from tropical countries, such as Indonesia and Malaysia, is not a viable option as they result in a high demand for the production of palm oil, which is 80 % of the global production (Korres et al. 2011). Consequently, deforestation is occurring at an annual rate of 1.5 % (Fargione et al. 2008). There are no net GHG emission savings with a change in land use (Reinhard and Zah 2009). According to Directive 2009/28/EC, palm oil biodiesel is not considered as biofuel because it needs to achieve GHG savings of 60 % by 2020 (EC 2009). The increase in palm oil production causes habitat loss, drainage of peatlands, and land conflicts (Colchester et al. 2006). Similar issues of deforestation, decarbonization, and soil degradation occur with the production of sugarcane ethanol (Goldemberg et al. 2008). According to Korres et al. (2010, 2011), biofuel in form of enriched biomethane produced from lignocellulosic biomass like grass silage is much better for Europe than rape seed biodiesel and wheat ethanol (Fig. 2). The low-input indigenous perennial grasses provide biofuel with more useable energy, GHG savings and less pollution related to agrochemical procedures than arable crops per hectare. The arable crops can be corn grain ethanol or soybean biodiesel (Tilman et al. 2006; Korres et al. 2011). The benefits of producing biomethane as a transport fuel from lignocellulosic biomass through the AD process are shown in Fig. 3.

3.2 GHG Emissions

Korres et al. (2010) assessed the GHG emissions of enriched biomethane as a transport biofuel produced by grass silage in place of diesel as 69.74 g CO_2 eq. MJ^{-1} energy replaced or 6,904 kg CO_2 eq. ha^{-1} yr^{-1}. This was

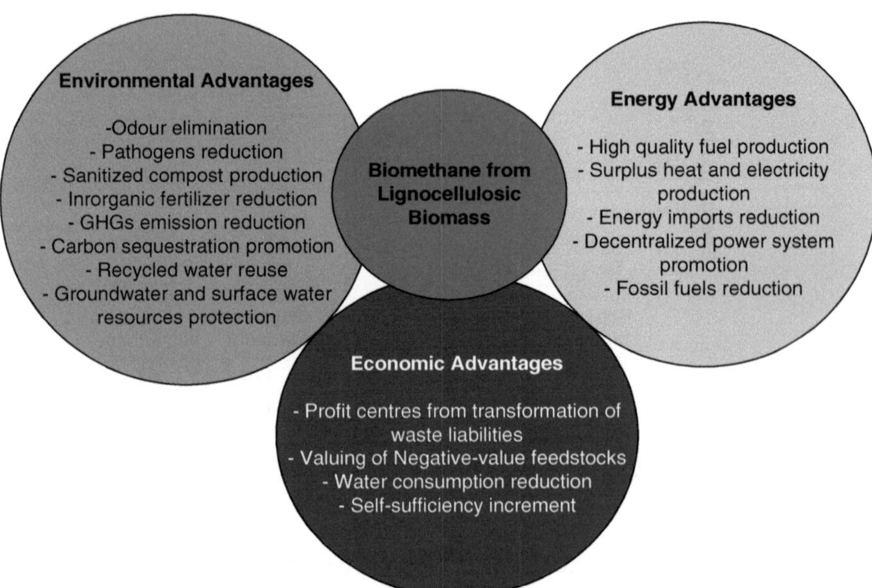

Fig. 3 Benefits of lignocellulosic biomethane production system

determined by considering different scenarios such as improved vehicle efficiency, electricity from wind, use of wood chips for AD heating requirements and carbon sequestration of 0.6 t C ha^{-1} yr^{-1}; a minimum value for most European permanent crops and grasslands. All of them results in GHG emission savings of up to 89.4 %. This achievement meets the EU directive 2009/28/EC requirements of 60 % GHG savings by 2018 (EC 2009). The crop production and AD process are the main GHG emissions contributors in grass biomethane (Fig. 4). Among the indirect GHG emissions, potassium and nitrogen fertilizers are the main contributors to agricultural emissions. The digester heating is the largest contributor in the biomethane production process (Korres et al. 2011).

The wheat ethanol, rapeseed biodiesel, and sunflower biodiesel do not meet the 60 % GHG emission savings in comparison with grass biomethane (Fig. 5). According to Thornley et al. (2009), issues of high nitrogen and pesticide requirements are associated with rape seed biodiesel, which impacts the GHG savings. Furthermore, the associated technology is poor. The low GHG savings with wheat ethanol is due to the emission of N_2O during cultivation and low biofuel yields (Smith et al. 2005; Börjesson 2009). Similarly, sunflower biodiesel only fulfills the conversion rate necessary to achieve 35 % GHG emission savings from 30 % of arable land (Ragaglini et al. 2010). Nevertheless, there are environmental benefits reported with sunflower biodiesel (Sanz-Requena et al. 2010). The biomethane production on farms from manure, which is an easily accessible substrate result in higher GHG emission savings (Korres et al. 2011). Nevertheless,

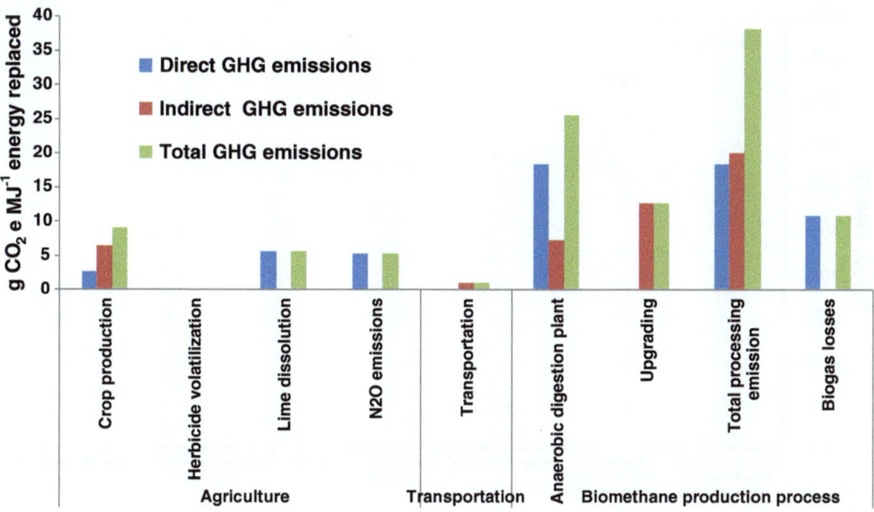

Fig. 4 The direct, indirect, and total GHG emissions from grass biomethane production (Korres et al. 2010, 2011)

the problem of high water contents, low biogas production rates, and high economies are barriers in AD of manure (Gerin et al. 2008). To overcome this problem, manure can be codigested with other lignocellulosic feedstocks. This results in higher biomethane production (Jagadabhi et al. 2008) with improved digester microbiology (Nizami and Murphy 2010).

3.3 Digestate: A Source of Fertilizer and Bioproducts

AD results in a residual digestate. This digestate can be a great source of commercial fertilizer. This additional environmental benefit included in the biofuel process chain lowers the production costs and loss to the environment and increases the process efficiency (Cherubini et al. 2009). The use of maize and grass silage as AD feedstocks and their digestate used as fertilizer have been studied by Gerin et al. (2008). Matsunaka et al. (2006) studied the Timothy grass for digestate purposes. They observed the benefit of nitrogen uptake by the grass digestate, especially during the spring. Liquid and fiber components are obtained from digestate (Salter and Banks 2009) and some of the liquid can be re-used to enhance the digestion process (Berglund and Börjesson 2006). The rest is processed into liquid biofertilizer or can be used for many practical purposes (Fig. 5). The solid digestate can be processed into either soil conditioner or high value insulating materials (Grass2004; Salter and Bank 2009). The concept of using biomethane as a biofuel and digestate for value-added products evolves into the concept of

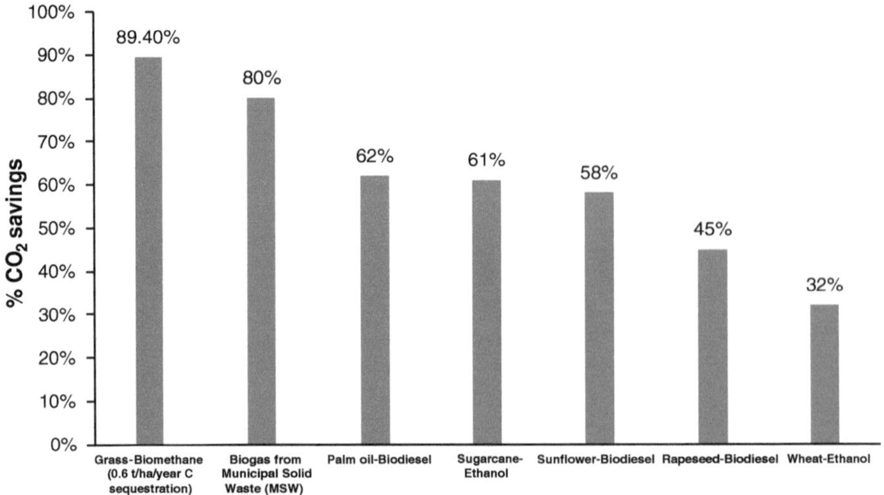

Fig. 5 The % CO_2 savings of different biofuel systems (Korres et al. 2010, 2011; Smyth et al. 2009)

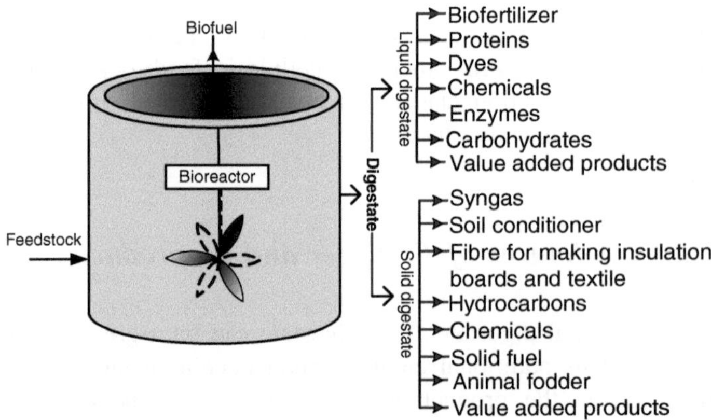

Fig. 6 Value-added products of digestate (Kamm and Kamm 2004; Korres et al. 2011)

biorefinery (Kamm et al. 1998; Nardoslawsky 1999; Kamm and Kamm 2004). According to Korres et al. (2010), these value-added products that emerge in addition to the biofuel (Fig. 6) will also help to reduce GHG emissions. However, calculation of these bioproducts GHG emissions is needed, as they are shaped into marketable products at additional energy and financial cost.

4 Scenarios to Increase Sustainability of Biomethane from Lignocellulosic Biomass

4.1 The Potential of CO_2 and GHG savings

A large portion of biogas consists of CO_2 (40–50 %), which is removed during biogas upgrading to achieve enriched biomethane as a transport fuel. The range of CO_2 removal during upgrading is 1.62–1.86 kg CO_2 m^{-3} (Power and Murphy 2009). This CO_2 removal is an additional source of GHG emission and thus can be minimized by its use in the AD (Fig. 7). Using CO_2 as a pretreatment option to accelerate the hydrolysis of cellulose (one of the major components in lignocellulosic biomass) is observed and described by Zheng et al. (1995), 1998 and Clark et al. (2006). The cellulose crystallinity, lignin sealing and cross-linkage of hemicellulose around cellulose are barriers in the attachment of enzymes and microbes to the cellulosic surfaces (Nizami et al. 2009; Fan et al. 1987). This is an issue that impacts the efficiency of lignocellulosic biomass undergoing cellulose hydrolysis (Kim and Hong 2001). The use of CO_2 as a pretreatment option in the AD process is preferred due to less expensive, clean, less energy demanding, easy to recover in a nontoxic manner and nonflammable properties in comparison with the physical, chemical, thermal, and steam explosion pretreatments (Chahal et al. 1981; Zheng et al. 1995; Kim and Hong 2000, 2001). The CO_2 can be used in two different forms: first in an explosive form at a high pressure where it disrupts the

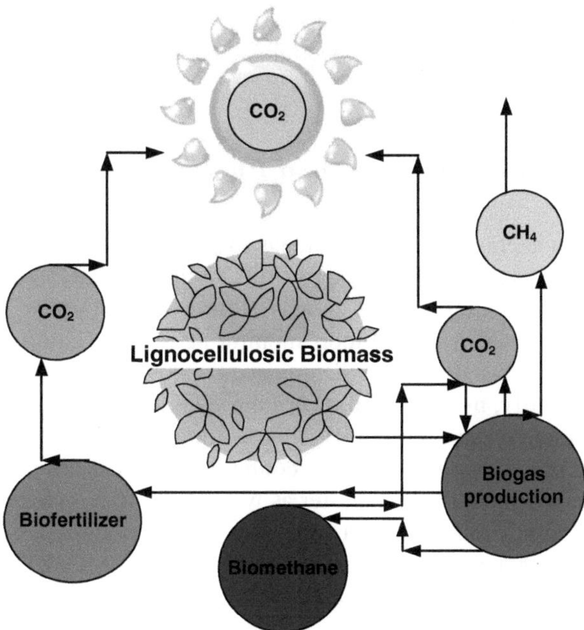

Fig. 7 The CO_2 movement through various subsystems involved in the lignocellulosic biomethane system

cellulose structure and second in a dissolved form where it forms carbonic acid. Carbonic acid is a weak acid that dissolves hemicellulose without toxicity and corrosivity to the AD process. These processes result in porous cellulosic surfaces, which are easily accessible to enzymatic and microbial activity (Zheng et al. 1998; Kim and Hong 1999, 2000, 2001). Thus, the use of CO_2 in biogas production has untapped potential that will not only enhance the efficiency of the process with reduced economic and energy requirements but will also decrease GHG emissions tremendously. Nevertheless, the use of CO_2 as a pretreatment option is limited to the ethanol industry in a supercritical CO_2 explosion form, where it increases glucose yield by 50 % and overall ethanol yield by 70 % (Zheng et al. 1998).

4.2 Digester Configurations and GHG Savings

Continuous stirred tank reactors (CSTR) are widely used to digest slurries and represent a simple and robust technology (Smyth et al. 2009; Mähnert et al. 2005). The addition of a separate preprocessing tank with chopper pump, screw-feeder, and flushing system (Weiland 2003) does increase the energy demand when using for lignocellulosic biomass. Therefore, the values for GHG emission of CSTRs will be higher than other digester configuration such as a dry batch digester or leach beds digester coupled with an up flow anaerobic sludge blanket (UASB) reactor (Nizami et al. 2009). In dry batch, leach beds, and UASB, there is less requirement for mechanical or electrical feeding and mixing (Köttner 2002; Nizami and Murphy 2010). This comparison of anaerobic digester configurations will assist developers and farmers in selecting digester types and digester processes suitable to digest lignocellulosic biomass with least GHG emissions.

4.3 Biogas Losses and Engine Efficiency

On average, the rate of biogas loss from AD to enriched biomethane production is 7.41 %, which accounts for indirect GHG emissions between 8.44 and 8.86 kg CO_2 m^{-3} of biogas (Power and Murphy 2009). Nizami et al. (2009), suggested a closed-loop monitoring system equipped with sensory devices for anaerobic digester (Nizami et al. 2009). The application of nanotechnology to identify, monitor, and record these losses using sensory chips and devices is at the infancy stage in the scientific community. Moreover, comparing the GHG emissions of various digester configurations will assist the development of different component of the digester as we attempt to reduce energy loss. Above all, vehicle engines must be improved, as existing engines are less efficient in utilizing biomethane and greater in their release of GHG (Power and Murphy 2009). According to Korres et al. (2010), an improvement of 18 % can be achieved by the improvements in engine efficiency to a similar km MJ^{-1} as diesel.

5 Conclusion

Lignocellulosic biomass is available in substantial quantities all over the globe and is a promising source of biofuel when digested anaerobically in a digester. LCA studies are important in analyzing biofuel sources from cradle to grave to determine optimal biofuel strategies. Grass and grass silage have recently been considered in many European countries as the crop of transport fuel. These feedstocks are perennial in nature and have high yields and volatile contents, which make them beneficial feedstocks for biomethane production. A GHG emission savings of 89 % is achieved by grass silage if digested anaerobically and biomethane is used as a transport fuel.

References

Bauen A (2005) Biomass in Europe. In: Silveira S (ed) Bioenergy-realizing the potential. Elsevier, Amsterdam, pp 19–30

Berglund M, Börjesson P (2006) Assessment of energy performance in the life cycle of biogas production. Biomass Bioenergy 30:254–266. doi:10.1016/j.biombioe.2005.11.011

IEA Bioenergy (2009) Sustainable bioenergy—a reliable energy option. Report prepared for IEA Bioenergy Implementing Agreement. www.ieabioenergy.org

Börjesson P (2009) Good or bad bioethanol from a greenhouse gas perspective-what determines this? Appl Energy 86:589–594. doi:10.1016/j.apenergy.2008.11.025

Browne JD, Murphy JD (2012) Assessment of the resource associated with biomethane from food waste. Appl Energy Accepted

Casey JW, Holden NM (2005) Analysis of greenhouse gas emissions from the average Irish milk production system. Agric Syst 86:97–114

Chahal DS, Mcguirc S, Pikor H, et al (1981) In: Proceedings of the 2nd world congress of chemical engineering. Montreal, Canada, October, 4–9

Chandra R, Takeuchi H, Hasegawa T (2012) Methane production from lignocellulosic agricultural crop wastes: A review in context to second generation of biofuel production. Renew Sustain Energy Rev 16:1462–1476

Cherubini F, Bird ND, Cowie A et al (2009) Energy-and greenhouse gas-based LCA of biofuel and bioenergy systems: key issues, ranges and recommendations. Resour Conserv Recycl 53:434–447. doi:10.1016/j.resconrec.2009.03.013

Cho JK, Park SC (1995) Biochemical methane potential and solid state anaerobic digestion of Korean food wastes'. Bioresour Technol 52(3):245–253

Clark JH, Budarin V, Deswarte FEI et al (2006) Green chemistry and the biorefinery: a partnership for a sustainable future. Green Chem 8:853–860

Colchester M, Jiwan N, Andiko et al (2006) Promised land: palm oil and land acquisition in Indonesia: implications for local communities and Indigenous Peoples. Forest Peoples Programme/Perkumpulan Sawit Watch, London/Bogor

Consoli F, Allen D, Boustead I, et al (1993) Guidelines for life-cycle assessment: a "Code of practice". In: Proceedings of society of environmental toxicology and chemistry (SETAC). SETAC Workshop, Sesimbra, 31 Mar–3 Apr

Dena HM, Kontges A, Rostek S (2009) Biogaspartner—a joint initiative. Biogas grid injection in Germany and Europe-Market, Technology and Players. German Energy Agency, HP Druck, Berlin

EC (2009) Directive 2009/28/EC of the European parliament and of the council of 23 April 2009 on the promotion of the use of energy from renewable sources and amending and subsequently repealing Directives 2001/77/EC and 2003/30/EC. Official J Eur Union, L140/16–62

Eisentraut A (2010) Sustainable production of second generation biofuels, information paper. International Energy Agency, Paris

Fan LT, Gharpuray MM, Lee Y-H (1987) Cellulose hydrolysis. Springer, Berlin

Fargione J, Hill J, Tilman D et al (2008) Land clearing and the biofuel carbon debt. Science 319:1235–1238. doi:10.1126/science.1152747

Farrell AE, Plevin RJ, Turner BT et al (2006) Ethanol can contribute to energy and environmental goals. Science 311:506–508

Gerin PA, Vliegen F, Jossart JM (2008) Energy and CO_2 balance of maize and grass as energy crops for anaerobic digestion. Bioresour Technol 99:2620–2627. doi:10.1016/j.biortech.2007.04.049

Goldemberg J, Teixeira CS, Guardabassi P (2008) The sustainability of ethanol production from sugarcane. Energy Policy 36:2086–2097. doi:10.1016/j.enpol.2008.02.028

Grass S (2004) Utilisation of grass for production of fibres, protein and energy. Biomass and agriculture: sustainability, markets and policies. Biomass Project Services, Switzerland, pp 169–177

ISO (2006) International Standardization Organization (ISO), Environmental management–life cycle assessment-principles and framework, ISO 14040

Jagadabhi PS, Lehtomäki A, Rintala J (2008) Co-digestion of grass silage and cow manure in a CSTR by re-circulation of alkali treated solids of the digestate. Environ Technol 29:1085–1093. doi:10.1080/09593330802180385

János T, Elza K (2008) Input materials of biogas production. Lecture on Environmental technology. Debreceni Egyetem a TÁMOP 4.1.2 pályázat keretein belül. Available from: http://www.tankonyvtar.hu/en/tartalom/tamop425/0032_kornyezettechnologia_en/ch01s02.html

Jha AK, Li J, Nies L et al (2011) Research advances in dry anaerobic digestion process of solid organic wastes. Afr J Biotechnol 10:14242–14253

Jørgensen H, Kristensen JB, Felby C (2007) Enzymatic conversion of lignocellulose into fermentable sugars: challanges and opportunities. Biofuels. Bioprod Biorefin 1:119–134

Juanga JP (2005) Optimizing dry anaerobic digestion of organic fraction of municipal solid waste. M.E. thesis, Asian Institute of Technology, Bankok

Kamm B, Kamm M (2004) Principles of biorefineries. Appl Microbiol Biotechnol 64:137–145

Kamm B, Kamm M, Soyez K (1998) The green biorefinery, concept of technology. In: Proceedings of (1st international symposium on green biorefinery) Neuruppin, Society of Ecological Technology and System Analysis, Berlin

Kim KH, Hong J (1999) Equilibrium solubilities of spearmint oil components in supercritical carbon dioxide. Fluid Phase Equil 164:107–115

Kim KH, Hong J (2000) Dynamic extraction of spearmint oil components by using supercritical CO2. Sep Sci Technol 35:315–322

Kim KH, Hong J (2001) Supercritical CO_2 pretreatment of lignocellulose enhances enzymatic cellulose hydrolysis. Bioresour Technol 77:139–144

Korres NE, Singh A, Nizami AS et al (2010) Is grass biomethane a sustainable transport biofuel? Biofuels Bioprod Bioref 4:310–325. doi:10.1002/bbb.228

Korres NE, Thamsiriroj T, Smyth BM et al (2011) Grass biomethane for agriculture and energy. In: Sustainable Agriculture Reviews, vol 7. Springer, London Ltd, pp 5–49, DOI: 10.1007/978-94-007-1521-9_2

Köttner M (2002) Biogas and fertilizer production from solid waste and biomass through dry fermentation in batch method. In: Wilderer, P, Moletta, R (eds) Anaerobic digestion of solid wastes III. IWA Publishing, London

Li D, Yuan Z, Sun Y (2010) Semi-dry mesophilic anaerobic digestion of water sorted organic fraction of municipal solid waste (WS-OFMSW). Bioresour Technol 101:2722–2728

Mähnert P, Heiermann M, Linke B (2005) Batch- and semi- continuous production from different grass species. Agric Eng Int: CIGRE E J V11, Manuscript EE 05 010

Matsunaka T, Sawamoto T, Ishimura H et al (2006) Efficient use of digested cattle slurry from biogas plant with respect to nitrogen recycling in grassland. Int Congr Ser 1293:242–252

Murphy JD, Braun R, Weiland P et al (2011) Biogas from crop digestion. IEA Bioenergy Task 37. Web access: http://www.iea-biogas.net/

Narodoslawsky M (1999) Green biorefinery. In: Proceedings of 2nd international symposium on green biorefinery, SUSTAIN, Feldbach

Nizami AS, Murphy JD (2010) What type of digester configurations should be employed to produce biomethane from grass silage? Renew Sustain Energy Rev 14:1558–1568

Nizami AS, Korres NE, Murphy JD (2009) A review of the integrated process for the production of grass biomethane. Environ Sci Technol 43:8496–8508

Payraudeau S, van der Werf HMG, Vertes F (2007) Analysis of the uncertainty associated with the estimation of nitrogen losses from farming systems. Agric Syst 94:416–430. doi:10.1016/j.agsy.2006.11.014

Peeters A (2009) Importance, evolution, environmental impact and future challenges of grasslands and grassland-based systems in Europe. Grassl Sci 55:113–125. doi:10.1111/j.1744-697X.2009.00154.x

Power MP, Murphy JD (2009) Which is the preferable transport fuel on a greenhouse gas basis; biomethane or ethanol? Biomass Bioenergy 33(10):1403–1412 doi: 10. 1016/j. biombioe. 2009. 06.044

Prasad S, Singh A, Joshi HC (2007) Ethanol as an alternative fuel from agricultural, industrial and urban residues. Resour Conserv Recycl 50:1–39

Ragaglini G, Triana F, Villani R et al (2010) Can sunflower provide biofuel for in-land demand? an integrated assessment of sustainability at regional scale. Energy. doi:10.1016/j.energy.2010.03.009

Reinhard J, Zah R (2009) Global environmental consequences of increased biodiesel consumption in Switzerland: consequential life cycle assessment. J Cleaner Prod 17(Supplement 1):S46–S56. doi:10.1016/j.jclepro.2009.05.003Salter

Salter A, Banks CJ (2009) Establishing an energy balance for crop-based digestion. Water Sci Technol IWA 59(6):1053

Sanz Requena JF, Guimaraes AC, Quiros Alpera S et al (2010) Life cycle assessment (LCA) of the biofuel production process from sunflower oil, rapeseed oil and soybean oil. Fuel Process Technol 92:190–199. doi:0.1016/j.fuproc.2010.03.004

Singh A, Pant D, Korres NE et al (2010a) Key issues in life cycle assessment of ethanol production from lignocellulosic biomass: challenges and perspectives. Bioresour Technol 101:5003–5012. doi:10.1016/j.biortech.2009.11.062

Singh A, Korres NE, Murphy JD (2010b) Grass and anaerobic digestion for biomethane production: a sustainable option. In: Grassland Science in Europe, the official peer-reviewed. Proceedings of European Grassland Federation, General Meeting, Kiel, 29 Aug–2 Sept 2010

Smith TC, Kindred DR, Brosnan JM et al (2005) Wheat as a feedstock for alcohol production. Research Review no. 61, Home-Grown Cereals Authority

Smyth B, Murphy JD, O'Brien C (2009) What is the energy balance of grass biomethane in Ireland and other temperate northern European climates? Renew Sustain Energy Rev 13(9):2349–2360 doi:10.1016/j.rser.2009.04.003

Thornley P, Upham P, Tomei J (2009) Sustainability constraints on UK bioenergy development. Energy Policy 37:5623–5635. doi:10.1016/j.enpol.2009.08.028

Tilman D, Hill J, Lehman C (2006) Carbon-negative biofuels from low-input high diversity grassland biomass. Science 314:1598–1600. doi:10.1126/science.1133306

Weiland P (2003) Production and energetic use of biogas from energy crops and wastes in Germany. Appl Biochem Biotechnol 109(1–3):263–274

Zheng Y, Lin HM, Wen J et al (1995) Supercritical carbon dioxide explosion as a pretreatment for cellulose hydrolysis. Biotechnol Lett 17:845–850

Zheng Y, Lin H-M, Tsao GT (1998) Pretreatment for cellulose hydrolysis by carbon dioxide explosion. Biotechnol Prog 14:890–896

Life Cycle Assessment of Biodiesel from Palm Oil

Keat Teong Lee and Cynthia Ofori-Boateng

Abstract Though the energy balance for the cultivation of oil palm biomass for biodiesel production is positive, current debate has been raised on its environmental sustainability due to the high consumption of fossil fuel, fertilizer, and pesticides. This chapter employs the well-to-wheel variant of life cycle analysis (LCA) to assess the various potential environmental impacts, energy and land use/conversion impacts associated with the production of biodiesel from palm oil. Eleven (11) main impact categories, namely land use, fossil fuel use, climate change, ozone layer depletion potential, minerals/heavy metals, acidification/eutrophication potential, ionizing radiation potential, ecotoxicity potentials, carcinogens, respiratory organics, and respiratory in organics based on Eco-Indicator 99, are analyzed and discussed. Excluding transportation impacts, the oil palm cultivation stage contributed the highest overall environmental impacts (44 % of the total impacts) compared to the other stages. On the other hand, fossil fuel consumption was highest (43 % of total impacts) in the transesterification unit exclusive of all impacts from transportation.

1 Introduction

The increasingly high cost, fast rate of exhaustion and negative impacts of fossil fuel's combustion on the environment have caused almost all economic sectors of the world to consider new lasting sources of energy to replace fossil fuels.

Biodiesel has recorded tremendous growth rate in its consumption and production over the past decade due to its positive environmental impacts (as well as other unique characteristics, e.g., biodegradability, non-toxicity, renewability) hence considered a feasible petro–diesel replacement (Vicente et al. 2004;

K. T. Lee (✉) · C. Ofori-Boateng
School of Chemical Engineering, Universiti Sains Malaysia (USM), 14300 Nibong Tebal, Pulau Pinang, Malaysia
e-mail: chktlee@eng.usm.my

A. Singh et al. (eds.), *Life Cycle Assessment of Renewable Energy Sources*, Green Energy and Technology, DOI: 10.1007/978-1-4471-5364-1_5,

Encinar et al. 2005). Moreover, biodiesel on combustion is reported to release insignificant amount of air emissions compared to fossil fuel (Antolin et al. 2002).

Biodiesel production processes utilize fossil-based fuels as main energy sources; thus, their emission effects add to the concentration of CO_2 in the atmosphere resulting in global warming. As reported by Roger et al. (2011), the production of 1 t of biodiesel from any feedstock averagely adds not less than 916 kg CO_2 to what is already in the atmosphere. Therefore, it can be inferred from these scenarios that energy use in biodiesel production processes is directly related to the emissions associated with its production thus needed to quantify these emissions and strategically allocate improvement measures. Life cycle assessment (LCA) presents a better assessment tool to quantify the environmental burdens associated with biodiesel over its life cycle. An LCA well-to-wheel assessment of palm oil biodiesel is discussed in this chapter.

LCA, also referred to as ecobalance analysis, is a technique used to quantify environmental impacts associated with the various stages of a product's life from raw material extraction through materials processing, distribution, use, repair, and maintenance, as well as waste management. The methodology of LCA brings out a wide outlook on the environmental burdens of a product because it considers a thorough inventory of energy, material inputs, and emissions; quantifies the potential environmental impacts associated with the specified inputs and emissions; and finally interprets the results which aid in policy making and implementation.

Currently, there has been a controversial debating issue on the environmental sustainability of biodiesel since its production makes use of great amount of fossil fuel which puts so much burden on the environment. The most important factor which affects the sustainability of biodiesel production is the choice of feedstock since each feedstock as well as its cultivation technology has its own specific ecological footprint. For instance, the environmental sustainability of palm oil production and subsequent conversion into biodiesel is characterized by land use, soil quality management, and genetic biodiversity (Parish et al. 2008). Biodiesel production from virgin feedstock such as palm oil is less sustainable than that from waste cooking oil (WCO) in terms of environmental impacts because the cultivation stage of WCO is eliminated from the life cycle assessment stage. Nonetheless, first generation biodiesel (FGB) feedstock, such as palm oil, soybean oil, rapeseed oil, has been pioneered and continued to saturate the biodiesel market until the commercial production of second and third generation feedstock become exceedingly sustainable over FGB feedstock.

1.1 Global Palm Oil Production Profile

Palm oil presents a better and attractive feedstock for biodiesel production compared to other first and second generation feedstock because of its high oil yield (averagely 8.6 t per hectare of land) which is almost three times more than that for

Table 1 Palm oil and palm biodiesel industries in major oil palm producing countries of the world

Country	Oil palm plantation, million ha	Number of palm oil mills	Number of CPO refineries	Number of palm kernel oil crushing units	Number of palm oil biodiesel production plants
Indonesia	6.170	405	32	21	–
Malaysia	4.890	416	57	46	25
Thailand	0.512	70	12	–	15
Nigeria	0.385	21	9	4	–
Colombia	0.209	7	–	–	6
Others	0.754	–	–	–	–
Total	12.900	–	–	–	81

Source MPOB (2010)

coconut, 12 times more than soybean oil yield, and seven times more than that for rapeseed (Schmidt 2007; Bockish 1998). Also, the per-unit production cost of palm oil is much lower compared to soybean oil which is 20 % higher. This makes it a better vegetable oil for biodiesel production.

Palm oil currently is the second largest edible oil source (after soybean oil) which forms approximately 34 % of the global oil supply (Schmidt 2007). In 2009, both palm oil and palm kernel oil accounted for 5 % of the total cultivated land for vegetable oil production globally. In 2010, the global palm oil production was 47.9 million tonnes of which 11 % were used for biodiesel production. For the production in 2010, Malaysia and Indonesia together contributed about 87 % of the total palm oil produced in the world with about 19.5 and 22.5 million tonnes, respectively (MPOB 2010). Projections for 2012 palm oil production growth in Malaysia and Indonesia indicate expansion of about 3.5 million tonnes. Though Malaysia has only about 12.5 % of its total landmass (i.e., 32 million hectares) under oil palm plantation (GOFBM 2009), it has been recognized as the world's largest producer of certified sustainable palm oil (CSPO) contributing over 50 % of total CSPO production (RSPO report 2011). Nigeria (who was the largest exporter of palm oil in 1934 but overtaken by Malaysia in 1966) remains the largest producer of palm oil in Africa and the world's fourth leading producer in 2010 with a total oil palm landmass of about 385, 000 hectares (RSPO report 2011). Other palm oil-exporting countries include Thailand, Colombia, Ecuador, Papua New Guinea, Ivory Coast, Brazil. Table 1 shows the number of palm oil and biodiesel industries in 2010 in major palm oil-producing countries of the world. These figures keep increasing from year to year thus needed to assess their environmental impacts and suggest measures for improvement.

In 2007, global biodiesel produced from palm oil recorded the highest production capacity of about 38 million tonnes followed by soy oil biodiesel (36 million tonnes), rapeseed oil biodiesel (16 million tonnes), and sunflower oil biodiesel (10 million tonnes). Presently, these capacities have increased still with palm oil biodiesel leading at 44 million tonnes (Biodiesel 2020). Figure 1 shows the production of biodiesel from palm oil in comparison with other seed oils

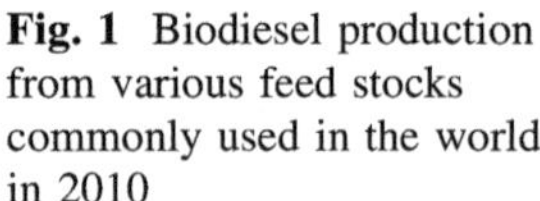

Fig. 1 Biodiesel production from various feed stocks commonly used in the world in 2010

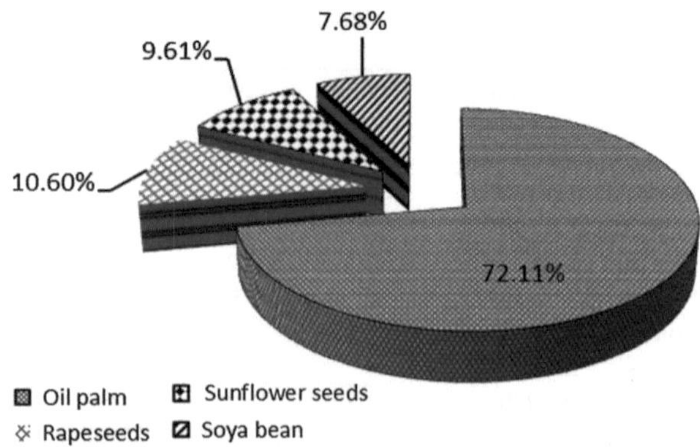

mainly used for biodiesel production in 2010. In 2010, Malaysia recorded the highest palm oil biodiesel installed capacity of 1.7 million tonnes. Between 2007 and 2008, Malaysia's biodiesel production gained a rise by 32 % from 129,715 to 171,700 t (Biodiesel 2020). The United States of America and the European Union were the main importers of biodiesel from Malaysia accounting for 39.2 and 38.6 % of the total biodiesel exports, respectively (Biodiesel 2020).

1.2 The Oil Palm

The oil palm is a perennial insect-pollinated plant which belongs to the family *Palmae* and genus *Elaeis* with many species including *guineensis, oleifera, kamerunicus*. *Elaeis guineensis Jacq.* has been the commonest species with an average generic life span of 150 years, an economic life of 20–25 years (11–16 months for nursery) and significantly high oil-to-bunch content (45–55 % oil) compared to the other species (Schmidt 2007). However, genus *oleifera* has been reported to have higher level of unsaturated fatty acids thus used for the production of interspecific hybrids with the genus *guineensis.*

The oil palm is cultivated in 45 countries in the world on a total land area of about 12.9 million hectares (GOFBM 2009). Oil palms are highly efficient producers of oil requiring less land than any other oil-producing crops. Only about 10 % of the oil palm produces the oil (which is extracted from the mesocarp or fleshy part of the fruits) and palm kernel oil (which is obtained from the kernel or seed in the fruit). The remaining 90 % is mainly the biomass comprising the empty fruit bunches (EFB), fibers, fronds, trunks, kernels, and mill effluent which are often disposed as wastes or used as mulch in the plantation.

After 24–30 months of planting a palm tree, it begins to bear fresh fruit bunches (FFB) and thus ready for harvest after some couple of months later. The normal frequency of harvesting is between 10 and 15 days (Xavier et al. 2008). The tree produces averagely 12 FFB annually with a bunch weighting 15–25 kg containing

1,000–1,300 fruitlets depending on the plantation management and establishment. On a per hectare basis, an oil palm plantation can yield averagely 35 t of FFB (from about 148 palm trees) and 8.6 t of palm oil (Henson 1990; Schmidt 2007). Generally, the extraction of 1 t of crude palm oil (CPO) requires 5 t of FFB which produces 1.15 t of EFB and 3.25 t of palm oil mill effluents (POME) as residues (Corley and Tinker 2003). The harvested FFB may contain around 20 % mesocarp oil, 25 % nuts (comprising 5 % kernels, 13 % fiber, and 7 % shell), and 23 % empty fruit bunches. The kernels also contain around 55 % oil and 8 % protein (Corley and Tinker 2003; Møller et al. 2000).

The oil palm industry now focuses on genetic means of improving the oil yield, palm disease tolerance, and the height of the tree (breeding dwarf palms in order to prolong the economic cropping cycle). Corley and Lee (1992) and Pushparajah (2002) have reported the possibility of commercializing genetically bred oil palms for the next 15–40 years. However, currently, transgenically high oleic acid palms have been field tested and proven to give high yields (Ravigadevi et al. 2002).

2 Biodiesel Production from Palm Oil: Process Description

2.1 Oil Palm Cultivation and Harvesting

The production of FFB involves six (6) main processes which are summarized in Fig. 2. The planning stage involves the feasibility studies of the proposed area for plantation. Usually, environment impact assessment (EIA) forms part of the planning stage and the implementation of management measures to assuage the adverse effects of some social and environmental practices are also considered. Oil palm nursery proceeds after confirmation of the suitability of area for plantation which is normally endorsed by respective bodies for development. The seedlings are raised in polybags as nursery for about a year with adequate irrigation with manuring, etc. The land for the oil palm plantation is then cleared of vegetation. Creation of road or paths, water drainage systems, and other soil conservation measures are put in place before the actual transplant. Most often the vegetation is cleared by burning which affects the environment negatively. In order to control soil erosion after the seedlings transplant, leguminous crops are interspersed with the oil palm trees which further fix nitrogen into the soil. Other field maintenance practices include pruning, pest and disease control, and mulching. After 24–30 months of transplanting depending on the nutritious value of the soil, harvesting of FFB may be due (Corley and Tinker 2003). Normally, harvesting is done manually with chisels and sickles mounted on bamboo or aluminum poles. The FFB are then transported to the oil mill for oil extraction. In order to ensure minimal amount of free fatty acid (FFA) content of the oil, handling of FFB after harvesting must be done in a way to reduce bruises on the fruits. Also, since the quality of the oil produced depends on the time interval between harvesting and

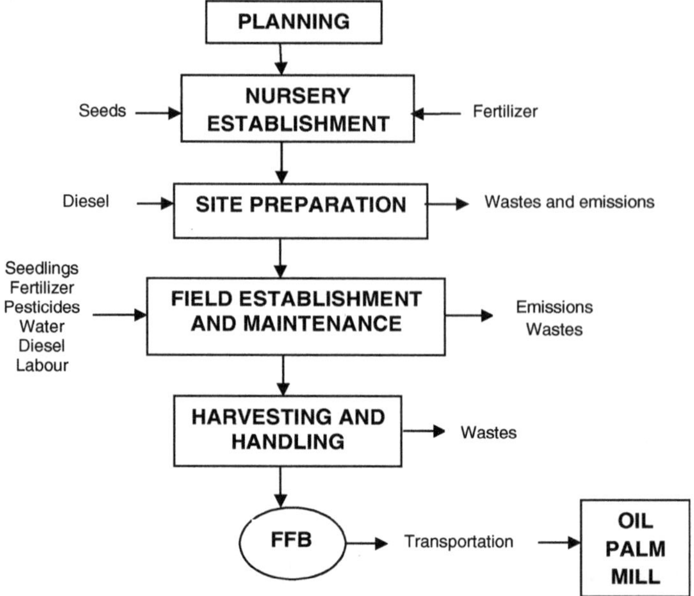

Fig. 2 Process flow diagram for oil palm cultivation

sterilization (the first stage of milling), FFB must be transported as soon as possible after harvesting and the distance from plantation to milling site must be close. Most oil mills are located near the plantation site to minimize the transportation distance and cost.

2.1.1 Environmental Interactions and Emissions from Palm Plantation

The cultivation and pretreatment of 1 tonne FFB emit approximately 285 kg of CO_2 eq. (Stichnothe and Schuchardt 2011). Oil palm cultivation has been reported as a major cause of substantial and irreversible damage to the natural environment (Schmidt 2007). Global warming potential (GWP) and eutrophication potential can be reduced by treating the palm oil mill effluents or co-composting the residues. In 2008, not less than 44 million tonnes of POME were generated in Malaysia which were and are still dumped in ponds releasing 5.5–9.0 kg of methane into the atmosphere for every tonne of FFB produced (Reijnders and Huijbregts 2008; Wu et al. 2010; Yacob et al. 2005).

Fertilizers applied to oil palm trees may be lost through volatilization and transformation to nitrous oxide (N_2O). Fertilizers may also contribute to nitrate and phosphate leakages to groundwater, hence causing water pollution. Paraquat (gramoxone) which is sprayed on the oil palm trees to kill herbs and weeds may leave about 11 mg (per kg body weight) of its content on the sprayer's skin after some few minutes (DEQ 1990).

Oil palm plantation has resulted in rampant deforestation, burning of forests, peat land degradation, and habitat loss of critically endangered species, and this direct land use has brought about significant emission of CO_2 and N_2O into the atmosphere. Forest fires used to clear vegetation for oil palm plantations are one source of CO_2. The smoke produced through forest burning can contribute to GWP as well as posing serious health threats to plantation workers and close neighbors. For instance, in Malaysia and Indonesia, 1997 recorded the highest CO_2 emissions resulting from bush burning since 1957 (Román-Cuesta et al. 2011). An estimated 0.81–2.57 gigatons of carbon was released into the atmosphere by the fires: 13–40 % of the mean annual global carbon emissions from fossil fuels in that year alone (Clay 2004). Again, in Indonesia and Malaysia, over 140 and 47 land mammalian species are endangered, respectively, as a result of oil palm plantation. Over 45 % of the total peat land has been converted to oil palm plantation due to the increase demand of palm oil, and this has put the leading producers of palm oil on top of major emitters of greenhouse gas (GHG) in the world. Currently, the annual cropland for oil palm plantation in Indonesia and Malaysia contributes about 2.63 t CO_2 eq. and 2.44 t CO_2 eq. per tonne FFB processed, respectively (Clay 2004). The situation becomes aggravated during deforestation and bogs draining which releases the peat bogs that store great quantities of carbon. Hence, appropriate management of plantation and the use of the biomass from the plantation as well as the processing residues from palm oil production (fibers, kernel shells, POME) for biofuel can have an effect on reducing GHG emissions.

2.2 Palm Oil Milling (Oil Extraction)

Figure 3 summarizes the main processes involved in the milling or extraction of palm oil from FFB. Sterilization of FFB is done in a steamer (pressurized cages) at about 2–3 bars to ameliorate the content of FFA which could reduce the quality of the oil. A rotation drum stripper is used to thresh the fruitlets from the sterilized bunches and the fruitlets sent to the digester. The EFB are also used as mulch in the oil palm plantation.

The digester then removes the fruits' mesocarp from the nuts by continuously heating the fruits with steam which helps to open the oil cells in the mesocarp for effective oil extraction. The oil extraction is done with the help of screw press where the press cake and nuts are conveyed to the palm kernel crushing (PKC) plant and the pressed liquor also sent to a vibrating screen where it is diluted. The oil is then clarified and purified to remove dirt and moisture before it is dried. The sludge (comprising mainly water soluble parts of the palm fruits and suspended materials like palm fibers) from the clarifier is desilted and further sent to the centrifuge to recover the excess oil which is recycled into the clarifier. The water–sludge mixture (palm oil mill effluent, POME) is then sent to the effluent treatment

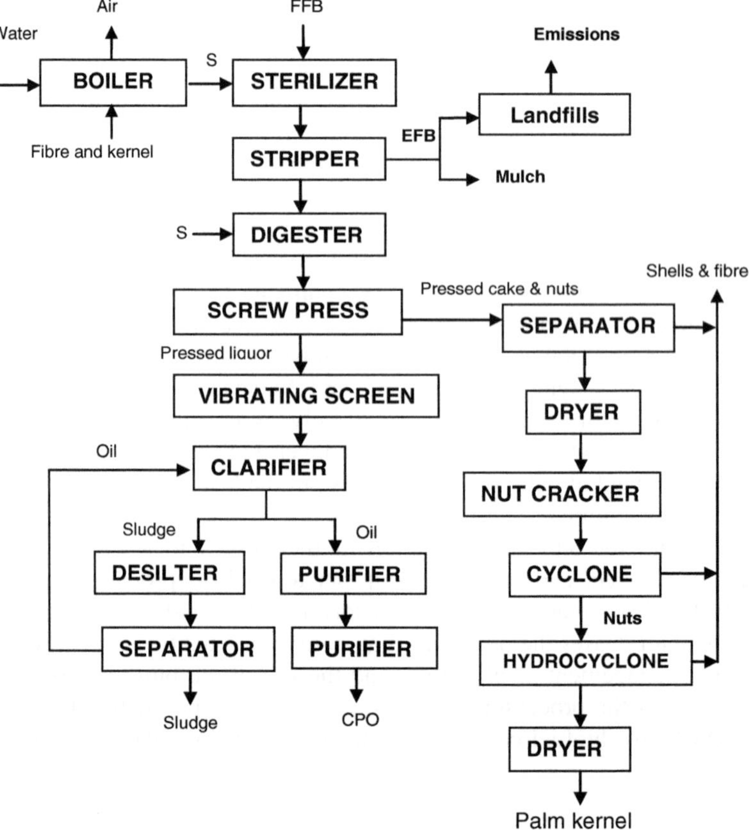

Fig. 3 Flow diagram of palm oil milling processes *S* represents steam

plant (ETP). The CPO produced is then stored and transported later to oil refinery. The palm kernel nuts are also cracked to separate the kernel from the shells. The oil palm fiber and kernel shells from the PKC plant are used as fuel in the boiler which generates steam for the oil milling processes.

2.2.1 Environmental Impacts Associated with CPO Production

The palm oil mill produces solid wastes such as palm pressed fiber (PPF), palm kernel shells (PKS), decanter cake, EFB, ash which are often damped without proper management or treatment. The EFB are also dumped in landfills or used as mulch in the oil palm plantation whose emissions also contribute to GWP. Due to some other problems associated with the use of EFB as mulch (including long decomposition period, high distribution and transportation cost), they can be used as fuel though it has a very small calorific value of 5 MJ/kg (Budiharjo 2010).

POME which are also a major liquid wastes from the palm oil mill are mostly mismanaged and disposed off wrongly. The direct rampant release of these effluents can cause water pollution which can affect downstream biodiversity and human beings. It has been reported that the average biochemical oxygen demand (BOD) of palm oil processing effluents is 25,000 ppm (Clay 2004). In Malaysia, for instance, effluents can legally be discarded into water bodies when their BOD levels are less than 100 ppm. However, the effluents also produce biogas mainly methane (Yusoff and Hansen 2007; Schmidt 2007) which can be tapped and used to generate electricity.

The biomass (mostly palm oil mills fruit fiber and kernel shells) powered combined heat and power (CHP) plants of the palm oil mills mostly operate without flue gas cleaning devices, hence causing the emissions of heavy metals and particulate matter which accounts for about 93 and 79 % of the human toxicity potential and heavy metals emissions to the air, respectively (Yusoff and Hansen 2007). Therefore, exhaust gas cleaning may help to reduce some of these environmental impacts drastically.

2.3 CPO Refinery

The refining of CPO helps to remove much FFA, odoriferous materials, phosphatides, waxes dirt, metal traces, etc., from the CPO. This process is achieved either through chemical or physical means. However, the physical process of CPO refining is the most commonly applied technology because of its simplicity, low capital cost, and high efficiency. Steam or physical refining involves degumming, bleaching, deodorizing, and fractionation into liquid olein and solid stearin fractions. CPO is acid treated in the degumming process to precipitate and separated out the gums or phosphatides. The oil is then bleached with activated clay or carbon under vacuum pressure to remove coloring pigment and metal ions. Deodorizing is carried out at high temperatures from 240 to 260 °C and pressure of 2–6 mmHg by injecting open steam which distills off the odoriferous matter present in the oil (Bockish 1998; Kheok and Lim 1982). The deodorized oil is then fractionated into palm olein and stearin by allowing the oil to crystalize under controlled temperature where the slurry passes through a filter press to obtain the stearin and olein fractions. The simple flow diagram of CPO treatment into refined palm oil (RPO) is shown in Fig. 4.

2.4 Transesterification of Palm Oil into Biodiesel

Biodiesel production from vegetable oil can be achieved through various means including pyrolysis, micro-emulsion, thermal cracking, transesterification. Transesterification of vegetable oil into biodiesel has been the most commonly

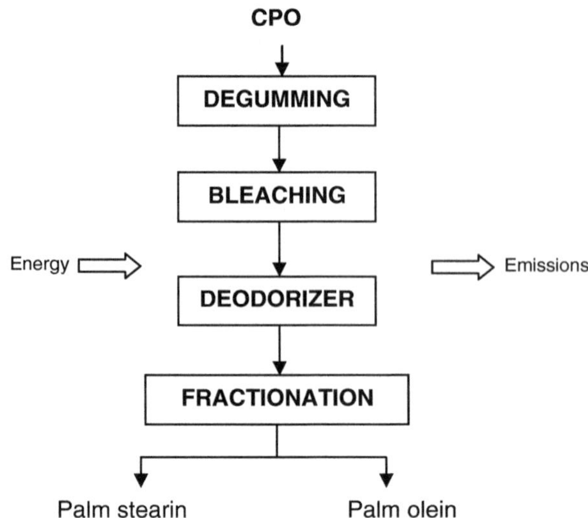

Fig. 4 Flow diagram of CPO refining processes

used technology due to its simplicity and environment-friendly processing. The catalyst (either sodium hydroxide, NaOH, or potassium hydroxide, KOH) is mixed with methanol in an agitator and the mixture made to react with the vegetable oil in a reactor at a temperature from 50 to 70 °C for 1–6 h. The resulting phases, i.e., glycerin and biodiesel phases usually containing some amount of methanol, are neutralized and then separated. Methanol is then recovered from the two phases with the help of distillation columns. The biodiesel is then purified by washing with warm water to remove soaps or excess catalysts, then dried and stored. Figure 5 shows the summary of flow diagram of transesterification processes of palm oil into biodiesel using alkaline catalyst.

2.4.1 Environmental Impacts Associated with the Transesterification of RPO into Biodiesel

In this chapter, homogenous base catalyst (NaOH) is used as the catalyst for transesterification thus there are bound to be the formation of soap together with the biodiesel, especially if the oil contains high amount of free fatty acids (FFA). The wastewater resulting from the washing of these soap stocks from the biodiesel is mostly released into water bodies untreated. Also, air emissions are released during the combustion of fossil fuel to produce energy to power the various unit operations within the plant. The transesterification unit is reported to contribute greatly to fossil fuel use compared to the other unit processes in the palm methyl ester (PME) production (Novizar and Dwi 2010).

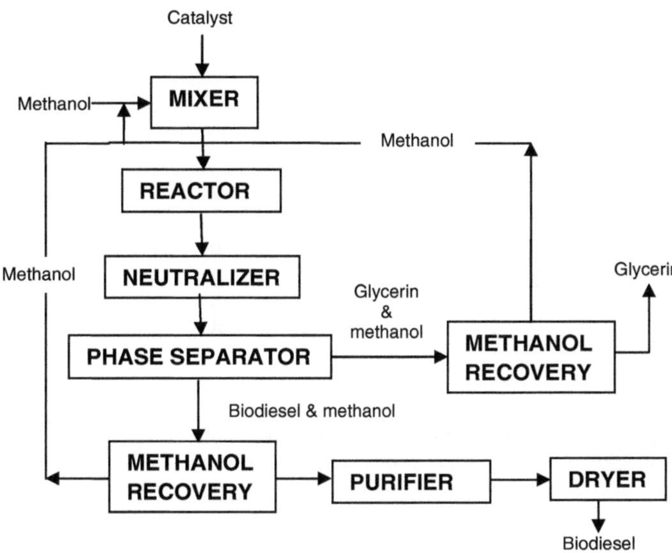

Fig. 5 Flow diagram of transesterification of palm oil into biodiesel

3 Tailpipe Emissions from Vehicles Using Biodiesel

Biodiesel is found to reduce tailpipe emissions from most vehicles compared to the conventional fuel such as petro-diesel. Tailpipe emissions such as hydrocarbon (HC), particulate matter (PM), carbon dioxide (CO_2), carbon monoxide (CO), sulfur dioxide (SO_2), nitrogen oxide (NO_x) are low with biodiesel use irrespective of the type of feedstock used.

Previous studies (Hitchcock et al. 1998; Turrio-Baldassarri et al. 2004) have carried out various investigations to compare the tailpipe emissions from conventional diesel and biodiesel. Figure 6 shows a summary of the contribution of tailpipe emissions from different vehicles that run on biodiesel and petro-diesel. It can be seen from Figure 6 that the emissions from biodiesel combustion are significantly lower than those for petroleum diesel. Nitrogen oxide emissions from biodiesel combustion, however, are slightly higher. Emissions of NO_x from biodiesel combustion can be reduced substantially by advancing the injection time. CO_2 emissions were also insignificant from almost all the vehicles running on biodiesel since CO_2 emitted during biodiesel combustion is recycled into the photosynthesis process in plants which is not so with CO_2 produced by fossil fuel combustion. Tailpipe emissions also differ in amount with different vehicles. The graph also shows that new buses (NB) and heavy goods vehicles (HG) running on biodiesel recorded a higher NO_x emissions compared to smaller cars and old cars.

Turrio-Baldassarri et al. (2004) reported the tailpipe emissions from buses which ran on biodiesel. Their results indicated that emissions of CO were 20 % lower than those for conventional diesel. SO_2 tailpipe emissions were also reduced

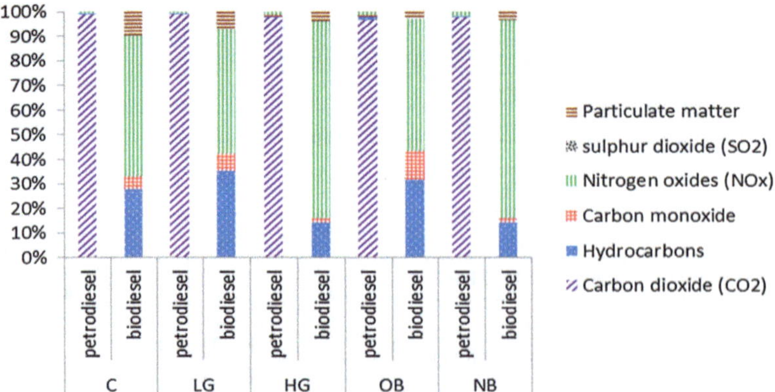

Fig. 6 Tailpipe emissions from vehicles using biodiesel and conventional diesel data source: Hitchcock et al. 1998. *C* Car, *LG* light goods vehicle, *HG* heavy goods vehicle, *OB* old bus, *NB* new bus

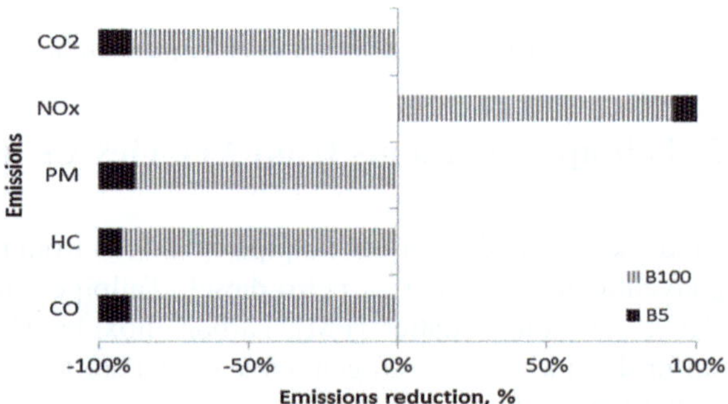

Fig. 7 Potential emission reduction from Tailpipe emissions from vehicles using biodiesel

by almost 100 %, while particulate matter was reduced by about 40 %. Figure 7 shows the potential reduction in tailpipe emissions from pure biodiesel (B100) and a diesel with 5 % biodiesel and 95 % petro-diesel (B5).

Engine modifications, efficient designs, and PM filters can also help to reduce tailpipe emissions from the use of biodiesel. Also, efficient recirculation of exhaust gas can help to reduce combustion temperature and pressure which leads to NO_x emissions reduction.

4 Life Cycle Assessment Methodology

LCA methodology used in this study followed the principles and framework of the International Organization for Standardization, ISO 14040 and 14044, which comprises four major steps that are summarized in Fig. 8. Some inventory data

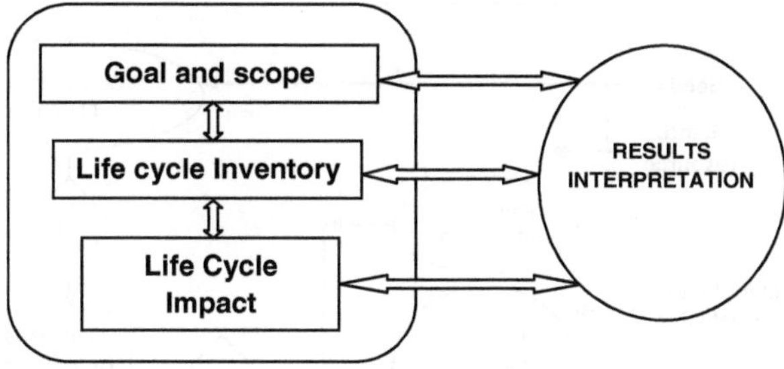

Fig. 8 Life cycle assessment methodology framework

were taken from GABi database and Ecoinvent 99 database version 2.1. GaBi 4.2 LCA software was used for the data analyses.

4.1 Goal and Scope Definitions

The goals of this chapter are (1) to come up with an inventory of resources associated with the production of palm oil biodiesel over its entire life cycle (from cradle to grave), (2) to identify and compare the major environmental impacts (damages) associated with each life cycle stage of the production of biodiesel from palm oil, (3) to identify the most important environmental loads on the production systems and suggest improvement measures. The focus of this chapter is geared toward the energy balance of biodiesel production from palm oil and its effect on the emissions of greenhouse gases, air, water, and solid waste pollutants on the environment. The assessment of the domestic economic importance of using the palm oil biodiesel does not fall within the scope of this chapter.

The scope according to ISO 14040 and 14044 includes the system boundary definition, the functional unit, allocation steps, temporal and geographical boundaries, data quality requirement, technology coverage, etc. This chapter dwells on palm oil production and conversion into biodiesel based on Malaysia's conditions (including plant location, feedstock origin, sources of electricity, and end-uses). Thus, geographical boundaries are not considered in this chapter as there are no assumptions of imports of biodiesel into the country.

4.1.1 System Boundary and Functional Unit

Figure 9 shows the system boundary for the production of biodiesel from palm oil considering the stages from cultivation of palm fruits to transesterification of palm oil into biodiesel. However, Figs. 2, 3, 4, and 5 detail the process flows of oil palm cultivation, palm oil milling, palm oil refining, and transesterification of palm oil into biodiesel, respectively. The major considerations within the system boundary for this chapter include the production of FFB (which comprises nursery stage and

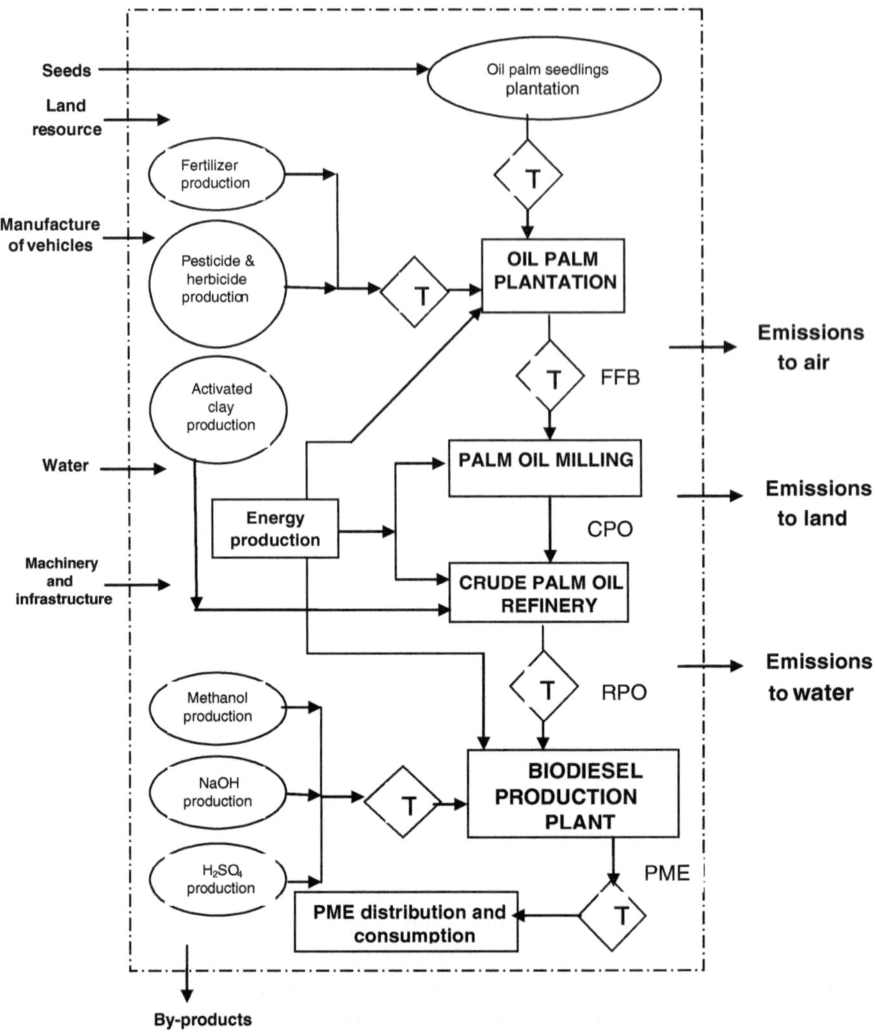

Fig. 9 LCA system boundary of biodiesel production from palm oil

oil palm plantation), transportation of FFB to crushing or milling facility, recovery of crude palm oil from the mill and refining of CPO, transportation of refined palm oil to biodiesel production plant and transesterification of RPO, transportation of palm oil methyl ester (PME) to consumers and finally the use of PME in diesel engines. Each of these stages has various substages which are detailed in Sect. 2.

Energy (such as electricity, fossil-based diesel) as well as the environmental inputs of the supply chain for the production of the raw materials used in each stage is also included in the system boundary. The life cycle environmental impacts associated with the production of machinery, infrastructure and land for the cultivation of FFB, palm oil milling and conversion into biodiesel are excluded from the system boundary for this chapter. This assumption is based on the results from previous studies (Schmidt 2007; Novizar and Dwi 2010) which report negligible contributions because on a basis of per kilogram inputs, small amount of energy is accounted for when the energy embodied in the machinery is distributed over the amount of outputs from the machine over its entire life cycle.

Again, the energy for constructing the biodiesel plant as well as the energy production facilities, such as hydropower plants, thermal plants, has equally negligible contributions of less than 1 % hence neglected in the system boundary for this chapter. This is because, on per kilogram basis of biodiesel, the plant would have very low energy and emission contributions since the energy embedded in fixed inputs would have to be distributed over the total biodiesel production during the lifespan of the plant (Schmidt 2007; Novizar and Dwi 2010). The production of seeds (for nursery) and organic fertilizer is also excluded from the system boundary. Organic fertilizers are assumed to be residues that are not produced specifically for oil palm cultivation.

The functional unit (FU) in LCA appends a reference to which the input and output resources are related. An FU of 1 kg of PME in Malaysia is chosen as the reference unit for all the input and output streams as well as the potential environmental effects.

4.1.2 Allocation Procedures

The choice of LCA allocation methods for multi-input/output process like biodiesel from palm oil is critical in quantifying the environmental burdens of the coproducts generated by the various unit operations (Tillman 2000) because they may have a significant impact on the final results (Bernesson et al. 2006). Since biodiesel production from palm oil generates many kinds of coproducts (by-products or wastes) such as oil palm fronds, empty fruit bunches, glycerin (depending on the raw material inputs and production processes employed), realistically the main product (biodiesel) should not carry all the environmental burdens. Allocations of such environmental burdens to the different coproducts are made based on ISO 14041 LCA allocation procedures. In this chapter, the main LCA allocation method used is the system expansion where no allocation is made ('avoiding allocation'). By this method, all the major unit processes to be allocated are divided into sub-unit processes. The PME is thus expanded to involve the other functions related to the coproducts, but PME is allocated the most share of the energy consumption within the process chain. This method has the advantage of modeling the indirect effects of the environmental burdens on the coproducts (Ekvall and Finnveden 2001). As reported by Bernessen et al. (2006) for systems whose coproducts can replace other products in later processes, the expansion method of LCA allocation is suitable for application in this chapter.

4.2 Life Cycle Inventory Analysis

4.2.1 Data Collection

Life cycle inventory (LIC) is a methodology for estimating the utilization and consumption of resources and the amount of waste streams and environmental emissions ascribed to a product's life cycle. The LCI analysis used in this chapter focused on materials and energy resource use, air emission, water emission, soil emissions, land use, and other wastes involved in the life cycle of biodiesel production based on 1 kg of 100 % PME. Data used for the LCI analysis in this chapter were obtained from plant reports, literature reviews (Yusoff and Hansen, 2005; Schmidt 2007), Ecoinvent database (GaBi Software and database for life cycle Engineering 4 2003), experimental results (Novizar and Dwi 2010; Choosak et al. 2009), and estimations based on Malaysia's oil palm industry (MPOB 2006). Table 2 shows the summary of materials and energy resources as inputs and outputs for producing 1 kg biodiesel from palm oil. The most important parameters associated with the key environmental impacts of biofuels were estimated for each impact category as detailed in Sect. 4.3 of this chapter.

4.2.2 Assumption and Limitations

For 1 ha of land, the oil palm plantation produces averagely 20 t of FFB annually which yields about 4.6 t of mesocarp oil (crude palm oil), and 0.750 t PKS which produces about 0.250 t of PKO, 0.500 t of kernel meal, and 300 t of POME. For the same size of oil palm plantation, about 18 t of oil palm fronds (OPF), 3 t of oil palm trunks (OPT), 15 t of EFB, and 3 t of palm pressed fiber are produced annually. In this chapter, methane from POME is assumed to be emitted into the air. However, efforts are currently being made to trap the methane as biogas for energy production. OPF and OPT are also assumed to be used as mulch in the oil palm plantation.

In Malaysia, the cultivation of 1 hectare land of oil palm requires about 191 kg nitrogen/year, 62 kg phosphorus oxide/year, 318 kg potassium oxide/year, and 98 kg magnesium oxide/year (Ng and Thamboo 1967). Comparing these fertilizer quantities to those applied in Nigeria, Malaysia's conditions require quite higher fertilizer quantities for the same hectare of land use in Nigeria. In Nigeria, 1 hectare of oil palm plantation requires about 149, 48, 236, and 93 kg nitrogen, phosphorus oxide, potassium oxide, and magnesium oxide per year, respectively (Tinker and Smilde 1963). Before field planting, the nursery also receives some amount of fertilizer in the form of nitrogen, phosphorus oxide, potassium oxide, and magnesium oxide. The first few weeks (from 8 to 24 weeks) may require little fertilizer (from 3.5–10 g fertilizer per seedling) bi-weekly. From the first year to the time of transplanting, in every 3 weeks, a seedling of oil palm may require about 12 kg nitrogen (from ammonium sulfate), 12 kg of phosphorus oxide (from

Table 2 Life cycle inventory of biodiesel production from palm oil

	Unit	Quantity	Energy coefficient (MJ/kg)	Total energy (MJ)
Nursery				
Input	No.	0.001336	33.6400	0.044943
Seeds				
Water	kg	0.698280	0.0042	0.002933
Fertilizer				
Nitrogen (N)	kg	9.259E−6	48.9000	0.004528
Urea	kg	1.346E−5	22.5000	3.029E−4
Phosphate (P_2O_5)	kg	2.934E−6	17.4300	5.114E−5
Potassium (K_2O)	kg	2.094E−5	10.3800	2.174E−4
Magnesium (MgO)	kg	5.011E−6	2.3200	1.163E−5
Boron (Borate)	kg	1.179E−6	32.2700	3.804E−5
Pesticides and herbicides				
Glyphosate	kg	0.000282	18.6200	0.005251
Paraquat	kg	0.000141	130.0000	0.018330
Furadan	kg	0.000211	13.1580	0.002776
Human energy	MJ			0.000037
Transportation of chemicals				
Diesel	kg	0.000604	48.1000	0.029052
Total energy input	MJ			1.08E−1
Output				
Oil palm seedlings	No.	0.001336	36.0400	0.048149
Emissions to soil				
Nitrogen (N)	kg	1.474E−6	48.9000	7.208E−5
Phosphate (P_2O_5)	kg	5.890E−7	17.4300	1.027E−5
Glyphosate	kg	0.000115	18.6200	0.002141
Paraquat	kg	0.000096	130.0000	0.012480
Furadan	kg	0.000104	13.1580	0.001368
Emissions to air				
NO_x	kg	1.474E−7	296.0000	4.363E−5
CO_2	kg	0.001657	32.1200	0.053222
SO_2	kg	5.895E−8	29.5000	1.739E−6
Total energy output	MJ			0.048149
Transportation of seedlings (T1)				
Diesel	kg	0.004692	48.1000	0.225685
Emissions to air				
CO	kg	0.014683	10.1100	0.148445
NO_x	kg	0.000054	296.0000	0.015984
Particulate matter	kg	0.000027	–	–
SO_2	kg	2.165E−5	29.5000	0.000639
TOC	kg	0.001457	–	–
Transplanting				
Input				
Seedlings	No.	0.001336	–	–

(continued)

Table 2 (continued)

	Unit	Quantity	Energy coefficient (MJ/kg)	Total energy (MJ)
Water	kg	2179.511	4.2E−3	9.153950
Fertilizer				
Nitrogen (N)	kg	0.02890	48.90	1.413210
Urea	kg	0.04200	22.50	0.945000
Phosphate (P_2O_5)	kg	0.00916	17.43	0.159659
Magnesium (MgO)	kg	0.01564	2.32	0.036285
Borate	kg	0.00368	32.27	0.118754
Potassium (K_2O)	kg	0.06536	10.38	0.678437
Pesticides				
Glyphosate	kg	0.00184	18.72	0.034445
Paraquat	kg	0.00070	130.00	0.091000
Furadan	kg	0.00299	13.16	0.039340
Herbicides	kg	0.00015	184.71	0.027707
Wood chippings	kg	3.54311	19.00	67.31909
Field establishment	kg	0.02659	48.1	1.278883
Diesel				
Field maintenance				
Diesel	kg	0.01329	48.1	0.639441
Electricity	MJ	1.23924	–	1.239240
Labor (harvest)	MJ	0.00600	–	0.006000
Total energy input	MJ			83.18044
Output				
FFB	kg	4.17000	43.33	180.6861
OPF	kg	1.50120	20.51	30.78961
OPT	kg	0.03336	16.88	0.563117
Emissions to soil/water				
Nitrogen (N)	kg	0.00460	48.90	0.224940
Phosphate (P_2O_5)	kg	0.00184	17.43	0.032071
Pesticides	kg	0.00037	245.57	0.090861
Emissions to air				
NOx	kg	0.00046	296.00	0.136160
SO_2	kg	0.00018	29.50	0.005310
Pesticides/herbicides	kg	0.00009	245.57	0.022592
CO_2	kg	0.01092	32.12	0.350706
Total energy output	MJ			212.9018
Transportation of raw materials to mill (T2)	kg	0.02346	48.10	1.12843
Diesel				
Emissions to air				
CO	kg	0.07342	10.11	0.74228
NOx	kg	0.00027	296.00	0.07992
Particulate matter	kg	0.00014	–	–
SO_2	kg	0.00011	29.50	0.00325
TOC	kg	0.00729	–	–

(continued)

Table 2 (continued)

	Unit	Quantity	Energy coefficient (MJ/kg)	Total energy (MJ)
Palm oil mill				
Inputs				
FFB	kg	4.17	43.33	180.686
Water	kg	6.58076	4.2E−3	0.02764
Energy consumption				
Electricity	MJ	–	–	0.29440
Diesel	MJ	0.00012	48.10	0.00578
Steam	MJ	5.47941	1.36	7.45200
Total energy input	MJ			188.4658
Output				
CPO	kg	0.94	31.40	29.5160
POME	kg	2.3	–	–
EFB	kg	1.0764	20.47	22.13391
PPF	kg	0.7176	19.22	13.79227
PKS	kg	0.6532	21.44	14.00461
Emissions to soil/air/water				
Steam	kg	5.479411	1.36	7.45200
NOx	kg	0.000588	296.00	0.17405
CO	kg	0.005188	10.11	0.05245
CO_2	kg	1.199897	32.12	38.5407
Particulate matter	kg	0.001269	–	–
SO_2	kg	0.000018	29.50	0.00053
TOC	kg	0.000046	–	–
VOC	kg	0.003607	–	–
Biogas from POME storage	MJ	0.0644	–	2.31800
BOD	kg	–		–
COD	kg	–		–
Nitrates	kg	0.000759	22.5	0.01708
Ash	kg	0.045	19.61	0.88245
Total energy output				128.884
CPO refining				
Inputs				
CPO	kg	0.94	41.86	39.3484
Activated clay	kg	0.00846	34.54	0.29222
Electricity	MJ	–	–	0.31505
Steam	MJ	–	–	4.96871
Total energy input	MJ	–	–	44.9244
Outputs				
RPO	kg	0.92500	31.40	29.0450
Emissions to air				
CO_2	kg	4.97541	32.12	159.810
SO_2	kg	0.00012	29.50	0.00345
NO_x	kg	0.00029	296.00	0.08622

(continued)

Table 2 (continued)

	Unit	Quantity	Energy coefficient (MJ/kg)	Total energy (MJ)
Total energy output	MJ			188.945
Transportation of RPO to biodiesel plant (T3)				
Diesel	kg	0.02531	48.10	1.22029
Emissions to air				
CO	kg	0.07920	10.11	0.80076
NOx	kg	0.00038	296.00	0.11248
Particulate matter	kg	0.00015	–	–
SO_2	kg	0.00017	29.50	0.00502
Transesterification				
Input				
RPO	kg	0.92500	31.40	29.04500
Methanol	kg	0.12278	19.70	2.418766
Sodium hydroxide (NaOH)	kg	0.01031	19.87	0.204879
Water	kg	0.29576	0.0042	0.001242
Phosphoric acid (H_3PO_4)	kg	0.00959	32.62	0.312826
Energy consumption				
Steam	MJ	0.21600	1.36	0.293760
Electricity	MJ	0.04214	–	0.042140
Total energy input	MJ			**32.31861**
Outputs				
Biodiesel	kg	1.00000	39.84	39.84000
Glycerol	kg	0.23860	18.05	4.306730
Wastewater	kg	0.05359	45.93	2.461388
Na_3PO_4	kg	0.00146	1.421	0.002075
Emissions to air				
CO_2	kg	2.55281	32.12	81.99626
Total energy output	MJ			128.6065
Transportation of biodiesel to consumer (T4)				
Diesel	kg	0.02490	48.10	1.1978
Emissions to air				
CO_2	kg	0.07792	32.12	2.5027
NO_x	kg	0.00028	296.00	0.0829
Particulate matter	kg	0.00014	–	–
SO_2	kg	0.00011	29.50	0.0032
TOC	kg	0.07732	–	–
Biodiesel use in diesel engine				
Biodiesel	kg	1.00000	39.84	39.840
Emissions to air				
CO_2	kg	−16.0890	32.12	−516.78
NO_x	kg	0.01360	296.00	4.0256
Particulate matter	kg	−0.00454	–	–

(continued)

Table 2 (continued)

	Unit	Quantity	Energy coefficient (MJ/kg)	Total energy (MJ)
SO_2	kg	−0.00454	29.50	−0.1339
Hydrocarbons (HC)	kg	−0.00454	–	
CO	kg	−0.05543	10.11	−0.5604

All quantities are wet weight averages from (MPOB 2010; Felda 2010; Schmidt 2007; Subramaniam et al. 2008; Yusoff 2006; Wicke et al. 2008)
Net energy value (NEV) = energy content of PME − net energy inputs = 20.28 MJ
NRnEV = energy content of PME − fossil energy inputs = 24.63 MJ
Net energy ratio (NER) = net energy outputs/net energy inputs = 1.4893

rock phosphate), 17 kg of potassium oxide, and 2 kg of magnesium oxide or kieserite (Von Uexkull and fairhurst 1991; Hartley 1988). At the early stage after transplanting, urea and limestone may be applied to the young palms at a rate of 6–8 kg/palm tree (Von Uexkull and fairhurst 1991). Borate is currently applied to oil palms up to the age of 6 years. Paraquat and glyphosate as herbicides are also applied twice or thrice a year at a rate of 0.625–1.25 kg/ha/time and 1.875–3.125 kg/ha/time, respectively (Von Uexkull and fairhurst 1991). Water requirement for the oil palm cultivation is assumed to be from rain water and irrigation at early stages of transplanting. The CHP system of the mill is assumed to utilize the PPF and palm kernel shells as fuel to produce heat for steam and electricity generation.

4.2.3 Transportation

The nursery field is assumed to be about 1.7 km away from the oil palm plantation field. The oil mill is assumed to be situated closer to the oil palm plantation; hence, the distance is negligible. CPO transportation to CPO refinery is assumed to be part of the biodiesel production plant which is estimated to be 296 km away from the oil mill.

4.2.4 Energy Analysis

For the production of 1 kg PME with energy content of 39.84 MJ, the NEV and NER of the whole life cycle of PME production are 20.28 MJ per kg of PME and 1.489 (without energy production from EFB, PPF, etc.), respectively. The NER would have been 4.81 if all EFB, PPF, PKS, etc., were considered as energy source (which is considered in this chapter). This clearly shows an energy profit for the PME production system. The total life cycle energy consumption of the PME product system is shown to be 20.28 MJ per kg PME.

The main energy supply to the palm oil mill includes diesel, electricity, and steam. PPF and PKS which are regarded as wastes from the palm oil mill are normally used to produce energy. Figures 10 and 11 summarize the energy contributions of various inputs into the various production stages excluding the transportation stages. Figure 12 shows the contribution of the total energy inputs by the four main transportation stages associated with the production of biodiesel from palm oil. Due to the conversion of solar radiation to biomass by means of photosynthesis, the chemical energy content of the harvested FFB and other biomass exceeds the energy input through the farming system. Oil palm is therefore regarded as a net source of useful energy (Corley and Tinker 2003). From Fig. 10, the consumption of water, fertilizer, pesticides, and chemicals as well as human work was highly recorded in the oil plantation stage. For chemical (including major input materials such as FFB for oil mill, wood chips for plantation) consumption, the energy inputs between FFB and wood chips were high which triggered the high values for the oil milling (57 % of total chemicals) and plantation (21 % of total chemicals) stages, respectively. The PPF and EFB (as part of chemical inputs in this chapter) from the mill with dry calorific values of 19.22 MJ/kg and 20.47 MJ/kg, respectively (Yusoff 2006), are used as fuel to produce steam and electricity for use within the mill.

Though the transesterification stage consumes many chemicals, the energy contents of these chemicals are quite low (10 % of total input chemicals), hence reducing the total energy consumption from chemicals in that stage. On the other hand, CPO refinery stage consumes activated clay with high heating value, hence increasing the energy contents of the chemical use (12 % of total energy of chemicals used) within that stage.

The oil mill, however, recorded the second highest (13 % of total input fuel excluding transportation) consumption of energy (including fossil and non-fossil fuel from EFB, PPF) compared to all the other stages. Since it was assumed in this chapter that 1 kg biodiesel was used in the "end use" or combustion stage, the

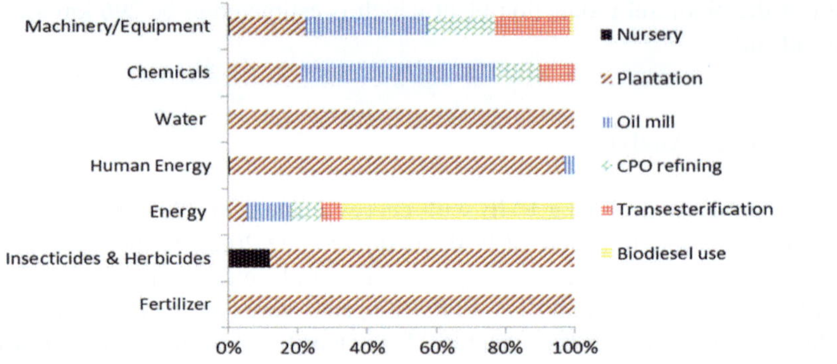

Fig. 10 Percent energy contribution of inputs into PME production by the various production stages (without transportation stages). Chemicals include catalyst, methanol, H_3PO_4, RPO, FFB, activated clay, etc. Energy includes diesel fuel, electricity, steam

highest energy consumption (67 % of total input fuel) was recorded in this stage. This shows positive environmental impacts because it releases insignificant air emissions upon combustion.

Plantation stage recorded highly significant human energy input (97 % of total human energy input) compared to other energy inputs (5 % of total energy input) such as diesel.

According to Henson (2004), palm oil mills are self-sufficient for electricity and heat. It has been reported (Husain et al. 2003) that the total heat and power generation for every tonne of FFB is about 1181 MJ (approximately 0.7 t steam). Within the mill, energy could be released as emissions into the atmosphere which is estimated to be 16 MJ per tonne FFB (Subramaniam 2006). It is assumed that the energy produced is more than the energy required by the mill; hence, the surplus is released into the atmosphere.

The highest water requirement for the whole production came from the oil palm plantation stage (contributing 99.6 % of the total water requirement).

Generally, according to Fig. 11, the highest energy consumption within the whole production cycle excluding transportations was obtained from the input chemicals/materials (which included EFB, PPF, etc.) with contribution of 81 % of total energy of inputs. Energy inputs from fossil and non-fossil fuel contributed about 16 % of the total energy inputs. Herbicide and pesticide use within both the nursery and plantation stages carried the least energy content of about 0.06 % of the total energy of inputs into the PME production.

The production of 1 kg of PME requires approximately 63.17 MJ of energy in the form of fuel (fossil and non-fossil fuel) and 396.67 MJ energy in the form of other raw materials, machinery, etc., including diesel consumption from transportation stages. The transportation of PME from biodiesel production plant to the consumer recorded the highest diesel consumption (32 %) compared to all the other transportation stages. This is attributed to the total distance covered by the truck delivering the raw materials.

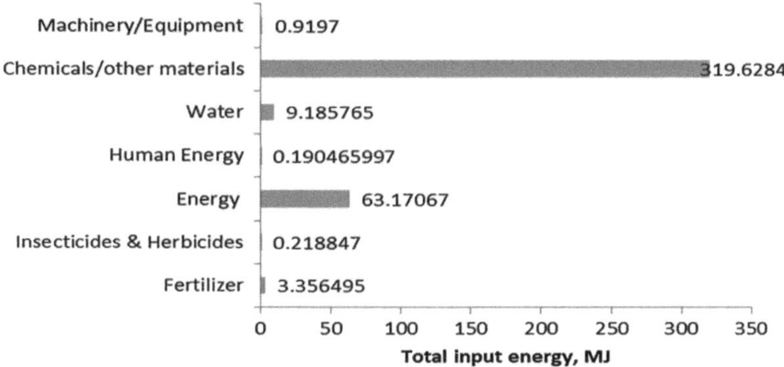

Fig. 11 Total energy inputs into PME production by all the stages within the system boundary

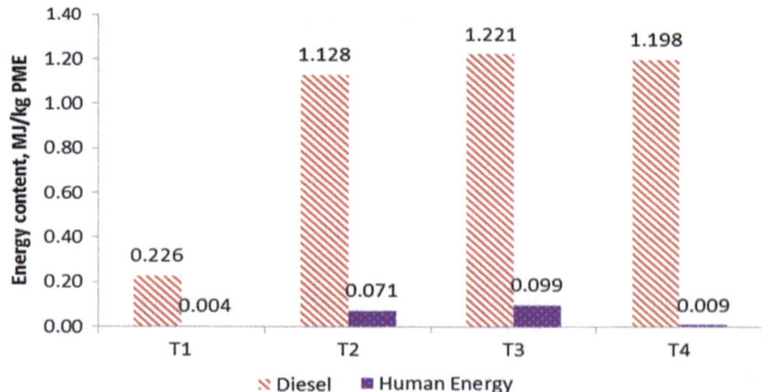

Fig. 12 Total energy inputs into PME production by the transportation stages. *T1* Transportation of seedlings to plantation site, *T2* transportation of FFB from plantation site after harvest to oil mill, *T3* transportation of RPO to biodiesel production plant, *T4* transportation of PME to consumer

4.3 Life Cycle Impact Assessment

Life cycle impact assessment (LCIA) is a major step in LCA which provides basic indicators for analyzing the potential environmental contributions of all the resource extractions including wastes and emissions. Eco-Indicator 99 (EI 99, Agalitarian Approach [AH]) was used to assess the environmental impacts associated with the life cycle of palm oil biodiesel. Standard LCIA comprises (1) impact categories selection and classification, (2) characterization, and (3) valuation steps.

4.3.1 Impact Category Selection and Classification

The potential environmental impact categories selected for this chapter according to EI99, AH method of LCIA, include land use/conversion (PDF*m^2*a)[1,2] acidification/nitrification potential (PDF*m^2*a), ecotoxicity potentials (PDF*m^2*a), fossil fuel use (MJ surplus energy), mineral resources (MJ surplus energy), climate change (DALY),[3] ozone layer depletion potential (DALY), radiation potential (DALY), carcinogenic effects (DALY), respiratory organics (DALY), and respiratory inorganics (DALY). The complete human health impact is obtained by adding up all the DALY values, while the ecosystem impacts and resource depletion are obtained by adding up the PDF and surplus energy, respectively. These categories were selected based on their relevance for assessing the

[1] PDF: Potentially disappeared fraction (plant species disappeared as a result of the impacts).

[2] a: year (annually).

[3] DALY: Disability adjusted life years (years of life lost due to the impacts).

environmental emissions associated with biofuel systems and those suggested by previous researches (Guinée et al. 2001; Edwards et al. 2007; Buchholz et al. 2009). These impact categories are evaluated using LCA software such as Simapro (developed by PRé Consultants), Gabi 4.2 (developed by PE International), Umberto (developed by IFU Hamburg GmbH), while the databases from Eco-Indicator 99 or 95, CML 2001 or 1996, Environmental Design of Industrial Products (EDIP 1997) or EDIP 2003, etc., are used to evaluate the final environmental impacts. Other LCIA methods that are implemented in Ecoinvent database include cumulative energy demand, ecological scarcity 1997, environmental priority strategies in product development (EPS 2000), IMPACT 2002, IPCC 2001 (climate change).

The main regulated pollutants evaluated in this chapter include CO, particulate matter, non-methane hydrocarbons, nitrogen oxides (NO_x), etc. Solid wastes, water, and CO_2 emissions as well as overall energy requirements are also evaluated. Each of these emissions and pollutants are classified into their main environmental impact category. For instance, CO_2 and CH_4 emissions were classified under climate change, NO_2 and SO_2, on the other hand, contribute to acid rain formation and some degree of direct effect on human health hence classified accordingly.

4.3.2 LCIA Characterization

This step involves the quantification of the extent to which each pollutant or emission contributes to different environmental impacts. Standard characterization factors conforming to Eco-Indicator 99 (EE99, EA) evaluation procedures are used in this stage. In this method, human health category is measured in DALY/kg FU, while ecosystem impacts and resource depletion are also measured in PDF*m2*a/kg FU and surplus energy/kg FU. These are available in Gabi 4.2. On the other hand, using CML 2001 database for the impacts evaluation generate different impact categories units. For instance, acidification potentials of NOx and SO_2 are based on proton formation potentials (PFP) (i.e., 0.7 for NOx and 1 for SO_2) expressed as SO_2 equivalent per FU. Thus, the total acidification potentials of 10 g NO_2 and 5 g SO_2 are given by $(10 \times 0.7) + (5 \times 1) = 12$ g SO_2 equivalent per FU. This is estimated by multiplying the amounts of the emissions by their proton formation factors and aggregating the results of these multiplications for each impact category.

Again, global warming potentials are based on CO_2 equivalent, while ozone layer depletion potentials are measured in CFC-11 equivalent. These characterization factors can be extended through normalization, grouping, and weighting. In normalization, the results of the impact categories are usually compared with the total impacts in the area of interest, for instance, in this chapter, Malaysia. Grouping also involves the sorting and ranking of the impact categories. In weighting, the different environmental impacts are weighted relative to each other,

Table 3 Characterization of air and water emissions in PME production

Type of emission	LCIA category
Emissions to air	*GWP (CO_2 eq./g)[a] – 100 years*
Carbon dioxide (CO_2)	1
Nitrous oxide (N_2O)	310
Carbon monoxide (CO)	3
Methane (CH_4)	21
Carbon tetra-fluoride (CF_4)	6,300
Sulfur hexafluoride (SF_6)	23,900
Hydro fluorocarbon (HFC)	140–12,100
Per-fluorocarbons (PFC)	6,500–9,200
	AP (SO_2 eq./g)[b]
Sulfur dioxide (SO_2)	1
Oxides of nitrogen (NO_x)	0.7
Hydrochloride acid (HCl)	0.88
Hydrogen fluoride (HF)	1.6
Nitrogen monoxide (NO)	1.07
Nitrogen dioxide (NO_2)	0.7
Ammonia (NH_3)	1.8
Emissions to water	*EP (PO_4 eq./g)[c]*
Phosphates (PO_4^{3-})	1
Nitrates (NO_3)	0.42
Nitrogen oxides (NO_x)	0.13
Ammonia (NH_3)	0.33

[a] U.S. DOE/EIA (1997)
[b] Heijungs (1992)
[c] Mark et al. (2001)

summed up to obtain a single number for the total environmental impact. Table 3 shows a summary of LCIA classification and characterization estimations associated with PME production.

4.3.3 Valuation

This step uses results from the LCIA to evaluate each process for improvements in the performance of every stage associated with the life cycle of palm oil biodiesel.

5 LCA Results and Interpretation

The major objective of this chapter's LCA interpretation is to detect or assess the points of potential environmental impacts which can lead to overall improvement of the performance of the palm oil biodiesel production industries in the world, especially in Malaysia. Figure 13 summarizes the environmental impact associated with each life cycle stage of PME production. Figure 14 shows the environmental

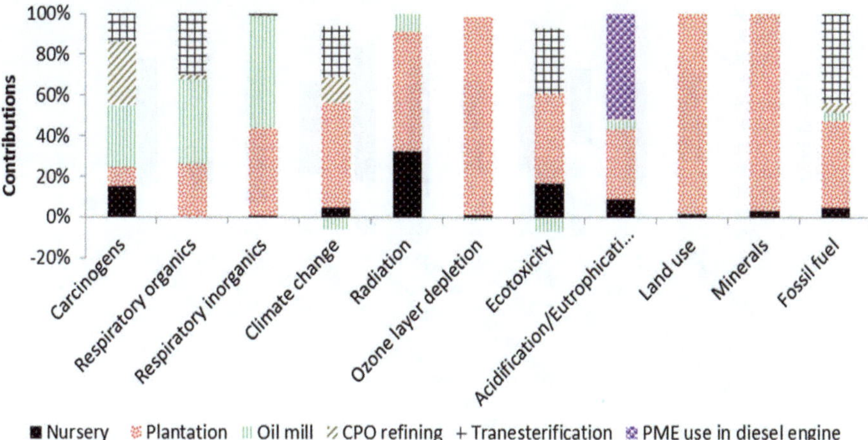

Fig. 13 Environmental impact potentials for 1 kg PME (excluding impacts from transportation)

impacts associated with the four main transportation stages within the PME production. Figure 15 summarizes the total environmental impacts (including transportation stages) within each of the production processes within the life cycle of palm oil biodiesel.

5.1 Emissions Associated with Oil Palm Nursery System

Out of the eleven impact categories, six of them, namely radiation (32.57 % of total radiation potentials), carcinogens (14.79 % of total carcinogens), ecotoxicity (19.11 % of total ecotoxicity), climate change (5.18 % of total climate change), land use (1.49 % of total land use), and acidification/eutrophication (8.81 % of total acidification) potentials, were highly significant (Fig. 13) excluding the impacts from transportation stages. The main emissions associated with these impacts include the use of herbicides and pesticides (radiation, ecotoxicity, carcinogens, etc.) and fertilizers (ecotoxicity, acidification, radiation). N-fertilizers emit N_2O into the air which contributes to the climate change effects. The use or spraying of herbicides and insecticides also emits particulate matter into the air.

Combining the effects on all impact categories as a single score, it can be seen that the nursery stage contributed only 1 % environmental impacts for 1 kg production of PME (Fig. 15). On the basis of human health, ecosystem depletion, and resource use, the nursery stage contributed 0.358, 0.192, and 0.188 %, respectively, for 1 kg PME.

In order to further reduce these impacts, the use of organic fertilizers can replace inorganic ones. Glyphosate and paraquat as pesticides and herbicides must

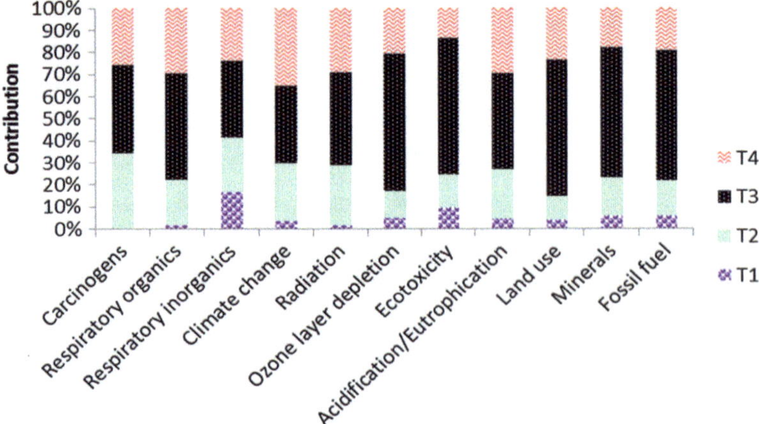

Fig. 14 Environmental impacts associated with transportation stages in the life cycle of PME. *T1* Transportation of oil palm seedlings to plantation site, *T2* transportation of FFB from palm plantation site to oil mill, *T3* transportation of RPO to transesterification unit, *T4* transportation of biodiesel from transesterification unit to consumer

be used in a minimal quantities, while efficient treatment of oil palm seeds are used for nursery.

5.2 Emissions Associated with Oil Palm Plantation

Emissions from the plantation are determined from material balance of the major substances such as N_2O, CO, CO_2, particulate matter into and out of the production stage. Since oil palm is a perennial crop, during the life cycle for the generation of FFB and uptake of nutrients, the harvesting and decomposition of biomass residues varies, hence making emissions data unavailable at early stages.

In the plantation stage, all the impact categories were significant compared to the other production stages but were higher for land use (98.51 % of total land use), minerals (96.75 % of total minerals), radiation (58.62 % of total radiation), climate change (58.42 % of total climate change), ecotoxicity (51.79 % of total ecotoxicity), respiratory inorganics (43.46 % of total respiratory inorganics), fossil fuel use (46.62 % of total fossil fuel use), and acidification (34.58 % of total acidification) (Fig. 13) excluding impacts from transportation stages. The use of fertilizers, herbicides, pesticides, and diesel use were the main sources of these emissions. Organic fertilizers could be used in place of inorganic ones in order to reduce some of these emissions. The commonly used herbicide, paraquat dichloride, is found to also emit quite substantial amount of minerals and metals into the soil. Table 4 shows the emissions of heavy metals/minerals associated with the production of 1 kg FFB from the plantation stage.

Table 4 Emissions of heavy metals from the production of 1 kg FFB

Heavy metal type	Emissions, mg/kg FFB
Arsenic	0.0285
Cadmium	0.05429
Chromium	1.26857
Cobalt	0.00571
Copper	0.0.331429
Mercury	0.000857
Molybdenum	0.002857
Nickel	0.145714
Lead	0.057143
Selenium	0.011429
Zinc	1.96286

The harvests of FFB, irrigation, etc., are done manually, some requiring the use of trucks and other machinery which utilizes fossil fuel. This can also result in greenhouse gas emissions contributing to high climate change effect. Land use/conversion effect was highest (98.51 %) for the plantation stage due to heavy land use. It is recommended that, new oil palm plantations should be cultivation on degraded land in order to reduce land conversion/use effects. On a percentage score with transportations impacts inclusive, the plantation stage alone contributed approximately 7 % of the total environmental impacts from the whole production stages (Fig. 15). On the basis of human health, ecosystem depletion, and resource use, the plantation stage had a percent contribution of 13.10, 7.26, and 1.70 %, respectively.

5.3 Emissions Associated with the Palm Oil Mill

The most significant impact categories in the oil milling stage are carcinogens, respiratory organics, respiratory inorganics, and ozone layer depletion with percent shares of 30.83, 41.44, 55.02, and 55.51 % with transportations excluded (Fig. 13). The major parameters resulting in high potentials of these impact categories are the POME and the boiler ash. POME is the wastewater generated from the clarification and other processing steps in the mill. This is normally treated in open ponds in order to reduce its biological oxygen demand. The EFB in this chapter is considered to be used for fuel production within the mill hence no emissions from dumping sites resulting in climate change effect. Climate change potential is insignificant due to the use of renewable fuels from PPF to EFB in the mill. The boiler ash also contributes to emissions into the soil. In this chapter, most of the wastes within the mill were considered to be recycled or treated before they were released into the environment.

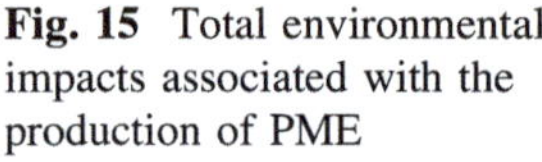

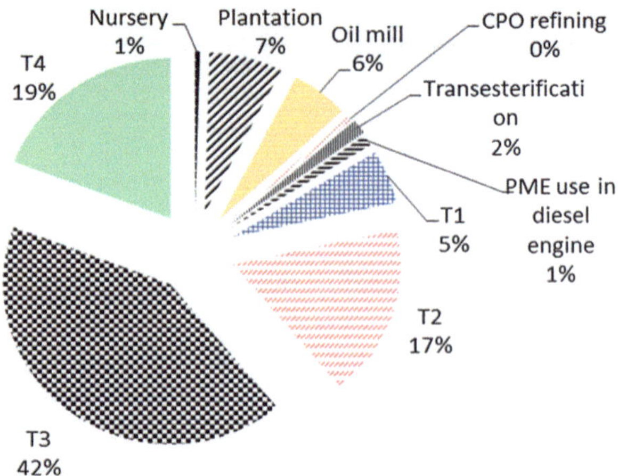

Fig. 15 Total environmental impacts associated with the production of PME

The oil milling stage contributed approximately 6 % of the total environmental impacts (Fig. 15). Human health, ecosystem depletion, and resource use for the oil milling stage recorded a percent environmental impact share of 13.01, 0.56, and 0.16 %, respectively.

5.4 Emissions Associated with CPO Refinery

The most dominant impact categories within the CPO refinery are carcinogens, climate change, fossil fuel use, and respiratory organics with contribution of 31.48, 14.34, 4.92, and 2.34 % of the total impacts for the PME production system, respectively (Fig. 13). The emissions from this stage come from the use of fossil fuel emitting N_2O, CO_2, CO, particulate matter, respiratory gases, etc., which result in these significant impact categories.

On a single score, the CPO refinery contributed approximately 0.002 % of the total impacts associated with PME production (Fig. 14). Human health, ecosystem depletion, and resource use were 0.84, 0.15, and 0.20 % of the total impacts, respectively.

5.5 Emissions Associated with Transesterification Stage

The significant impact potentials within the transesterification stage are carcinogens, respiratory organics, climate change effect, ecotoxicity, and fossil fuel use with contributions of 13.64, 30.31, 28.77, 37.19, and 43.59 %, respectively (Fig. 13). These are due to the emissions resulting from the use of fossil fuel, sodium hydroxide, phosphoric acid as well as wastewater containing soap stocks which are not transformed into any useful products. Emissions such as CO_2, N_2O, particulate matter are predominant in the transesterification stage.

On the whole, 2 % (Fig. 15) of the total environmental impacts were associated with the transesterification stage with 1.36, 0.80, and 1.73 % of the total impacts being assigned to human health, ecosystem depletion, and resource use, respectively.

PME uses in diesel engine showed significance in acidification/eutrophication potential due to the emission of CO_2, N_2O, CO, and other particulate matter into the environment. This stage offsets some of the environmental impacts by negating them. Hence, the use of biodiesel contributed to the reduction of most of the environmental impacts.

5.6 Emissions Associated with Transportation

Four main transportation stages were considered in this chapter, namely transportation of oil palm seedlings to plantation site (T1), transportation of FFB from plantation site to oil mill (T2), transportation of RPO to transesterification unit (T3), and transportation of biodiesel to the consumer (T4). These stages contributed the most impacts (83 %) (Fig. 15) due to the use of fossil fuel which emitted greenhouse gases and other particulate matter into the environment upon combustion. The most significant impacts category associated with transportation is climate change (Fig. 14). T1, T2, T3, and T4 contributed 5, 17, 43, and 19 % of the total impacts associated with PME production, respectively. On the whole, all the transportation stages contributed 71.33, 80.45, and 96.02 % to human health, ecosystem depletion, and resource use, respectively.

From Fig. 16, resource use was the major environmental concern (41 %) followed by ecosystem depletion. This means that fossil fuel use and mineral/metal emissions were high in the PME production. Ecotoxicity, acidification, and land use potentials (ecosystem depletion) were high at 37 % of the total impacts

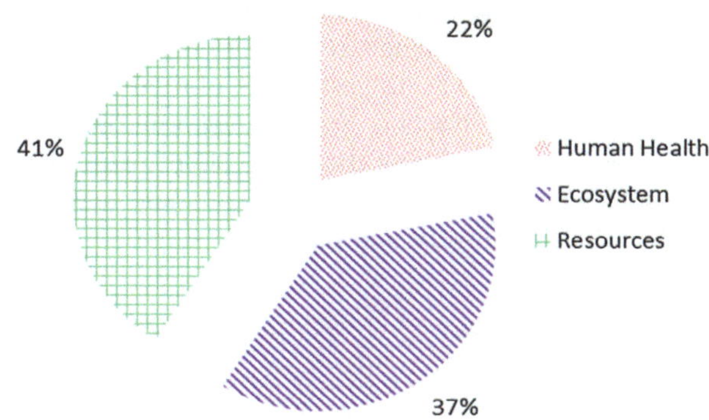

Fig. 16 Environmental impacts associated with the life cycle of biodiesel from palm oil

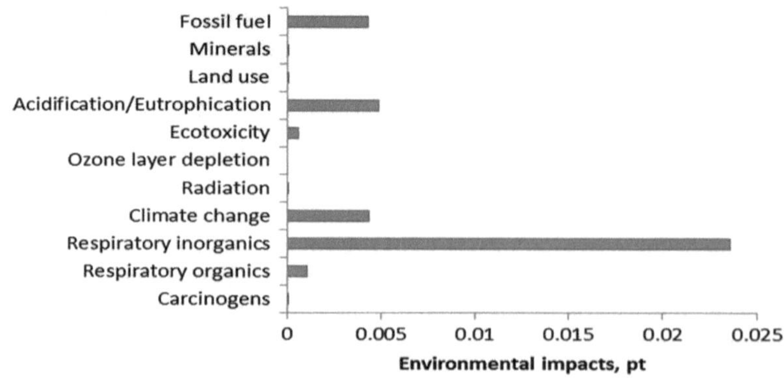

Fig. 17 Environmental impacts associated with biodiesel from palm oil. Pt: Eco-Indicator 99 (EI99, EA) points. 1 pt ~ impacts from one-thousandth person per year

associated with PME production. Figure 17 summarizes the total environmental impacts associated with the production of PME over its life cycle.

The most significant environmental damage or impacts were caused by respiratory inorganics which are caused by emissions from fossil fuel combustion and other chemical use such as sodium hydroxide. Fossil fuel, acidification/eutrophication, ecotoxicity, climate change, respiratory organics, and respiratory inorganics were also significant at 11.07, 12.58, 1.49, 11.17, 3.08, and 60.61 % of the total environmental impacts associated with 1 kg PME production, respectively.

6 Conclusion

Environmental impacts associated with the life cycle of palm oil biodiesel were assessed in this chapter using the well-to-wheel variant. The potentials of eleven main impact categories were considered for oil palm nursery, plantation, oil mill, CPO refinery, transesterification, biodiesel use as well as four transportation stages. Results from the analysis indicated that fossil fuel use was high in the plantation, transesterification, and transportation stages which further led to increase in climate change, respiratory organics, and acidification potentials. Fossil fuel consumption was highest (43 % of total impacts) in the transesterification unit exclusive of all impacts from transportation. Emissions from diesel use and transportation throughout the life cycle of palm biodiesel were more that 90 % of the total impacts. The use of fertilizers and herbicides also increased the overall impacts in the nursery and plantation stages.

7 Acknowledgment

The authors would like to thank Universiti Sains Malaysia for the financial support given through their Research University Grant No. 854002 and USM Fellowship.

References

Antolin G, Tinaut FV, Briceno Y et al (2002) Optimization of biodiesel production of sunflower oil transesterification. Bioresour. Technol. 83:111–114

Bernesson S et al (2006) A limited LCA comparing large and small scale production of ethanol for heavy engines under Swedish conditions. Biomass Bioenergy 30:46–57

Biodiesel 2020 (2008) A Global market survey, 2nd edn. Available via http://www.emerging-markets.com/biodiesel/. Cited 13th Nov 2011

Bockish M (1998) Fats and oils handbook. AOCS Press, Champaign

Buchholz T, Luzadis VA, Volk TA (2009) Sustainability criteria for bioenergy systems: results from an expert survey. J Clean Prod 17:S86–S98

Budiharjo D (2010) Palm oil mill waste utilisation. Climate change conference -technology cooperation and economic benefit of reduction of GHG emissions in Indonesia, Hamburg, Germany, 1–2 Nov 2010

Choosak K, Chanporn T, Pornpote P (2009) LCA studies comparing biodiesel synthesized by conventional and supercritical methanol methods. J Clean Prod 17:143–153

Clay JW (2004) World agriculture and environment. Island Press. ISBN 978-1559633703, Washington DC. Available via http://books.google.com/books?id=_RU8D9kB714C. Cited 14 Nov 2011

Corley RHV, Lee CH (1992) The physiological basis for genetic improvement of oil palm in Malaysia. Euphytica J 60:179–184

Corley RHV, Tinker PB (2003) The oil palm, 4th edn. Blackwell Publishing, Berlin

DEQ (Division of Environmental Quality) (1990) Paraquat. In: Annual conference 1990, toxicological association Thailand, chemical in agriculture: prevention and preservation, faculty of medicine, Ramathibodi Hospital, Mahidol University, Thailand

Ekvall T, Finnveden G (2001) Allocation in ISO 14041—a critical review. J Clean Prod 9:197–208

Encinar JM, Gonzalez JF, Rodriguez-Reinares A (2005) Biodiesel from used frying oil: variables affecting the yields and characteristics of the biodiesel. Ind. Eng. Chem. Res. 44:5491–5499

Felda report (2010) Waste product statistics for Malaysian palm oil producer Felda Palm industries Sdn. Bhd. with 349,000 ha oil palm plantations and 70 palm oil mills, Presented in October 2010

Gabi 4 Software and database for life cycle Engineering (2003) PE Europe GmbH and IKP University of Stuttgart

GOFBM (Global Oils and Fats business magazine) (2009) Oil palm-the backbone of economic growth 6(2):6–8, (April–June)

Guinée JB, Gorrée M, Heijungs R, Huppes G, Kleijn R, et al (2001) Life cycle assessment; an operational guide to the ISO standards; Part 1 and 2: Scientific Background. Ministry of Housing, Spatial Planning and Environment (VROM) and Centre of Environmental Science (CML), Den Haag and Leiden, The Netherlands

Hartley CWS (1988) The oil palm, 3rd edn. Longman, London

Heijungs R (1992) Environmental life cycle assessment of products, guidelines and backgrounds, center of environmental science (CML), Leiden, The Netherlands, Leiden University

Henson IE (1999) Comparative ecophysiology of oil palm and tropical rain forest. In: Gurmit S, Lim KH, Teo L, Lee K. (eds) Oil palm and the environment, Malaysian Oil Palm Growers' Council, Kuala Lumpur, pp 9-39

Henson IE (2004) Modelling carbon sequestration and emissions related to oil palm cultivation and associated land use change in Malaysia, MPOB technology, Malaysian Palm Oil Board, Kuala Lumpur Malaysia. Available at http://www.mpob.gov.my

Hitchcock GS, Lewis CA, Moon DP et al (1998) Alternative road transport fuels—UK field trials, Volume 1 analysis report, Energy Technology Support Unit, Harwell

Husain Z, Zainal ZA, Abdullah MZ (2003) Analysis of biomass-residue-based cogeneration system in palm oil mills. Biomass Bioenergy 24:117–124

Kheok SC, Lim EE (1982) Mechanism of palm oil bleaching by montmorillonite clay activated at various acid concentrations. J. Am. Oil Chem. Soc. 59:129–131

Edwards R, Larive, JF, Mahieu V, Rouveirolles P (2007) Well-to-wheel analysis of future automotive fuels and powertrains in the European context, Version 2c. Report from Concawe, EUcar and Joint Research Centre of the European Commission, Brussels

Mark A, Huijbregts J, Seppälä J (2001) Life cycle impact assessment of pollutants causing aquatic eutrophication. LCA Methodol. doi:0.1065/Ica2001.08.060

Møller JR,Thøgersen AM, Kjeldsen MR et al. (2000) Feeding component table—Composition and feeding value of feeding components for cattle), Report No. 91 National Committee for Cattle, Denmark

MPOB (2006) Review of the Malaysian oil palm industry 2005. Economics and industry development division, Malaysian Palm Oil Board, Kelana Jaya, Malaysia

MPOB (2010) Review of the Malaysian oil palm industry. Selangor, Malaysian Palm Oil Board (MPOB), Malaysia

Ng SK, Thamboo S (1967) Nutrient contents of oil palms in Malaya. I. Nutrients required for reproduction: fruit bunches and male inflorescence. Malay Agric J 46:3–45

Novizar N, Dwi S (2010) Life cycle assessment of biodiesel production from palm oil and Jatropha oil in Indonesia. Biomass Asia workshop, Nov 29–Dec 01, 2010, Jakarta, Indonesia

Parish F, Sirin A, Charman D et al (2008) Assessment on peatlands, biodiversity and climate change. Global Environment Centre, Kuala Lumpur and Wetlands, Wageningen

Pushparajah E (2002) Cultivation of oil palms in marginal areas—A revisit. In: Plantation management: back to basics. Proceedings of national ISP Sem. 2002, Kuching, Sarawak, ISP, Kuala Lumpur, pp 69–82

Ravigadevi S, Siti Nor AA, Ahmad Parveez GK (2002) Genetic manipulation of the oil palm-challenges and prospects. Planter Kuala Lumpur 78:547–562

Reijnders L, Huijbregts MAJ (2008) Palm oil and the emission of carbon-based greenhouse gases. J Cleaner Prod 16:477–482

Roger S, Jennifer C et al (2011) The role of life cycle assessment in identifying and reducing environmental impacts of CCS. Lawrence Berkeley National Laboratory report, University of California, CA

Román-Cuesta RM, Asbjornsen H, Salinas N et al (2011) Implications of fires on carbon budgets in Andean cloud montane forest: The importance of peat soils and tree resprouting. For Ecol Mngt 261:1987–1997

RSPO (The Roundtable on Sustainable Palm oil report) (2011) EU media market communications. Available via http://www.rspo.eu/. Cited 13 Nov 2011

Schmidt JH (2007) Life assessment of rapeseed oil and palm oil. Ph.D. Thesis, Part 3: lifecycle inventory of rapeseed oil and palm oil, Dissertation. Department of Development and Planning, Aalborg University, Aalborg

Stichnothe H, Schuchardt F (2011) Life cycle assessment of two palm oil production systems. Biomass Bioenergy 35:3976–3984

Subramaniam V (2006) Life cycle inventory for palm kernel crushing. Unpublished data by research officer, Vijaya Subramaniam, Engineering and processing division, energy and environment Unit, Malaysian Palm Oil Board (MPOB), Kajang

Subramaniam V, Ma AN, Choo YM, Nik MNS (2008) Environmental performance of the milling process of Malaysian palm oil using the life cycle assessment approach. Ame J Env Sci 4(4):310–315

Tillman AM (2000) Significance of decision-making for LCA methodology. Env Imp Ass Rev 20:113–123

Tinker PBH, Smilde KW (1963) Dry matter production and nutrient content of plantation oil palms in Nigeria. II. Nutrient content. Plant Soil 19:350–363

Turrio-Baldassarri L, Battistelli CL, Conti L, Crebelli R, De Berardis B, Iamiceli AL (2004) Emission comparison of urban bus engine fueled with diesel oil and 'biodiesel' blend. Sci Total Environ 327:147–162

United States Department of Energy/Energy Information Administration (US DOE/EIA) (1997) Emissions of greenhouse gases in the United States. DOE/EIA-0573(97), Distribution Category UC-950, United States

Vicente G, Martinez M, Aracil J (2004) Integrated biodiesel production: a comparison of different homogeneous catalysts systems. Bioresour. Technol. 92:297–305

Von Uexkull HR, Fairhurst T (1991) The oil palm: fertilizer management for high yield. International Potash Institute, Berne

Wicke B, Dornburg V, Junqinger M, Faaij A (2008) Different palm oil production systems for energy purposes and their greenhouse gas implications. Biomass Bioenergy 32:1322–1337

Wu TY, Mohammad AW, Jahim JM, Anuar N (2010) Pollution control technologies for the treatment of palm oil mill effluent (POME) through end-of-pipe processes. J. Environ. Manage. 91:1467–1490

Xavier A, Ho SH, Vijiandran JR, Gurmit S (2008) Managingcoastal and alluvial soils under oil palm. In: Act 2008: agronomic principles and practices of oil palm cultivation, 13–16 Oct 2008, Sibu, Sarawak, pp 415–452

Yacob S, Hassan MA, Shirai Y et al (2005) Baseline study of methane emission from open digesting tanks of palm oil mill effluent treatment. Chemospher 59:1575–1581

Yusoff S (2006) Renewable energy from palm oil—innovation on effective utilization of waste. J Clean Prod 14:87–93

Yusoff S, Hansen SB (2007) Feasibility study of performing a life cycle assessment on crude palm oil production in Malaysia. Int J Life Cycl Ass 12:50–58

Environmental Sustainability Assessment of Ethanol from Cassava and Sugarcane Molasses in a Life Cycle Perspective

Shabbir H. Gheewala

Abstract Liquid transportation fuels derived from biomass (biofuels) are widely promoted inter alia due to their perceived environmental benefits. The environmental sustainability of biofuels, however, must be rigorously tested using scientific tools. Several such tools, namely net energy ratio (NER), net energy balance (NEB), renewability, and life cycle assessment (LCA), are defined and then used to analyze the environmental sustainability of ethanol produced from cassava and sugarcane molasses in Thailand. These studies show the utility of such tools in the evaluation and also point various areas of improvement. Use of renewable energy sources in the supply chain, utilization of co-products as well as waste products, and good management practices at the farm such as use of organic fertilizers are some of the options that can help improve the environmental benefits of biofuels.

1 Background/Context

Liquid transportation fuels derived from biomass feedstocks, popularly referred to as "biofuels", seem to be an attractive substitute for oil-based gasoline and diesel for several reasons. Firstly, their similarity in physical properties facilitates the use of the existing infrastructure designed for their oil-based counterparts. The similarity in chemical properties induces minimal modification requirements of the car engines, especially for biodiesel. Being of biomass origin, the carbon dioxide

S. H. Gheewala (✉)
The Joint Graduate School of Energy and Environment, King Mongkut's University of Technology Thonburi, Bangkok, Thailand
e-mail: shabbir_g@jgsee.kmutt.ac.th

S. H. Gheewala
Center for Energy Technology and Environment, Ministry of Education, Bangkok, Thailand

A. Singh et al. (eds.), *Life Cycle Assessment of Renewable Energy Sources*,
Green Energy and Technology, DOI: 10.1007/978-1-4471-5364-1_6,

emissions from the combustion of biofuels in car engines are counterbalanced by the atmospheric carbon dioxide sequestered during growth of biomass; this is supposed to yield substantial benefits toward climate change mitigation. Also, their biomass origin makes them renewable (if sustainably managed), which is a major advantage over the non-renewable oil-based fuels. The substitution of oil-based fuels by biofuels is of particular interest to a large number of countries that do not have large resources of the former and thus have to rely on expensive imports from the few countries that do. Energy security is a strong driving force for the promotion of biofuels, particularly in the developing world. Socioeconomic benefits, particularly for the agricultural community, are also an important consideration (Daniel et al. 2010; ERIA 2008).

This chapter focuses particularly on the environmental sustainability assessment of biofuels. As mentioned earlier, there are perceived environmental benefits due to the "biogenic" greenhouse gas (GHG) emissions during use phase as well as the renewable nature of biofuels. However, environmental sustainability of biofuels cannot be evaluated only on the basis of carbon dioxide emissions in the use phase. A broader perspective based on the entire life cycle of the biofuel is imperative. In this case then, the carbon neutrality of biofuels does not hold as there are GHG emissions associated with the cultivation as well as processing of feedstock which are not balanced by the uptake of atmospheric carbon dioxide during plant growth. Rigorous tools, based on the entire life cycle, are thus required for the proper assessment of the environmental sustainability of biofuels (Nguyen and Gheewala 2008a, b).

2 Tools for Environmental Sustainability Assessment

As mentioned earlier, the most commonly used tools are based on the entire life cycle of the biofuels. They are distinguished into three broad categories: (1) net energy balance and ratio, (2) renewability, and (3) life cycle assessment. The first two are based on energy (particularly the first law of thermodynamics), and the third one is for evaluating potential environmental impacts.

2.1 Net Energy Balance and Net Energy Ratio

As biofuels are energy carriers, two indicators that are absolutely essential in their initial evaluation are the net energy balance and net energy ratio (Shapouri et al. 2006; Nguyen et al. 2007; Prueksakorn and Gheewala 2008). These are preliminary indicators based on the first law of thermodynamics. The net energy balance or NEB is the difference of the total energy output and the total energy input over the entire life cycle of the biofuel. Intuitively, the NEB of the system must be positive or there must be a net energy gain; else, it does not make sense to produce the biofuel. Of course, the first law of thermodynamics does not allow for "creation of energy." The positive NEB is possible for biofuels only because the solar energy input during biomass growth is not accounted for in the calculation. Similar in approach to NEB, the net energy ratio or NER is the ratio of the total energy output to the total energy input over the entire life cycle of the biofuel. This ratio must be greater than one for the biofuel production to be meaningful. It must be emphasized that having a positive NEB or an NER greater than one is not sufficient to establish the environmental sustainability of biofuels. However, they give a very good first check; if the NEB of a biofuel is negative or NER is less than one, then it probably does not make sense to produce the biofuel as we are investing more energy into the production of the fuel than we actually get back from its use. Of course in certain special circumstances, we may still go ahead with the biofuel production even in such a case if the energy carriers used as input to the biofuel system are cheap and easily available, but cannot themselves be used as transportation fuel substitutes.

2.2 Renewability

Another energy-based indicator, similar to NER, is the renewability ratio which is defined as the ratio of the total energy output to the total *fossil* energy input. This ratio distinguishes between the types of energy carriers that are input to the biofuel cycle; if there is more (renewable) energy output than the (non-renewable) fossil energy input, i.e., renewability ratio higher than one, it indicates that the investment of fossil energy into biofuel system has yielded a higher amount of renewable energy.

2.3 Life Cycle Assessment

Life cycle assessment (LCA) is a tool for environmental sustainability assessment that is particularly suited and hence widely used for assessment of biofuels. This tool evaluates the environmental impacts of a product (or service) over the entire period of its life, starting from raw materials extraction (or production) and including materials processing, distribution, use, and waste disposal at the end of life. The principles and criteria of LCA are covered by the ISO 14040:2006 and ISO 14044:2006 standards. The environmental impact categories commonly of interest for evaluating biofuel systems are climate change (global warming), acidification, nutrient enrichment (eutrophication), human and ecotoxicity, land use, and biodiversity. Very often, biofuel studies are limited to climate change partly due to its global nature and importance in many scientific discussions and partly because of the perceived benefits of biofuels toward reducing GHG emissions as compared to their fossil counterparts which must be scientifically verified.

3 System Boundaries

For biofuels, the system boundaries usually cover cultivation of the feedstock (including agrochemical production and application), processing of the feedstock for conversion to the biofuel, use of the biofuel, and transportation in all the intermediate stages (Fig. 1) (Gheewala 2011). In recent years, the system boundaries of biofuels have been expanded to include the pre-cultivation stage which is land use change. This particular stage has significant implications on the emissions of GHGs and on biodiversity, especially when forests and other high conservation value lands are converted to agriculture for cultivation of biofuel feedstocks (Danielson et al. 2008; Fargione et al. 2008).

4 Data Sources

For conducting environmental sustainability assessment, data are required at every stage of the life cycle outlined in Fig. 1. If land use change from a natural system to an agricultural one has taken place in the recent past (say, less than 20 years ago), then data on the type of land that existed need to be known and, if possible, the carbon stock in the soil as well as above and below ground biomass. In the absence of very detailed data, default values from the Intergovernmental Panel on Climate Change methodology could also be used (IPCC 2006).

At the cultivation stage, detailed data need to be collected on the use of agrochemicals (herbicides, pesticides, and fertilizers), their rates, and frequency of application throughout the cultivation period. Labor requirements, if necessary, also need to be assessed. Use of irrigation water and agricultural machinery (particularly fuel use) needs to be recorded as also the yield of the main product and by-products. These data are usually from primary sources at the farm level though in the case of national studies, national statistics from agricultural organizations may also provide useful information.

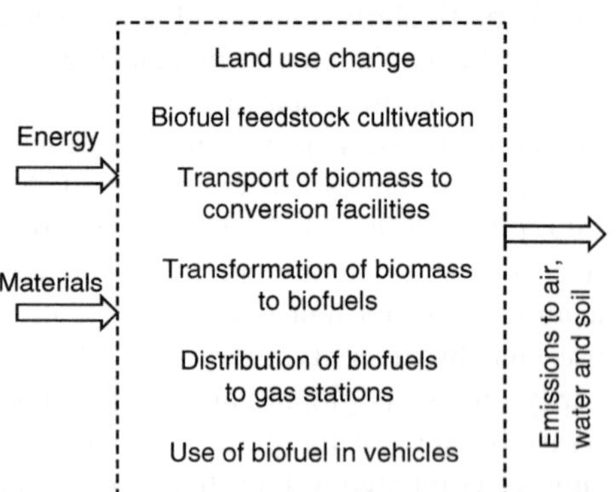

Fig. 1 Generic system boundaries of biofuels

Feedstock processing entails the use of various chemicals and energy carriers (boiler fuels, electricity, etc.). These data are usually primary data collected from plant records through production data of chemicals and energy carriers, which are usually from literature or national databases. This stage usually involves the production of co-products, which play an important role in the assessment and must be carefully noted. As environmental burdens must be shared between the co-products, additional information must be collected for this based on the methodology to be used for allocation. For example, if mass or energy allocation is to be used, data on the mass and energy content of the co-products should be assessed. If economic allocation is to be used, then data on the economic value of all the outputs must be collected; this is done either at the company level or a more average level, depending on the scope of the study. Usually, an average over several years is considered because economic values can vary substantially over time. Wastewater emissions are particularly critical as they usually have high organic content; the type of treatment can significantly affect environmental emissions. Treatment of wastewater in ponds can result in high methane emissions, whereas high-rate anaerobic processes could result in the collection and subsequent utilization of the generated methane for energy. These data are usually based on standard calculations.

Emissions from the use of biofuels in vehicles are usually measured based on chassis dynamometer tests where specific engine types are operated using a particular fuel under standardized driving conditions.

Data required for transportation of intermediate and final products are either directly the fuel used or data on vehicle type, capacity, loading, and transportation distances. These are usually primary data collected from the transportation companies except for transportation distances which may be estimated from secondary sources if the source and destination of the transported product are known.

In all the steps, information should be collected on the data uncertainty to ensure the meaningfulness of the interpretation of results, especially for comparative purposes.

5 Environmental Sustainability Assessment

To illustrate the use of the tools for biofuels environmental sustainability assessment introduced in the preceding section, the cases of ethanol from cassava and sugarcane molasses in Thailand are used (Silalertruksa and Gheewala 2009). Energy balances, renewability, and various environmental impact categories will be evaluated using LCA. The system boundaries are "cradle-to-gate" ending at ethanol production as the ethanol produced is then blended with gasoline before use in vehicles (a 10 % blend of ethanol with 90 % gasoline is most commonly used in Thailand); thus, the study was limited to neat ethanol to facilitate direct comparison with gasoline. For the impact assessment, the CML2 methdology, a problem-oriented (midpoint) approach, has been used (Guinée et al. 2002). Thus, for example, greenhouse gases are aggregated into global warming potential represented in terms of kg CO_2eq instead of evaluating the final damage due to climate change. The midpoint approach reduces the uncertainty introduced from complex modeling approaches as well as forecasting and effect modeling (Blottnitz and Curran 2007). Similar to the case for global warming, the other impact categories of interest are acidification (kg SO_2eq), eutrophication (kg PO_4^{3-}eq), and human toxicity (kg 1,4 DCBeq).

5.1 *Ethanol from Cassava*

The life cycle diagram of ethanol production from cassava is shown in Fig. 2. The first step in the life cycle is the cultivation of cassava that includes land preparation, planting, farming (including agricultural activities such as fertilizer application), and finally harvesting. Manual labor is used for most activities. The fresh cassava roots are transported by trucks or pickup trucks to the ethanol plants where they are either used in fresh form or sun-dried into cassava chips (and stored for later use). The roots then undergo liquefaction, fermentation, and distillation, followed by molecular sieve dehydration to produce 99.5 % ethanol. These processes, particularly distillation and dehydration, are energy intensive and have a substantial effect on the energy and environmental performance of the system. The type of fuel used in the boilers and the treatment method of the wastewater are crucial issues affecting performance.

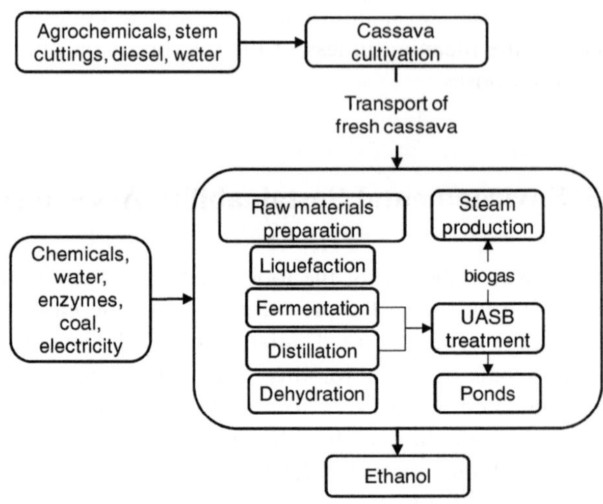

Fig. 2 Life cycle diagram of cassava ethanol (cradle-to-gate)

5.1.1 NER, NEB and Renewability of Cassava Ethanol

The proportions of energy used in cassava farming, transportation, and ethanol production are shown in Fig. 3a. As anticipated, ethanol production contributes almost three-fourths of the total energy use, followed by cassava cultivation contributing almost one-fifth. Steam production has the highest contribution; the fuel used for this step could thus have a large contribution also to the environmental emissions. The net energy ratio of the ethanol works out to 1.19; this is higher than one, indicating an energy gain. The NEB for 1,000 L of ethanol is 3,827 MJ: a positive value once again confirming a net energy gain. However, it is difficult to decide on the basis of the NER and NEB alone whether the gain is "enough" to justify production.

The next step in the evaluation is to calculate how much renewable energy is obtained with the investment of a unit of fossil energy; the proportions of only fossil energy use are presented in Fig. 3b. When the non-fossil energy sources (e.g., biogas and human labor) are removed, the contribution of steam is even more pronounced, contributing more than one-half of the total. The renewability of the ethanol from cassava works out to 1.38, which is marginally better than the NER. The very small difference between the NER and renewability is because very limited amount of energy is from renewable resources.

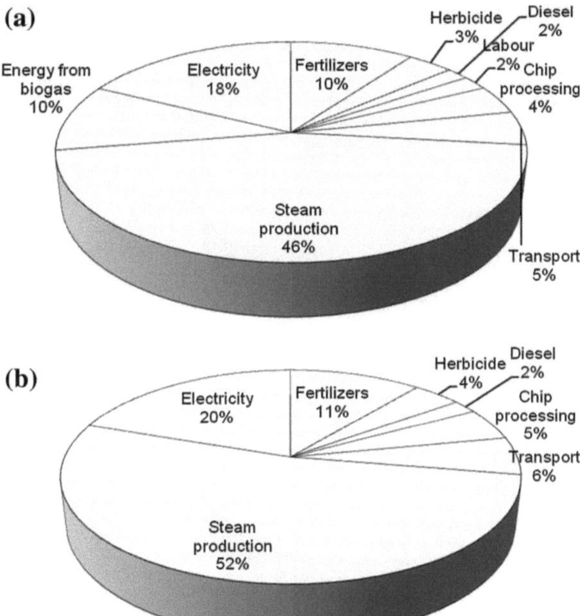

Fig. 3 Proportion of energy inputs in the production of cassava ethanol (cradle-to-gate). **a** Total energy. **b** Fossil energy

5.1.2 LCA of Cassava Ethanol

The impact assessment results from cassava ethanol production are presented in Fig. 4. As anticipated from the energy balance evaluation, the ethanol conversion step is the major contributor to all the impact categories considered. During this step, steam is produced in a boiler fired by sub-bituminous coal and is responsible for emissions of CO_2, SO_2, NOx, CO, and particulates; it thus contributes 52, 51, and 43 % to global warming, acidification, and human toxicity impacts, respectively. The other big contributor is of course the electricity, which is largely produced from natural gas and coal. Eutrophication is contributed largely by fertilizer use in the cultivation stage and the wastewater discharged from the upflow anaerobic sludge blanket (UASB) reactor after biogas recovery.

5.2 Ethanol from Sugarcane Molasses

The life cycle diagram of ethanol production from sugarcane molasses is shown in Fig. 5. The first step in the life cycle is sugarcane cultivation, which consists of planting stem cuttings in the initial year, followed by three ratoons. Manual labor is used for land preparation, planting, farming, and harvesting. The sugarcane is transported to the sugar mill in trucks and trailers. The next step is at the sugar mill where the sugarcane is crushed to extract sugarcane juice; the residue remaining

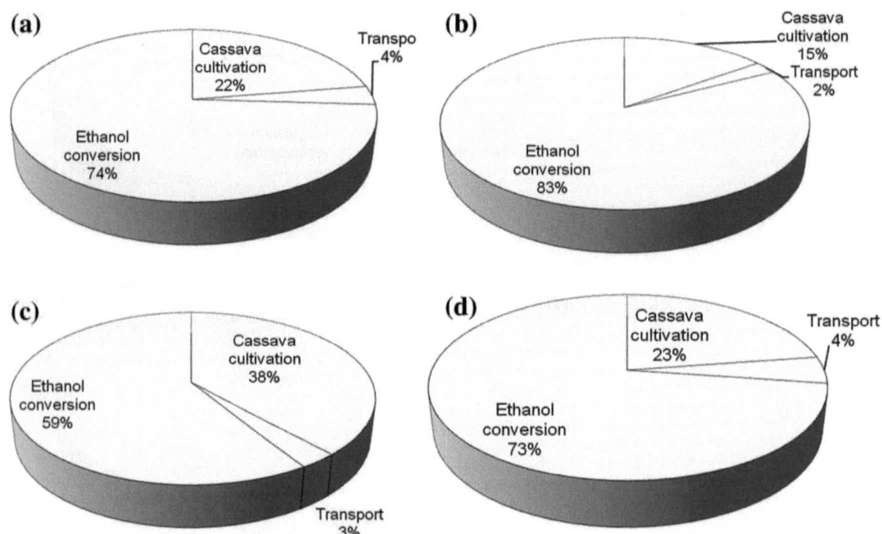

Fig. 4 Potential environmental impacts of 1,000 L cassava ethanol production. **a** Global warming (1,922 kg CO_2eq). **b** Acidification (16 kg SO_2eq). **c** Eutrophication (2.79 kg PO_4^{3-}eq). **d** Human toxicity (18.53 kg 1,4 DCBeq)

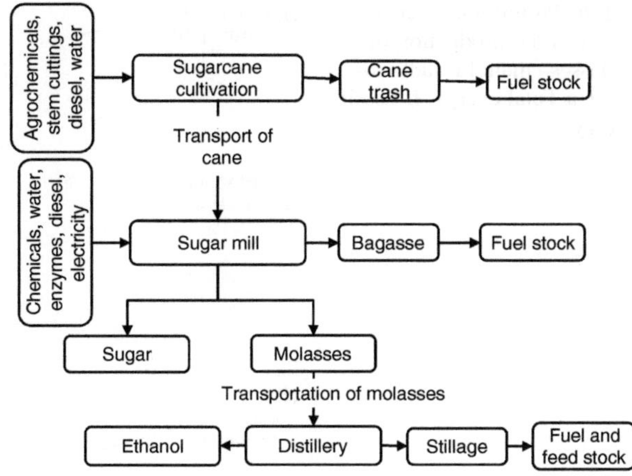

Fig. 5 Life cycle diagram of molasses ethanol (cradle-to-gate)

after the juice is extracted, called bagasse, is used as an energy source to produce steam and electricity both for the sugar milling process and for export. The sugarcane juice is concentrated by boiling, and sugar is extracted in several steps by crystallization. After the final economic extraction of sugar, the remaining molasses (which still contains a substantial amount of sugar) is used for producing ethanol. The sugar content of the molasses is the key quality criterion as it is converted to ethanol. Energy-based allocation is used to share the environmental burdens of sugarcane cultivation and sugar milling between sugar and molasses. As for the bagasse, one part is used within the sugar mill itself and is treated as internal recycling. The excess electricity produced from the bagasse is exported to the grid, and credits are provided to the sugar milling system from the avoided conventional electricity production. Molasses is transported to the ethanol plants via trucks or through pipelines. The ethanol production process is similar to that for cassava except that the liquefaction step is not required as molasses can be directly fermented, followed by distillation and dehydration to produce 99.5 % ethanol.

5.2.1 NER, NEB, and Renewability of Molasses Ethanol

The proportions of the energy used in sugarcane cultivation, sugarcane transportation, molasses transportation, and ethanol production are shown in Fig. 6a. The sugar milling step is not included in the figure as this step actually yields energy in excess of that being used by the process from the conversion of one of the co-products, bagasse, which is used for steam and electricity production. As in the case of cassava ethanol, the ethanol production step contributes the major share, almost two-thirds of the total. In the case of molasses ethanol, the energy outputs are not only from the ethanol but also from the exported bagasse electricity from the sugar milling step. The net energy ratio is 1.12; though higher than one it is

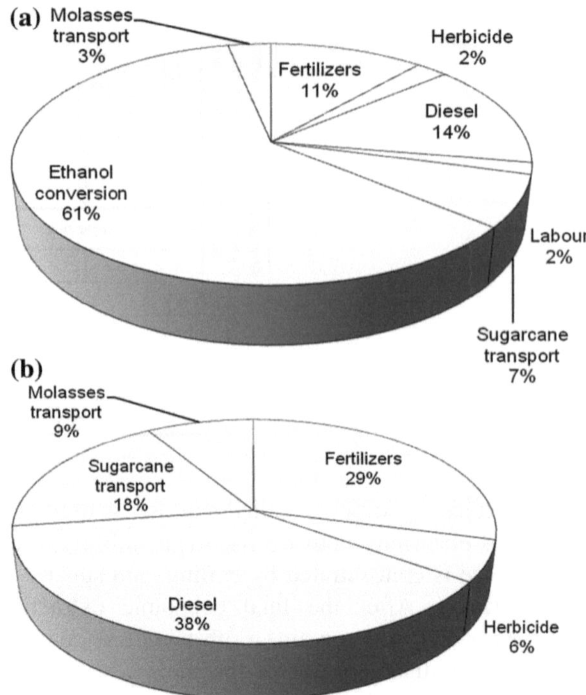

Fig. 6 Proportion of energy inputs in the production of molasses ethanol (cradle-to-gate). **a** Total energy. **b** Fossil energy

even more modest than that of the cassava ethanol. The NEB for 1,000 L of ethanol is 3,350 MJ: once again positive but lower than cassava ethanol. Until this stage, the molasses ethanol seems to be doing slightly worse than that produced from cassava.

However, the calculation of renewability shows a significantly different picture. The results of only fossil energy input are presented in Fig. 6b. Here, it can be seen that the major contributor to energy use, ethanol conversion, is absent because all the energy in this step is provided by rice husk and recovered biogas (from wastewater treatment). This has a significant effect on the renewability which amounts to 3.02, substantially higher than the NER of molasses ethanol and even much higher than the renewability of cassava ethanol. This in fact goes on to show the importance of the use of renewable energy sources in the life cycle, particularly in an energy-intensive step like ethanol conversion.

5.2.2 LCA of Molasses Ethanol

Figure 7 shows the contributions of the various life cycle stages to the potential environmental impacts of ethanol production from sugarcane molasses. Global warming, acidification, eutrophication, and human toxicity are 685 kg CO_2eq, 12.5 kg SO_2eq, 19.55 kg PO_4^{3-}eq, and 19.11 kg 1,4 DCBeq, respectively. The

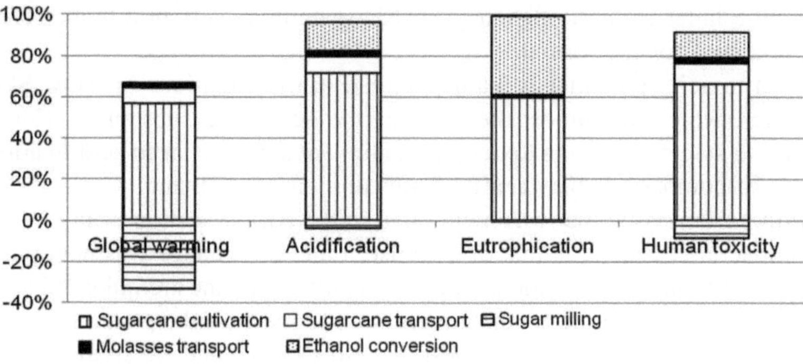

Fig. 7 Potential environmental impacts of molasses ethanol production

sugarcane cultivation stage has an outstanding contribution to all the impact categories, which is different from the case of cassava ethanol where ethanol conversion was the major contributor. This is because of the electricity production from bagasse, which brings credits to the molasses, the feedstock for ethanol production; this can also be seen clearly in the Fig. 7 where sugar milling contributes to reducing impacts (just as it also helped improving the energy balance).

5.3 Comparison of LCA Results

The global warming potential of molasses ethanol is significantly lower than that of cassava ethanol. This is largely due to the use of renewable energy carriers (corncob, bagasse, and rice husk) in the ethanol conversion from molasses, whereas coal is used in the cassava ethanol factory. The other reason is the credits from bagasse electricity exported to the grid. Burning of cane trash in the sugarcane field before harvesting contributes greenhouse gases to the molasses ethanol cycle. Avoidance of this burning and possible utilization of energy could have double benefits though these must be balanced with retaining these residues on the field for improving soil fertility.

Acidification potential for cassava ethanol is higher than molasses ethanol largely due to the use of sub-bituminous coal in the cassava ethanol plant. For molasses ethanol, acidifying substances are released mainly in feedstock production and cane trash burning. Avoiding burning of cane trash and utilizing it for energy would also help further reduce acidification potential of molasses ethanol.

Feedstock production and ethanol conversion are the large contributors to eutrophication potential, which is higher for molasses ethanol than cassava ethanol. Wastewater released during molasses ethanol production has much higher organic matter than that from cassava ethanol production; this combined with the situation that wastewater from the former is treated in less efficient pond systems, whereas the latter in high-efficiency UASB systems results in a higher eutrophication potential for the former.

Human toxicity potential is mainly related to the emissions of NOx, SO_2, and particulates, which are almost the same as the acidifying substances. Hence, the contributing stages of human toxicity potential are also similar to acidification as is the trend of comparison between molasses and cassava ethanol.

One important issue that was mentioned in the earlier sections, but not covered in the impact assessment, is that of land use. Land use change was not considered because all the current plantations of cassava and sugarcane are quite old and there is no plan in the near future by the government to increase the plantation area. The policy focus is on increasing the yield of both cassava and sugarcane. Nevertheless, land is a scarce resource and it is interesting to evaluate its utilization in terms of land occupation. The evaluation shows that to produce 1,000 L of cassava ethanol, 0.37 ha.y of land is required, which is almost the same for molasses ethanol at 0.39 ha.y (after accounting for sugar based on energy allocation). Increasing the yields of cassava and sugarcane as well as better utilization of co-products would help reduce this.

6 Key Messages

The energy balance and LCA studies helped evaluate the environmental sustainability of biofuel systems as illustrated for the examples of ethanol production from cassava and sugarcane molasses. A combination of NER and renewability was useful for evaluating the energy performance and utility of using renewable energy sources to replace fossil energy. LCA was useful for evaluating the environmental and health impacts. For the studied cases, the following improvement options were identified:

(a) Optimum utilization of the land resource could be achieved by improving the yields of sugarcane as well as cassava. This could be obtained by improving soil fertility through utilization of organic fertilizers or animal waste and reducing chemical fertilizer use. This would also result in reduced eutrophication, which originates mainly from the use of chemical fertilizers.
(b) Air emissions from the sugarcane cultivation stage can be reduced by avoiding the burning of cane trash. This would result in the reduction in all the potential environmental impacts too.
(c) Effective waste management would go a long way in enhancing the efficiency of the system. Thus, biogas recovery from wastewater, organic fertilizer, and distiller's dried grains with solubles (DDGS) production would yield valuable products and reduce the environmental burdens.
(d) Use of renewable energy sources especially in the ethanol conversion stage would reduce the emission of greenhouse gases as well as improve the renewability of the system.

References

Blottnitz HV, Curran MA (2007) A review of assessments conducted on bio-ethanol as a transportation fuel from a net energy, greenhouse gas, and environmental life cycle perspective. J Clean Prod 15(7):607–619

Daniel R, Lebel L, Gheewala SH (2010) Agrofuels in Thailand: policies, practices and prospects. In: Lebel L et al (eds) Sustainable production consumption systems: knowledge, engagement and practice. Springer, London

Danielsen F, Beukema H, Burgess ND, Parish F, Brühl CA, Donald PF, Murdiyarso D, Phalan B, Reijnders L, Struebig M, Fitzherbert EB (2008) Biofuel plantations on forested lands: double jeopardy for biodiversity and climate. Conserv Biol 23(2):348–358

Fargione J, Hill J, Tilman D, Polasky S, Hawthorne P (2008) Land clearing and the biofuel carbon debt. Science 319:1235–1238

Guinée JB, Gorrée M, Heijungs R, Huppes G, Kleijn R, de Koning A, van Oers L, Sleeswijk AW, Suh S, Udo de Haes HA, de Bruijn H, van Duin R, Huijbregts MAJ (2002) Life cycle assessment: an operational guide to the ISO standards. Kluwer Academic Publishers, Dordrecht

Gheewala SH (2011) Life cycle assessment (LCA) to evaluate environmental impacts of bioenergy projects. J Sustain Energy Environ 2(1):35–38

IPCC (2006) 2006 IPCC Guidelines for National greenhouse gas inventories Vol. 4-Agriculture, Forestry and Other Land Use (AFOLU), Institute for Global Environmental Strategies (IGES), Hayama

Nguyen TLT, Gheewala SH (2008a) Life cycle assessment of fuel ethanol from cassava in Thailand. Int J LCA 13(2):147–154

Nguyen TLT, Gheewala SH (2008b) Life cycle assessment of fuel ethanol from cane molasses in Thailand. Int J LCA 13(4):301–311

Nguyen TLT, Gheewala SH, Garivait S (2007) Full chain energy analysis of fuel ethanol from cassava in Thailand. Environ Sci Technol 41(11):4135–4142

Prueksakorn K, Gheewala SH (2008) Full chain energy analysis of biodiesel from *Jatropha curcas* L. Thailand. Environ Sci Technol 42(9):3388–3393

ERIA (2008) Sustainable biomass utilisation vision in East Asia. ERIA Research Project Report 2007, No. 6-3, ERIA, Jakarta

Shapouri H, Wang M, Duffield JA (2006) Net energy balancing and fuel-cycle analysis. In: Dewulf J, van Langenhove H (eds) Renewables-based technology: sustainability assessment, Wiley, West Sussex

Silalertruksa T, Gheewala SH (2009) Environmental sustainability assessment of bio-ethanol production in Thailand. Energy 34(11):1933–1946

Comparison of Algal Biodiesel Production Pathways Using Life Cycle Assessment Tool

Anoop Singh and Stig Irving Olsen

Abstract The consideration of algal biomass in biodiesel production increased very rapidly in the last decade. A life cycle assessment (LCA) study is presented to compare six different biodiesel production pathways (three different harvesting techniques, i.e., aluminum as flocculent, lime flocculent, and centrifugation, and two different oil extraction methods, i.e., supercritical CO_2 (sCO_2) and press and co-solvent extraction). The cultivation of *Nannochloropsis* sp. considered in a flat-panel photobioreactor (FPPBR). These algal biodiesel production systems were compared with the conventional diesel in a EURO 5 passenger car used for transport purpose (functional unit 1 person km (pkm). The algal biodiesel production systems provide lesser impact (22–105 %) in comparison with conventional diesel. Impacts of algal biodiesel on climate change were far better than conventional diesel, but impacts on human health, ecosystem quality, and resources were higher than the conventional diesel. This study recommends more practical data at pilot-scale production plant with maximum utilization of by-products generated during the production to produce a sustainable algal biodiesel.

1 Introduction

The production and utilization of algal biomass for biodiesel production is a rapidly growing industry all over the world due to increasing crude oil prices, import reliance, depletion of domestic petroleum resources, environmental

A. Singh (✉) · S. I. Olsen
Department of Scientific and Industrial Research (DSIR), Ministry of Science and Technology, Technology Bhawan, New Mehrauli Road, New Delhi 110016, India
e-mail: apsinghenv@gmail.com

S. I. Olsen
e-mail: siol@dtu.dk

A. Singh et al. (eds.), *Life Cycle Assessment of Renewable Energy Sources*,
Green Energy and Technology, DOI: 10.1007/978-1-4471-5364-1_7,

disasters, and national security concerns (Brentner et al. 2011). Alga is a very promising source of biomass for bioenergy production as it sequesters a significant quantity of carbon from atmosphere and industrial gases and is also very efficient in utilizing the nutrients from industrial effluents and municipal wastewater (Singh and Olsen 2011a; Singh et al. 2011a, b).

Algae represent a vast variety of photosynthetic species populating in diverse environments (Nigam and Singh 2011; Mata et al. 2010), and they might be autotrophic or heterotrophic in nature (John et al. 2011). Using only sunlight and abundant and freely available raw materials (e.g., CO_2 from atmosphere/flue gas and nutrients from wastewater), algae can synthesize and accumulate large quantities of lipids and carbohydrates along with other valuable co-products (e.g., astaxanthin, omega-3 fatty acids) (Subhadra and Edwards 2010; Singh et al. 2011c, 2012). Like other biomass, algal biomass is also a carbon-neutral source for the production of bioenergy (Singh and Olsen 2011b). Thus, algae can play a major role in the treatment/utilization of wastewater and reduce the environmental impact and disposal problems. They can also be grown on saline/coastal seawater and on non-agricultural lands (Hu et al. 2008; Melis and Happe 2001). Recent research initiatives have proven that microalgae biomass appears to be one of the promising sources of renewable biodiesel, which is capable of meeting the global energy demand and displaces the fossil diesel without compromising with production of food, fodder, and other products derived from crops (Singh et al. 2011b). Microalgae grow extremely rapidly and many are exceedingly rich in oil. Microalgae commonly double their biomass within 24 h. Biomass doubling times during exponential growth are commonly as short as 3.5 h (Chisti 2007). Oil content in microalgae can exceed 80 % by weight of dry biomass (Metting 1996; Spolaore et al. 2006).

Life cycle assessment (LCA) has in recent years been the method of choice for environmental assessment of various kinds of new technologies for bioenergy and carbon sequestration. LCA is a universally accepted approach of determining the environmental consequences of a particular product over its entire production cycle (Korres et al. 2010; Pant et al. 2011). The LCA of biofuels is the key to observe their sustainability (Singh and Olsen 2012). Yang et al. (2011) examined the life cycle of water and nutrients usage of microalgae-based biodiesel production. This study quantified the water footprint and nutrient usages during microalgae biodiesel production. The results indicated that using seawater or wastewater can reduce the life cycle of freshwater usage by as much as 90 %. They also reported that utilization of sea/wastewater for algal culture can reduce nitrogen usage by 94 % and eliminate the need for potassium, magnesium, and sulfur. An analysis of the energy life cycle for production of microalgal biomass of *Nannochloropsis* sp. was performed by Jorquera et al. (2010), which included raceway ponds, tubular and flat-plate photobioreactors (PBRs) for algal cultivation. They concluded that net energy ratio (NER) for ponds and flat-plate PBRs could be raised significantly by selecting algal strains having higher lipid content. Clarens et al. (2010) demonstrated the benefits of algae production coupled with wastewater treatment and concluded that the use of wastewater effluent as pond medium could significantly reduce not only the need for chemical fertilizers and their associated life cycle burdens but also the use of freshwater during algae cultivation. Lardon et al. (2009) conducted an LCA study to analyze the potential environmental impacts of biodiesel production from microalgae, and the outcome of this study confirms the potential of microalgae as an energy source but highlights the imperative necessity of decreasing the energy and fertilizer consumption.

The present study is an effort to compare three harvesting technologies and two oil extraction technologies involved in the production of algal biodiesel and analyze the sustainability to use produced algal biodiesel in passenger cars in comparison with fossil diesel.

2 Methodology

The study is conducted considering Danish conditions to cultivate algal biomass and further processed for algal biodiesel production. In this study, six scenarios were studied for *Nannochloropsis* sp. (Table 1).

2.1 *Biodiesel Production Steps*

The production process is divided into four steps, i.e., reactor design, biomass cultivation, harvesting, and oil extraction and trans-esterification.

2.1.1 Bioreactor Design

The FPPBR is made up of LDPE sheet, and steel is considered for cultivation of algal biomass as Ugwu et al. (2008) reviewed various algal cultivation systems and

Table 1 Description of different algal biodiesel production scenarios compared

Algal biodiesel production scenarios	Abbreviation
Algal biodiesel production using Al flocculent and sCO_2 extraction and esterification method	AB1
Algal biodiesel production using Al flocculent and press and co-solvent oil extraction and esterification method	AB2
Algal biodiesel production using centrifugation and sCO_2 extraction and esterification method	AB3
Algal biodiesel production using centrifugation and press and co-solvent oil extraction and esterification method	AB4
Algal biodiesel production using lime flocculent and sCO_2 extraction and esterification method	AB5
Algal biodiesel production using lime flocculent and press and co-solvent oil extraction and esterification method	AB6

found that FPPBR is the most advantageous system compared to others with lesser limitations. Wegeberg and Felby (2010) from University of Copenhagen in a report made for DONG Energy also stated that FPPBR is the most suitable photobioreactor for Danish conditions. The size of bioreactor considered was 2.5 × 1.5 × 0.07 m, similar to Brentner et al. (Brentner et al. (2011)), 90 % volume of the reactor was considered as working volume, and 1.5 m distance between one row of FPPBR and another was considered.

2.1.2 Cultivation

The seawater and CO_2 from cement industry were considered for cultivation of algae. The transportation of seawater and CO_2 was deemed by pipelines. Nitrogen and phosphorus were supplied by ammonium nitrate and monocalcium phosphate, respectively. Transport of material (100 km) to the field was considered by 40 t truck, and electricity was supplied from the grid. The productivity of *Nannochloropsis* sp. was reported in the range of 0.24–1.7 g/l/d (Jorquera et al. 2010; Chiu et al. 2009; Richmond and Cheng-Wu 2001; Zou and Richmond 1990; Zittelli et al. 1999). Biomass productivity in this study was considered as 0.73 g/l/d, lower than the average of the range of biomass productivity (0.97 g/l/d) and similar to outdoor productivity (0.73 g/l/d) reported by Zittelli et al. (1999).

2.1.3 Harvesting

Harvesting of algal biomass was considered by three different techniques, i.e., centrifugation, flocculation with lime, and flocculation with aluminum, in different scenarios. The efficiency of all techniques was assumed to be 95 %, similar to Brentner et al. (2011). Grid electricity was taken for electricity supply.

2.1.4 Oil Extraction and Trans-esterification

Oil extraction from algal biomass is the least developed area in the algal biodiesel production technology. In the present study, the biodiesel conversion from the biomass was considered by two different techniques. One was pressing followed by solvent extraction (using recovered and recycled hexane), followed by trans-esterification. Another technique was supercritical CO_2 (sCO_2) extraction, which avoids the use of organic solvents, followed by trans-esterification. Danish average grid electricity and district heat were used to supply electricity and heat. Oil content of *Nannochloropsis* sp. was reported in the range of 20–60 % dry mass basis (Sierra et al. 2008; Rodolfi et al. 2009; Chiu et al. 2009; Jorquera et al., 2010). 29.6 % oil content was considered in this study because most of the researcher reported the same (Jorquera et al. 2010; Chiu et al. 2009; Sierra et al. 2008).

2.2 Life Cycle Assessment

The LCA was conducted by using SimaPro 7.2.3 software and ecoinvent 2.0 database. The cradle-to-grave system was adopted to analyze different scenarios of algal biodiesel production system.

2.2.1 Goal and Scope

The goal of the study was to evaluate options for harvesting and oil extraction of algal biomass in the biodiesel production and to identify the most sustainable algal biodiesel production pathway. The system was modeled using literature data and communications with scientists working in the area of algal biomass cultivation and biodiesel production. Robust operational data were not included as there are relatively very few facilities (even at pilot scale) available round the world. Thus, documentation of most of the modeled process reflects laboratory-scale operational data. The key assumptions used in this study are presented in the Table 2.

2.2.2 Functional Unit

The functional unit used in the study is 1 person km (pkm). During the different steps, reference flows change between 1 t algal biomass during cultivation step, 1 t biodiesel during biodiesel step, and 1 pkm for comparing algal biodiesel with fossil diesel.

2.2.3 Reference System

The EURO 5 passenger car running on diesel is used as reference system to compare the impacts of algal biodiesel use for transport with EURO 5 passenger car.

2.2.4 System Boundary

The algal biodiesel production system considered in this study included biomass cultivation, harvesting, oil extraction, trans-esterification, and biodiesel use in EURO 5 passenger car. This system does not include the potential use of co-product and/or by-product. The material used in the construction of FPPBR was considered inside the system boundaries, while the productions of other machinery/buildings were out of the system boundaries. Allocation of impacts on co-product/by-products was also not considered; hence, all impacts were allocated to the biodiesel.

Table 2 Key assumptions used in the algal biodiesel LCA

Parameter	Value considered	Calculated value	Reference
Biomass productivity	0.73 kg/m^3/y	0.73 kg/m^3/y	Zittelli et al. (1999)
Lipid content	29.6 %	29.6 %	Jorquera et al. (2010)
FPPBR	2.5 × 1.5 × 0.07 m	2.5 × 1.5 × 0.07 m	Brentner et al. (2011)
Working volume		90 %	
Productivity days		200	
Cultivation			
Land		0.01 ha/t algal biomass	
CO_2	1.50 % CO_2 0.40 l/min/l culture	7.98 t/t algal biomass	Cheng-Wu et al. (2001)
Solar energy		405.5 GJ/t algal biomass	
Water		474.2 m^3/t algal biomass	
Electricity	Aeration 10 W/m^3 10 h/day Pumping water and outflow 53 W/m^3	3009.2 kWh/t algal biomass	Sierra et al. (2008), Jorquera et al. (2010)
LDPE sheet	7.6 kg/FPPBR	4.4 kg/t algal biomass	Brentner et al. (2011)
Steel	0.6 kg/FPPBR	0.3 kg/t algal biomass	Brentner et al. (2011)
Ammonium nitrate	7.4 g/l biodiesel	7.2 kg/t algal biomass	Yang et al. (2011)
Monocalcium phosphate	27.70 g/l biodiesel	71.6 kg/t algal biomass	Yang et al. (2011)
Harvesting			
Option 1: Centrifugation			
Electricity	1 kWh/m^3	499.1 kWh/t algal biomass	Brentner et al. (2011)
Transport		8.4 t km/t algal biomass	
Option 2: Flocculation with lime			
Lime	0.316 kg/m^3	149.7 kg/t algal biomass	Brentner et al. (2011)
Electricity	0.1 kWh/m^3	49.9 kWh/t algal biomass	Brentner et al. (2011)
Transport		23.3 t km/t algal biomass	

(continued)

Table 2 (continued)

Parameter	Value considered	Calculated value	Reference
Option 3: Flocculation with aluminum			
Aluminum	0.074 kg/m^3	34.9 kg/t algal biomass	Brentner et al. (2011)
Electricity	0.1 kWh/m^3	49.9 kWh/t algal biomass	Brentner et al. (2011)
Transport		11.9 tkm/t algal biomass	
Oil extraction and trans-esterification			
Option 1: Press, co-solvent, and esterification			
Algal biomass		3.79 t/t algal biodiesel	
Extraction efficiency	91 %	91 %	Brentner et al. (2011)
Conversion efficiency	98 %	98 %	Brentner et al. (2011)
Electricity	69 kWh/10,000 MJ algal biodiesel	271.53 kWh/t algal biodiesel	Brentner et al. (2011)
Heat	17,585 MJ/10,000 MJ algal biodiesel	69200.2 MJ/t algal biodiesel	Brentner et al. (2011)
HCl (30 % v/v)	1.1 kg/10,000 MJ algal biodiesel	4.33 kg/t algal biodiesel	Brentner et al. (2011)
H_3PO_4 (85 % v/v)	2.8 kg/10,000 MJ algal biodiesel	11.02 kg/t algal biodiesel	Brentner et al. (2011)
Option 2: Supercritical CO_2 extraction and esterification			
Algal biomass		3.63 t/t algal biodiesel	
Extraction efficiency	95 %	95 %	Brentner et al. (2011)
Conversion efficiency	98 %	98 %	Brentner et al. (2011)
Electricity	1,840 kWh/10,000 MJ algal biodiesel	7240.74 kWh/t algal biodiesel	Brentner et al. (2011)
Heat	225 MJ/10,000 MJ algal biodiesel	885.12 MJ/t algal biodiesel	Brentner et al. (2011)
HCl (30 % v/v)	1.1 kg/10,000 MJ algal biodiesel	4.33 kg/t algal biodiesel	Brentner et al. (2011)
H_3PO_4 (85 % v/v)	2.8 kg/10,000 MJ algal biodiesel	11.02 kg/t algal biodiesel	Brentner et al. (2011)

2.2.5 Sensitivity Analysis

The sensitivity analysis was conducted with higher oil content of algal biomass (i.e., 60 %), reported by Rodolfi et al. (2009), and this oil content can be obtained in *Nannochloropsis* sp. by creating N stress at the adequate time of biomass cultivation.

3 Results and Discussion

Impact assessment of all scenarios was made using Impact 2002+ method in SimaPro 7.3.2. The impact categories considered in the present study include human health, ecosystem quality, climate change, and resources. These impact categories are further subdivided into carcinogens, non-carcinogens, respiratory inorganics, ionizing radiation, ozone layer depletion, aquatic ecotoxicity, terrestrial ecotoxicity, respiratory organics, terrestrial acid/nutrient, land occupation, global warming, non-renewable energy, and mineral extraction. The network of all scenarios is presented in Figs. 1, 2, 3, 4, 5, and 6. The scenarios considering the press and co-solvent oil extraction perform better than sCO_2 for oil extraction. Among the harvesting techniques, lime flocculation showed best results, followed by centrifugation and aluminum flocculent techniques. The press and co-solvent method might be better than sCO_2 due to lesser requirement of electricity in comparison with sCO_2 technique. Centrifugation technique required about 10 times higher electricity demand than the flocculation techniques. Aluminum production might involve higher energy input than the lime production, which resulted in a higher impact harvesting technique.

The comparison results of passenger car running with algal biodiesel produced in different scenarios and fossil diesel are presented in Figs. 7, 8, 9, 10, and 11. The characterization results depicted that algal biodiesel provides very high savings of GHG in comparison with diesel, while other impacts (carcinogens, non-carcinogens, respiratory inorganics and organics, ionizing radiation, mineral extraction, etc.) were higher in algal biodiesel produced in all scenarios in comparison with fossil diesel (Fig. 7). The higher reduction in global warming caused by the uptake of CO_2 during growth of algae makes algal biodiesel superior to the fossil diesel. Impact on mineral extraction was very high in the scenarios of aluminum flocculent used for harvesting (Fig. 7), because of the use of aluminum in the scenarios that require mineral extraction. Impact on non-renewable energy and aquatic eutrophication was also high in algal biodiesel scenarios than in fossil diesel ones (Fig. 7) and this might be due to higher chemical, electricity and heat demand during cultivation and conversion steps of algal biodiesel production.

Damage assessment of the study showed savings in climate change impact with the algal biodiesel in comparison with the fossil diesel. Human health, ecosystem quality, and resources get higher impacts with the use of algal biodiesel in

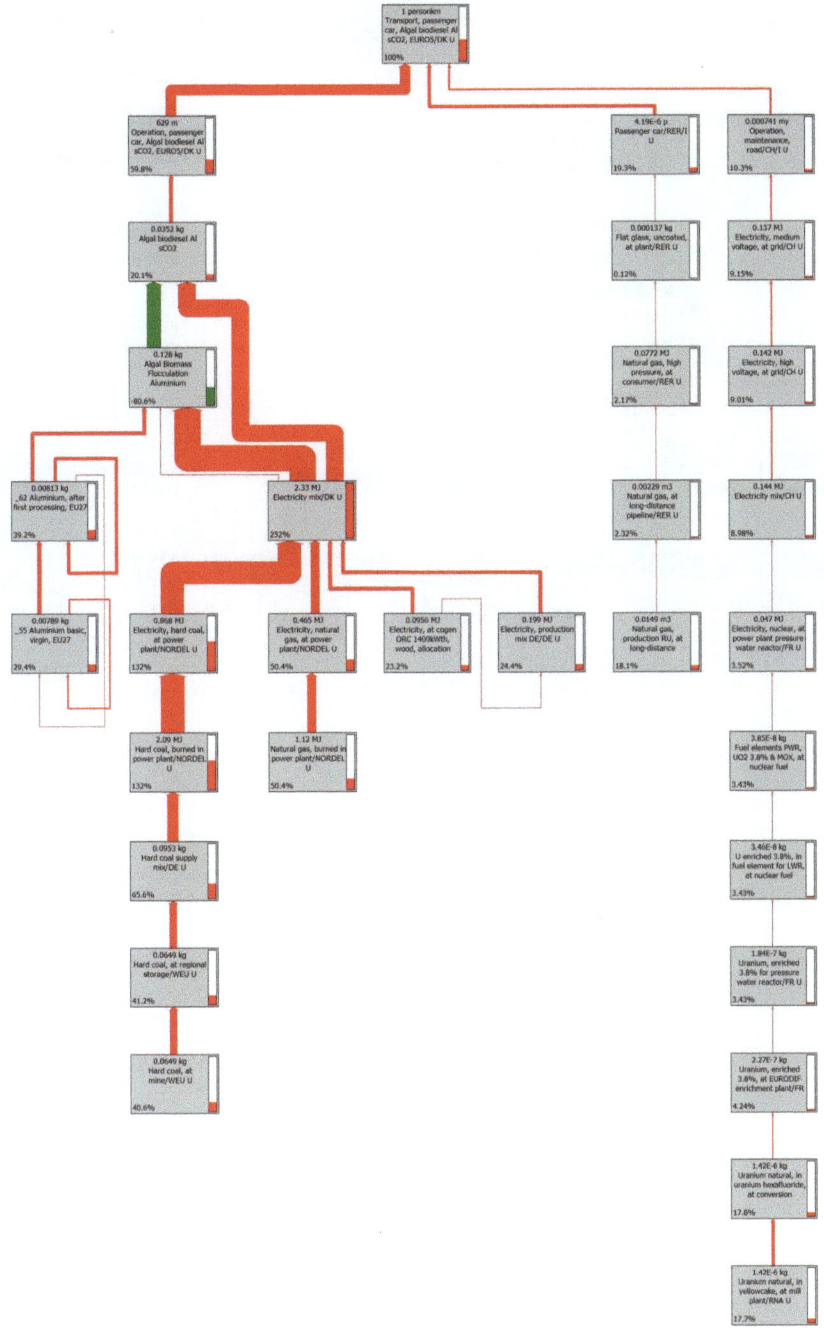

Fig. 1 Network of algal biodiesel production using aluminum flocculent for harvesting and sCO_2 for oil extraction (scenario AB1) for single score impact assessment (*Green sankey* represents positive impact, and *Red sankey* represents negative impact)

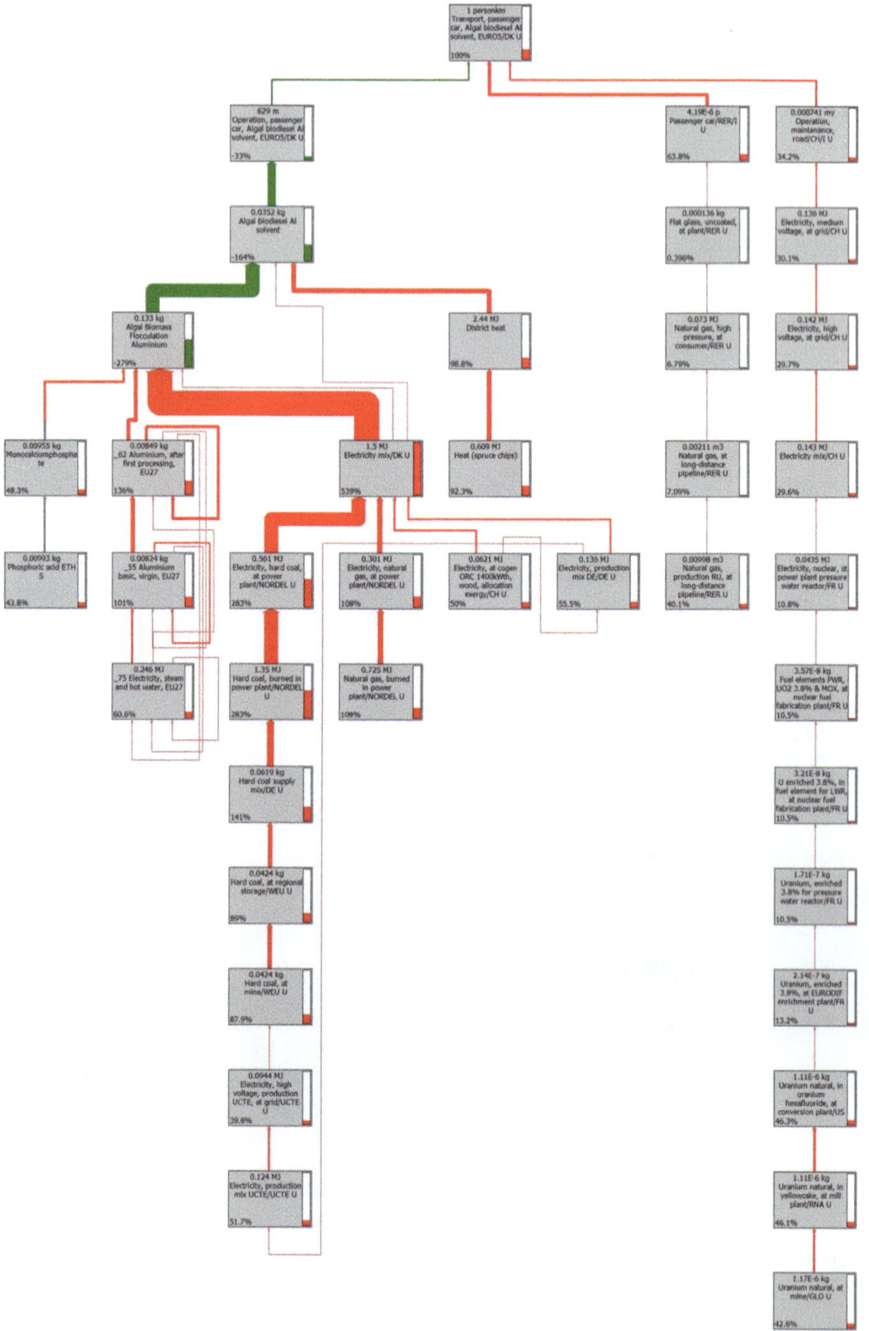

Fig. 2 Network of algal biodiesel production using aluminum flocculent for harvesting and press and co-solvent for oil extraction (scenario AB2) for single score impact assessment (*Green sankey* represents positive impact, and *Red sankey* represents negative impact)

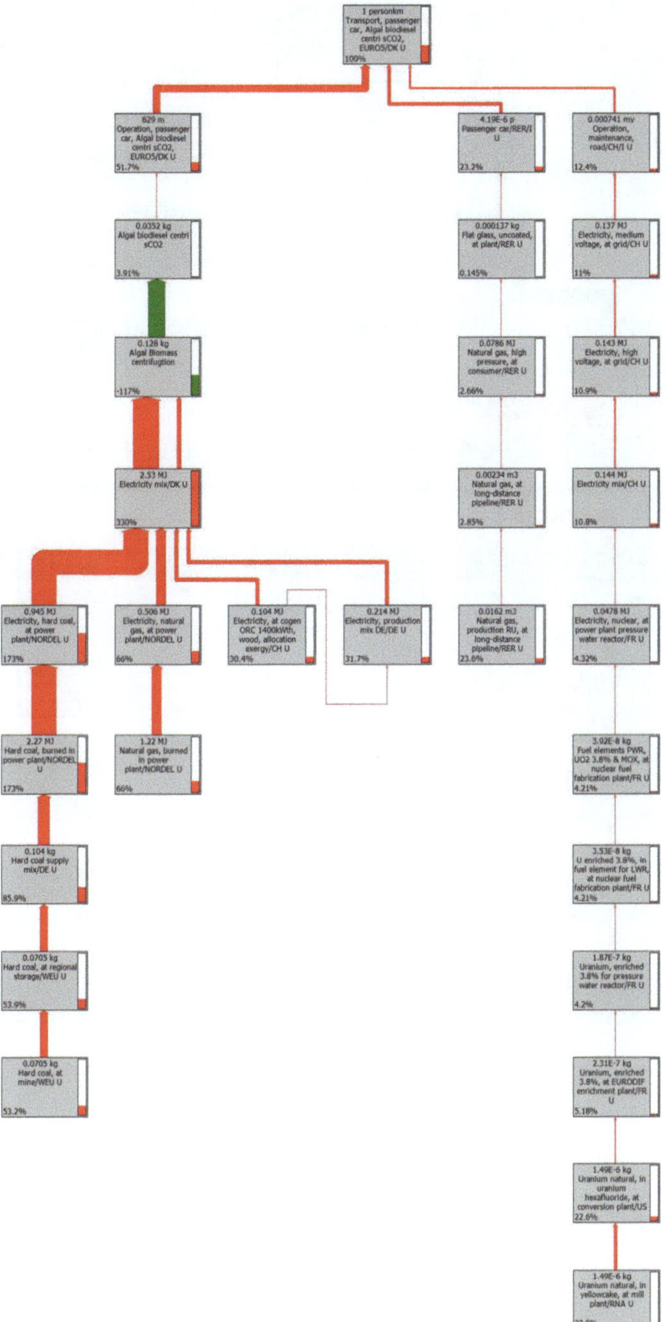

Fig. 3 Network of algal biodiesel production using centrifugation for harvesting and sCO_2 for oil extraction (scenario AB3) for single score impact assessment (*Green sankey* represents positive impact, and *Red sankey* represents negative impact)

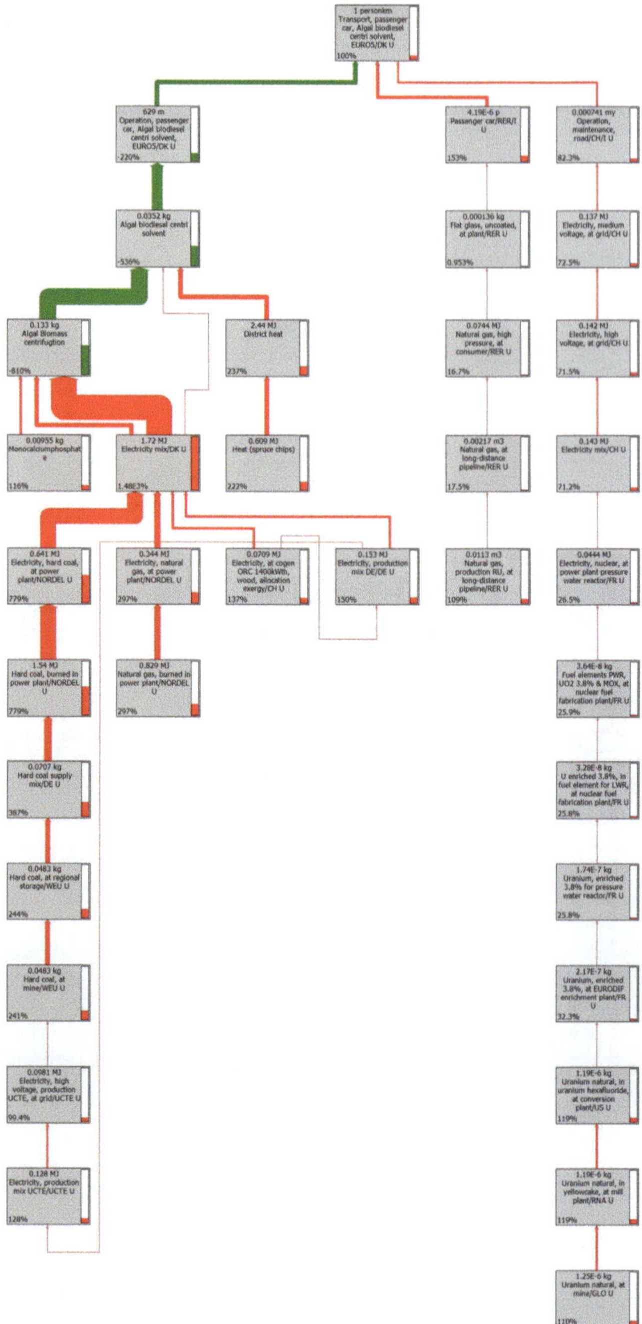

Fig. 4 Network of algal biodiesel production using centrifugation for harvesting and press and co-solvent oil extraction (scenario AB4) for single score impact assessment (*Green sankey* represents positive impact, and *Red sankey* represents negative impact)

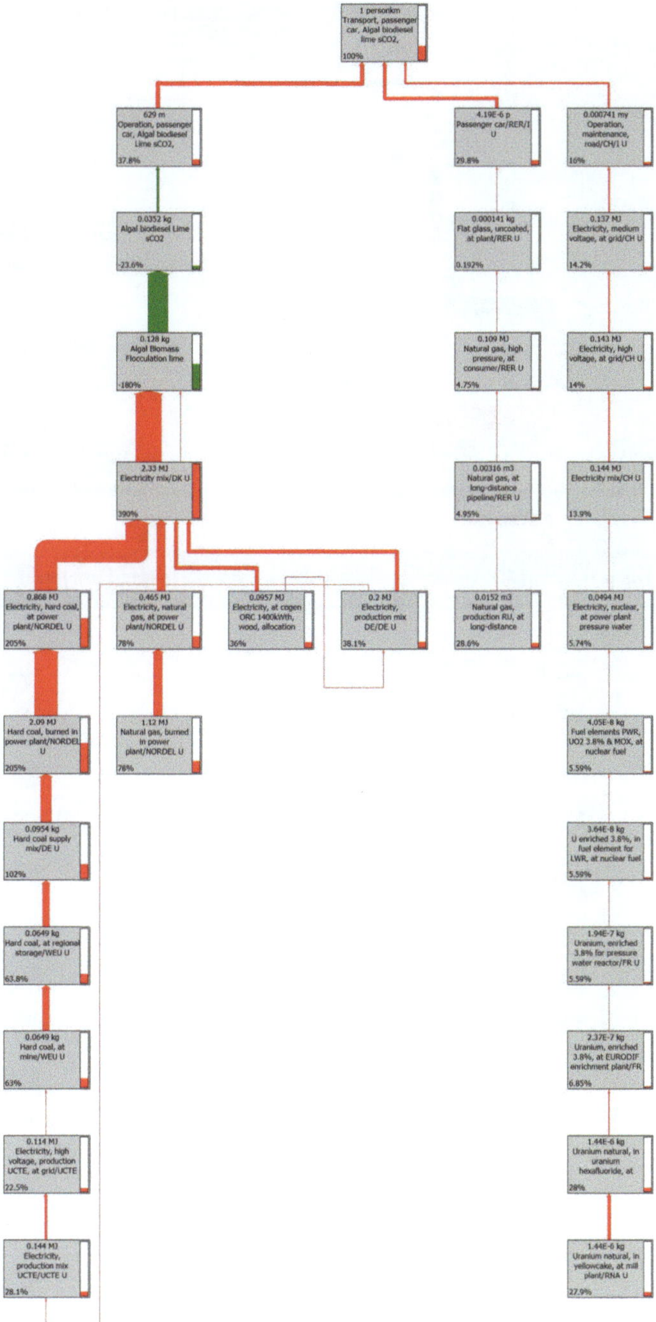

Fig. 5 Network of algal biodiesel production using lime flocculent for harvesting and sCO_2 for oil extraction (scenario AB5) for single score impact assessment (*Green sankey* represents positive impact, and *Red sankey* represents negative impact)

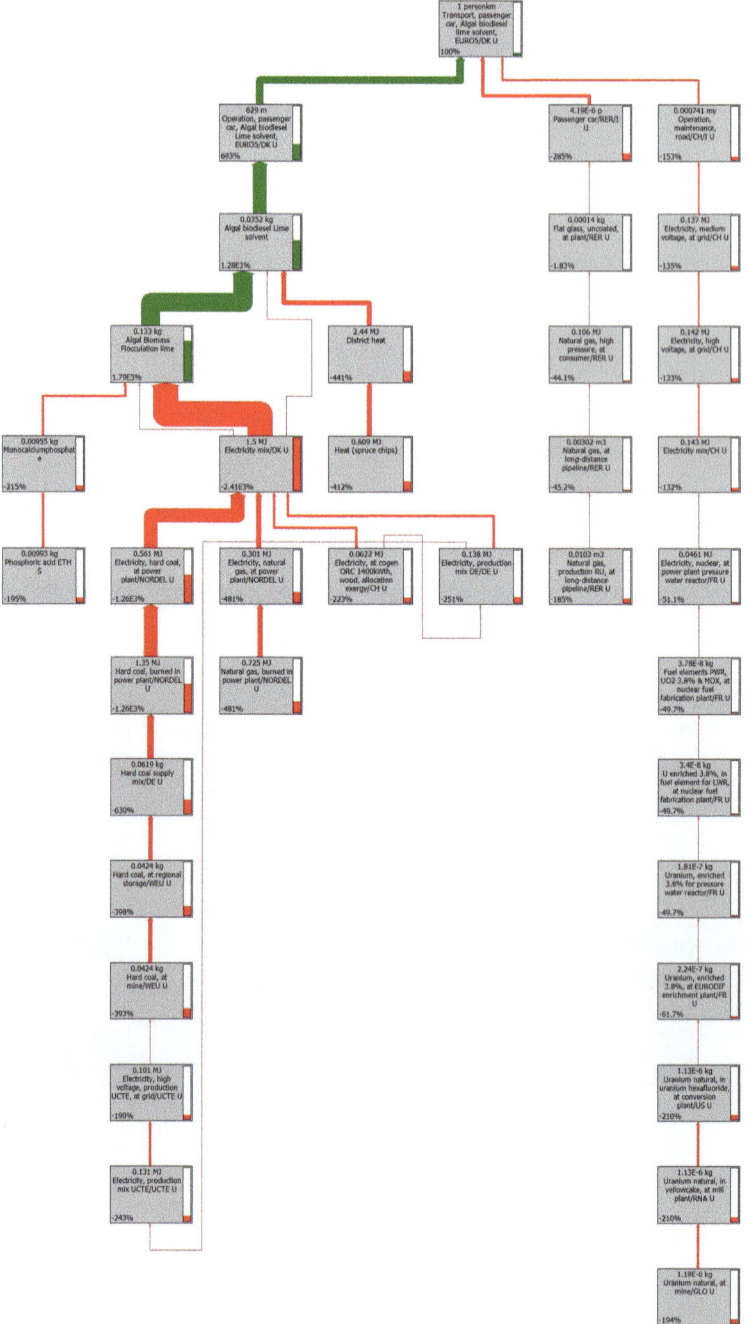

Fig. 6 Network of algal biodiesel production using lime flocculent for harvesting and press and co-solvent oil extraction (scenario AB6) for single score impact assessment (*Green sankey* represents positive impact, and *Red sankey* represents negative impact)

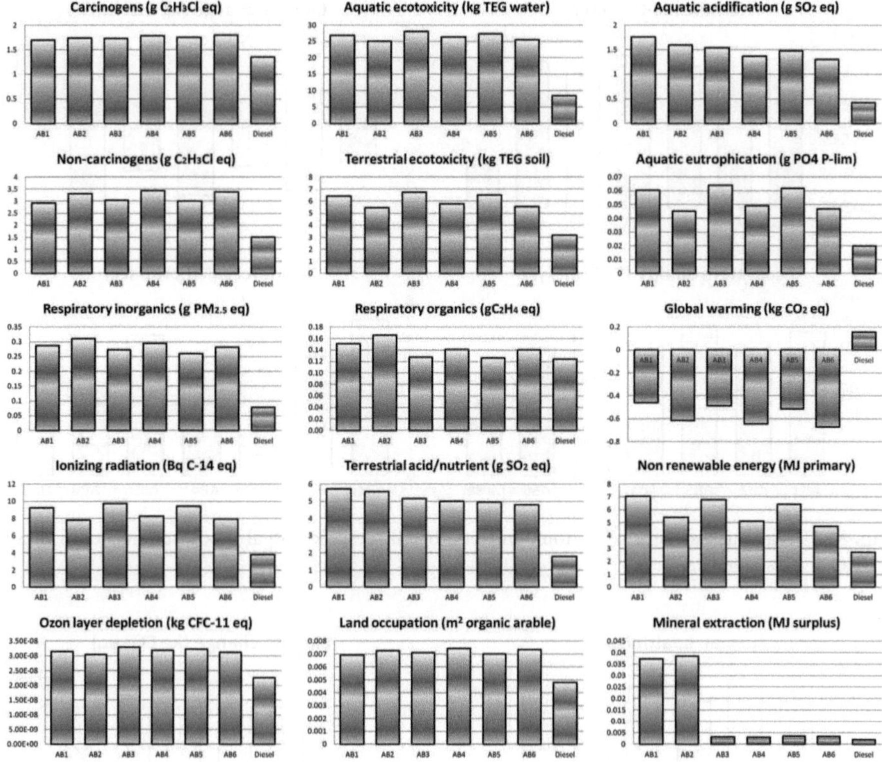

Fig. 7 Characterization of various impacts for all scenarios algal biodiesel and fossil diesel EURO 5 passenger car

passenger car than fossil ones (Fig. 8). Similar results were recorded for normalization (Fig. 9).

The results of single score per impact category are presented in Fig. 10. The comparison results clearly showed a positive impact of algal biodiesel use in passenger car on global warming as it provides savings of about 397–532 % in comparison with fossil diesel. About 18 times higher impacts were recorded on mineral extraction in case of aluminum flocculation harvesting scenarios than fossil diesel and about 1.5 times in centrifugation harvesting scenarios and about 1.7 times in lime flocculation scenarios. The different oil extraction techniques had not showed any significant difference in impacts on mineral extraction. Impacts on land occupation were about 44–54 % higher in algal biodiesel scenarios in comparison with fossil diesel. The total impact of all categories was positive for the use of algal biodiesel in passenger car. Results showed about 22–105 % reduction in total impacts by using algal biodiesel in passenger car in comparison with fossil diesel (Fig. 10).

Scenarios with press and co-solvent oil extraction showed better savings than the sCO_2 oil extraction scenarios. Lime flocculation harvesting with press and

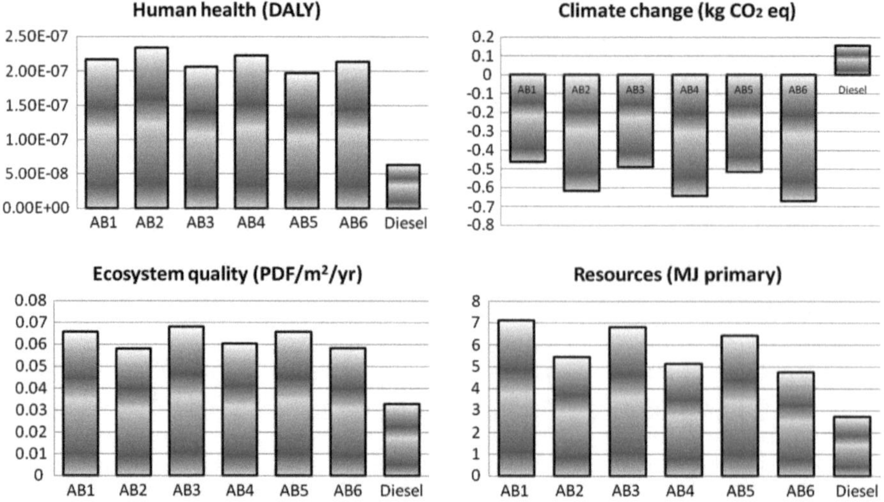

Fig. 8 Damage assessment of various impacts for all scenarios algal biodiesel and fossil diesel EURO 5 passenger car

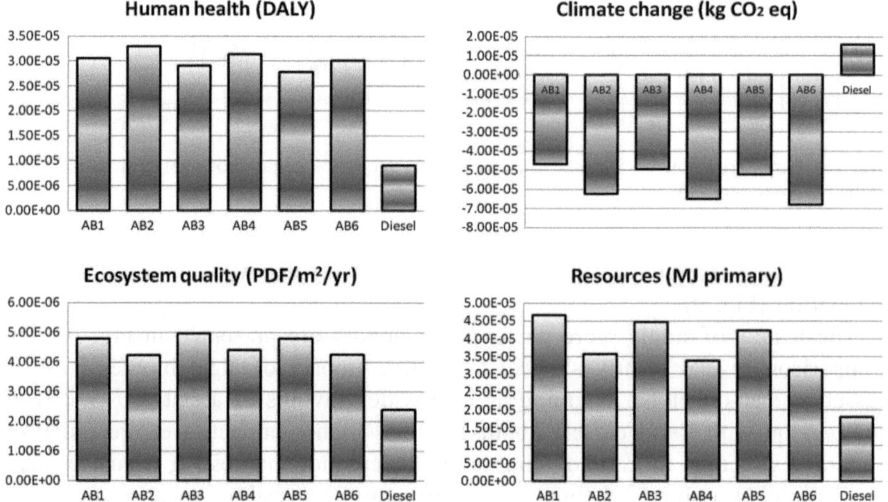

Fig. 9 Normalization of various impact categories for all scenarios algal biodiesel and fossil diesel EURO 5 passenger car

co-solvent oil extraction provides maximum saving of total impacts compared to other algal biodiesel scenarios (Fig. 10). Single score impacts on human health, ecosystem quality, and resources were higher in algal biofuel scenarios, while impacts on climate change were negative in all algal biodiesel scenarios in comparison with fossil diesel (Fig. 11). Impacts on human health were higher in

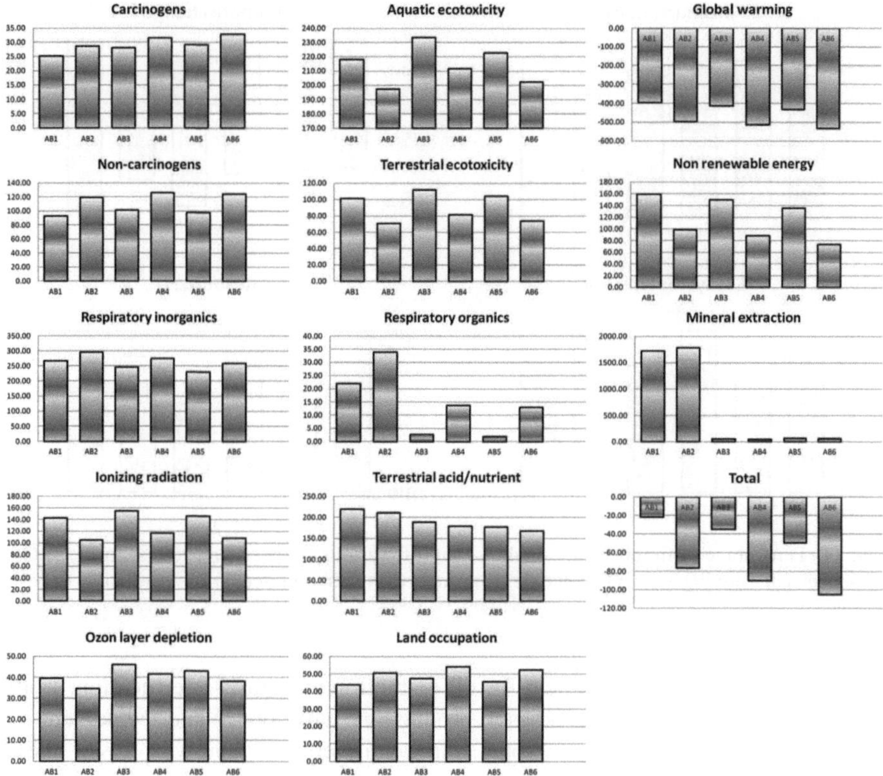

Fig. 10 Percentage difference of single score impact per impact category for EURO 5 passenger car running on algal biodiesel produced in different scenarios in reference to fossil diesel

scenarios with press and co-solvent oil extraction than sCO_2 oil extraction, while impacts on ecosystem quality, and resources were higher in sCO_2 oil extraction process in comparison with press and co-solvent oil extraction scenarios. The higher impacts on human health in press and co-solvent oil extraction scenarios might be due to the use of solvent in oil extraction. Impacts of different harvesting techniques showed similar trend on human health and resources, maximum with aluminum flocculation, followed by centrifugation and lime flocculation. Centrifugation harvesting technique provides higher impact on ecosystem quality, while aluminum and lime flocculation techniques provide about similar impacts (Fig. 11).

The sensitivity analysis was conducted by increasing the oil content in algal biomass from 29.6 to 60 %, and results of sensitivity analysis are presented in Tables 3, 4, 5, and 6. The increase in oil content reduces the impacts on various categories and decreases the savings of global warming potential of algal biodiesel (Table 3). The savings on climate change decreased to 257 from 531 %. This might be due to less biomass necessitates less uptake of CO_2, higher consumption

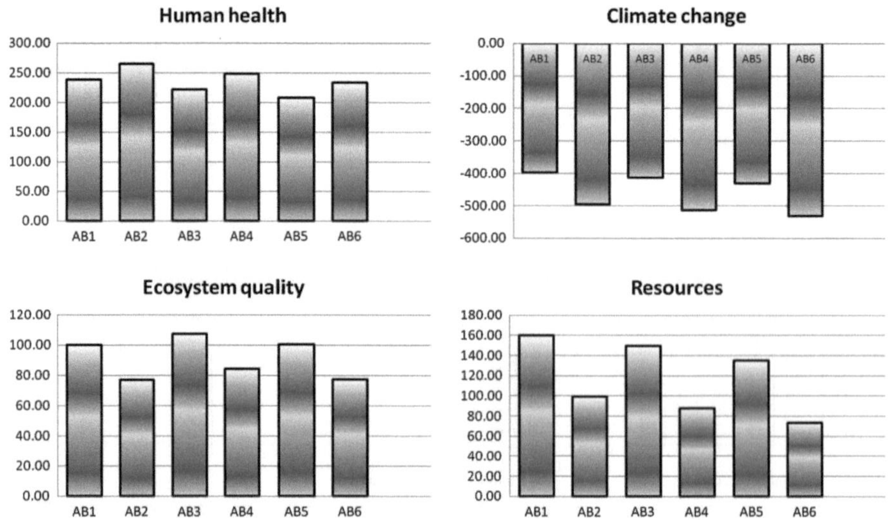

Fig. 11 Percentage difference of single score of various impact categories for EURO 5 passenger car running with algal biodiesel produced in different scenarios in relation to fossil diesel

of chemicals, electricity and heat for oil extraction and trans-esterification. In the sensitivity analysis, algal biodiesel production with press and co-solvent oil extraction showed savings on total impacts in comparison with fossil diesel (Table 6).

The results of present investigation showed that harvesting with lime flocculation and press and co-solvent oil extraction scenarios of algal biodiesel production provides maximum savings on total impacts. Frank et al. (2012) also reported a saving on GHG emissions from the use of algal biodiesel in comparison with low-sulfur petroleum diesel. Brentner et al. (2011) concluded that cultivation of algal biomass in FPPBR, harvesting with chitosan flocculation, and supercritical methanol process for oil extraction and trans-esterification along with anaerobic digestion and nutrient recycling reduce energy demand and eutrophication along with reductions in water and land use. Batan et al. (2010) also reported a saving of 75 g CO_2eq emissions per MJ of energy produced in a well-to-tank study of microalgal biodiesel production. The scenarios in this study also provide GHG savings over fossil diesel, but they are not better in terms of impacts on human health, ecosystem quality, and resources. To make the model more feasible there is need of more realistic commercial scale data and need to compare more scenarios of different cultivation, harvesting, oil extraction and trans-esterification techniques also need to expand the system boundaries to include utilization of co-products and by-products.

Table 3 Characterization of various impact categories for all scenarios algal biodiesel considering oil content 60 % and fossil diesel EURO 5 passenger car 60 (the values in parenthesis represent percent change from the corresponding value for diesel)

Impact category	AB1	AB2	AB3	AB4	AB5	AB6	Diesel
Carcinogens (kg C_2H_3Cl eq)	0.0016 (15.1)	0.0016 (18.2)	0.0016 (16.6)	0.0016 (19.6)	0.0016 (17.1)	0.0016 (20.2)	0.0013
Non-carcinogens (kg C_2H_3Cl eq)	0.0025 (63.4)	0.0028 (87.9)	0.0025 (67.2)	0.0029 (91.8)	0.0025 (66.0)	0.0029 (90.5)	0.0015
Respiratory inorganics (kg $PM_{2.5}$ eq)	0.00022 (178.3)	0.00024 (203.7)	0.00021 (168.5)	0.00023 (193.4)	0.00020 (160.8)	0.00022 (185.4)	0.00008
Ionizing radiation (Bq C-14 eq)	7.69 (102.3)	6.15 (61.8)	7.91 (108.2)	6.38 (67.9)	7.76 (104.1)	6.22 (63.6)	3.80
Ozone layer depletion (kg CFC-11 eq)	2.61E − 08 (15.7)	2.48E − 08 (9.9)	2.68E − 08 (18.9)	2.55E − 08 (13.1)	2.65E − 08 (17.4)	2.52E − 08 (11.6)	2.25E − 08
Respiratory organics (kg C_2H_4 eq)	0.00013 (3.0)	0.00014 (14.0)	0.00012 (−6.4)	0.00013 (4.2)	0.00012 (−6.8)	0.00013 (3.8)	0.00012
Aquatic ecotoxicity (kg TEG water)	22.44 (166.0)	20.49 (143.0)	23.01 (172.8)	21.09 (150.0)	22.63 (168.3)	20.70 (145.4)	8.43
Terrestrial ecotoxicity (kg TEG soil)	5.25 (64.6)	4.22 (32.4)	5.41 (69.8)	4.39 (37.8)	5.30 (66.2)	4.28 (34.1)	3.19
Terrestrial acid/nutria (kg SO_2 eq)	0.0043 (142.5)	0.0041 (131.4)	0.0041 (127.5)	0.0039 (115.7)	0.0040 (121.9)	0.0038 (109.9)	0.0018
Land occupation (m^2org.arable)	0.0063 (31.0)	0.0066 (37.2)	0.0064 (32.8)	0.0067 (39.0)	0.0063 (31.9)	0.0066 (38.1)	0.0048
Aquatic acidification (kg SO_2 eq)	0.0013 (209.7)	0.0011 (166.9)	0.0012 (184.2)	0.0010 (140.3)	0.0012 (176.6)	0.0010 (132.4)	0.0004
Aquatic eutrophication (kg PO_4 P-lim)	4.71E − 05 (138.0)	3.16E − 05 (59.9)	4.90E − 05 (147.4)	3.36E − 05 (69.7)	4.79E − 05 (142.0)	3.25E − 05 (64.0)	1.98E − 05
Global warming (kg CO_2 eq)	−0.080 (−151.4)	−0.217 (−239.5)	−0.093 (−159.9)	−0.231 (−248.4)	−0.106 (−168.3)	−0.245 (−257.1)	0.156
Non-renewable energy (MJ primary)	5.097 (86.9)	3.353 (22.9)	4.969 (82.2)	3.219 (18.0)	4.782 (75.3)	3.024 (10.9)	2.727
Mineral extraction (MJ surplus)	0.020 (864.6)	0.020 (890.6)	0.003 (40.1)	0.003 (29.7)	0.003 (47.1)	0.003 (36.9)	0.002

Table 4 Damage assessment of various impact categories for all scenarios algal biodiesel considering oil content 60 % and fossil diesel EURO 5 passenger car 60 (the values in parenthesis represent percent change from the corresponding value for diesel)

Damage category	AB1	AB2	AB3	AB4	AB5	AB6	Diesel
Human health (DALY)	1.66E − 07 (159.3)	1.81E − 07 (182.4)	1.61E − 07 (151.3)	1.75E − 07 (174.0)	1.56E − 07 (144.6)	1.71E − 07 (167.0)	6.40E − 08
Ecosystem quality (PDF*m^2*yr)	0.054 (65.0)	0.046 (40.2)	0.055 (68.4)	0.047 (43.9)	0.054 (65.2)	0.046 (40.5)	0.033
Climate change (kg CO_2 eq)	−0.080 (−151.4)	−0.217 (−239.5)	−0.093 (−159.9)	−0.231 (−248.4)	−0.106 (−168.3)	−0.245 (−257.1)	0.156
Resources (MJ primary)	5.12 (87.4)	3.37 (23.6)	4.97 (82.1)	3.22 (18.0)	4.79 (75.3)	3.03 (10.9)	2.73

Table 5 Normalization of various impact categories for all scenarios algal biodiesel considering oil content 60 % and fossil diesel EURO 5 passenger car (the values in parenthesis represent percent change from the corresponding value for diesel)

Damage category	AB1	AB2	AB3	AB4	AB5	AB6	Diesel
Human health	2.34E − 05 (159.3)	2.55E − 05 (182.4)	2.27E − 05 (151.3)	2.47E − 05 (174.0)	2.21E − 05 (144.6)	2.41E − 05 (167.0)	9.02E − 06
Ecosystem quality	3.95E − 06 (65.0)	3.35E − 06 (40.2)	4.03E − 06 (68.4)	3.44E − 06 (43.9)	3.95E − 06 (65.2)	3.36E − 06 (40.5)	2.39E − 06
Climate change	−8.09E − 06 (−151.4)	−2.20E − 05 (−239.5)	−9.43E − 06 (−159.9)	−2.34E − 05 (−248.4)	−1.07E − 05 (−168.3)	−2.47E − 05 (−257.1)	1.57E − 05
Resources	3.37E − 05 (87.4)	2.22E − 05 (23.6)	3.27E − 05 (82.1)	2.12E − 05 (18.0)	3.15E − 05 (75.3)	1.99E − 05 (10.9)	1.80E − 05

Table 6 Single score of various impact categories for all scenarios algal biodiesel considering oil content 60 % and fossil diesel EURO 5 passenger car 60 (the values in parenthesis represent percent change from the corresponding value for diesel)

Damage category	AB1	AB2	AB3	AB4	AB5	AB6	Diesel
Total (Pt)	5.29E − 05 (17.3)	2.91E − 05 (−35.6)	5.00E − 05 (10.8)	2.60E − 05 (−42.4)	4.68E − 05 (3.6)	2.26E − 05 (−49.8)	4.51E − 05
Human health (Pt)	2.34E − 05 (159.3)	2.55E − 05 (182.4)	2.27E − 05 (151.3)	2.47E − 05 (174.0)	2.21E − 05 (144.6)	2.41E − 05 (167.0)	9.02E − 06
Ecosystem quality (Pt)	3.95E − 06 (65.0)	3.35E − 06 (40.2)	4.03E − 06 (68.4)	3.44E − 06 (43.9)	3.95E − 06 (65.2)	3.36E − 06 (40.5)	2.39E − 06
Climate change (Pt)	−8.09E − 06 (−151.4)	−2.20E − 05 (−239.5)	−9.43E − 06 (−159.9)	−2.34E − 05 (−248.4)	−1.07E − 05 (−168.3)	−2.47E − 05 (-−257.1)	1.57E − 05
Resources (Pt)	3.37E − 05 (87.4)	2.22E − 05 (23.6)	3.27E − 05 (82.1)	2.12E − 05 (18.0)	3.15E − 05 (75.3)	1.99E − 05 (10.9)	1.80E − 05

4 Conclusion

The algal biodiesel could be a better option to replace the fossil diesel, but to make any policy for the algal biodiesel, there is a need for more in-depth analysis with multiple scenarios and more realistic data, at least pilot-scale data required to draw any solid recommendation. The impacts of algal biodiesel production could be minimized by utilization of co-products and by-products generated during various steps, within the system.

References

Batan L, Quinn J, Willson B, Bradley T (2010) Net energy and greenhouse gas emission evaluation of biodiesel derived from microalgae. Environ Sci Technol 44:7975–7980

Brentner LB, Eckelman MJ, Zimmerman JB (2011) Combinatorial life cycle assessment to inform process design of industrial production of algal biodiesel. Environ Sci Technol 45(16):7060–7067

Cheng-Wu Z, Zmora O, Kopel R, Richmond A (2001) An industrial-size flate plate glass reactor for mass production of *Nannochloropsis* sp. (Eustigmatophyceae). Aquaculture 195:35–49

Chisti Y (2007) Biodiesel from microalgae. Biotechnol Adv 25:294–306

Chiu SY, Kao SY, Tasi MT, Ong SC, Chen CH, Lin CS (2009) Lipid accumulation and CO_2 utilization in *Nannochloropsis oculata* in response to CO_2 aeration. Bioresour Technol 100:833–838

Clarens AF, Resurreccion EP, White MA, Colosi LA (2010) Environmental life cycle comparison of algae to other bioenergy feedstocks. Environ Sci Technol 44:1813–1819

Frank ED, Han J, Palou-Rivera I, Elgowainy A, Wang MQ (2012) Methane and nitrous oxide emissions affect the life cycle analysis of algal biofuels. Environ Res Lett 7:014030

Hu Q, Sommerfeld M, Jarvis E, Ghirardi M, Posewitz M, Seibert M, Darzins A (2008) Microalgal triacylglycerols as feedstocks for biofuel production: perspectives and advances. Plant J 54:621–639

John RP, Anisha GS, Nampoothiri KM, Pandey A (2011) Micro and macroalgal biomass: a renewable source for bioethanol. Bioresour Technol 102:186–193

Jorquera O, Kiperstok A, Sales EA, Embirucu M, Ghirardi ML (2010) Comparative energy life-cycle analyses of microalgal biomass production in open ponds and photobioreactors. Bioresour Technol 101:1406–1413

Korres NE, Singh A, Nizami AS, Murphy JD (2010) Is grass biomethane a sustainable transport biofuel? Biofuels Bioproducts Biorefining 4:310–325

Lardon L, Hélias A, Sialve B, Steyer JP, Bernard O (2009) Life-cycle assessment of biodiesel production from microalgae. Environ Sci Technol 43(17):6475–6481

Mata TM, Martins AA, Caetano NS (2010) Microalgae for biodiesel production and other applications: a review. Renew Sustain Energy Rev 14:217–232

Melis A, Happe T (2001) Hydrogen production: green algae as a source of energy. Plant Physiol Biochem 127:740–748

Metting FB (1996) Biodiversity and application of microalgae. J Ind Microbiol Biotechnol 17:477–489

Nigam P, Singh A (2011) Production of liquid biofuels from renewable resources. Prog Energy Combust Sci 37:52–68

Pant D, Singh A, Bogaert GV, Diels L, Vanbroekhoven K (2011) An introduction to the life cycle assessment (LCA) of bioelectrochemical systems (BES) for sustainable energy and product generation: relevance and key aspects. Renew Sustain Energ Rev 15:1305–1313

Richmond A, Cheng-Wu Z (2001) Optimization of a plate glass reactor for mass production of *Nannochloropsis* sp. outdoors. J Biotechnol 85:259–269

Rodolfi L, Chini Zittelli G, Bassi N, Padovani G, Biondi N, Bonini G, Tredici M (2009) Microalgae for oil: strain selection, induction of lipid synthesis and outdoor mass cultivation in a low-cost photobioreactor. Biotechnol Bioeng 102:100–112

Sierra E, Acién FG, Fernández JM, García JL, González C, Molina E (2008) Characterization of flat plate photobioreactors for the production of microalgae. Chem Eng J 138:136–147

Singh A, Nigam PS, Murphy JD (2011a) Renewable fuels from algae: an answer to debatable land based fuels. Bioresour Technol 102:10–16

Singh A, Nigam PS, Murphy JD (2011b) Mechanism and challenges in commercialisation of algal biofuels. Bioresour Technol 102:26–34

Singh A, Olsen SI (2011a) Critical analysis of biochemical conversion, sustainability and life cycle assessment of algal biofuels. Appl Energy 88:3548–3555

Singh A, Olsen SI (2011b) Algal biofuels: key issues, sustainability and life cycle assessment. In: Petersen LS, Larsen H (eds) Energy systems and technologies for the coming century. Proceedings, Risø international energy conference, 10–12 May 2011, pp 275–282

Singh A, Olsen SI (2012) Key issues in life cycle assessment of biofuels. In: Gopalakrishnan K et al (eds) Sustainable bioenergy and bioproducts, green energy and technology. Springer, London, pp 213–228

Singh A, Olsen SI, Nigam PS (2011c) A viable technology to generate third generation biofuel. J Chem Tech Biotech 86:1349–1353

Singh A, Pant D, Olsen SI, Nigam PS (2012) Key issues to consider in microalgae based biodiesel production. Energy Educ Sci Technol Part A: Energy Sci Res 29(1):687–700

Spolaore P, Joannis-Cassan C, Duran E, Isambert A (2006) Commercial applications of microalgae. J Biosci Bioeng 101:87–96

Subhadra B, Edwards M (2010) An integrated renewable energy park approach for algal biofuel production in United States. Energ Policy 38:4897–4902

Ugwu CU, Aoyagi H, Uchiyama H (2008) Photobioreactors for mass cultivation of algae. Bioresour Technol 99:4021–4028

Wegeberg S, Felby C (2010) Algae biomass for bioenergy in Denmark biological/technical challenges and opportunities. University of Copenhagen, Copenhagen

Yang J, Xu M, Zhang X, Hu Q, Sommerfeld M, Chen Y (2011) Life-cycle analysis on biodiesel production from microalgae: water footprint and nutrients balance. Bioresour Technol 102:159–165

Zittelli GC, Lavista F, Bastianini A, Rodolfi L, Vincenzini M, Tredici MR (1999) Production of eicosapentaenoic acid by *Nannochloropsis* sp. cultures in outdoor tubular photobioreactors. J Biotechnol 70:299–312

Zou N, Richmond A (1990) Effect of light-path length in outdoor flat plate reactors on output rate of cell mass and of EPA in *Nannochloropsis* sp. J Biotechnol 70:351–356

Sustainability of (H_2 + CH_4) by Anaerobic Digestion via EROI Approach and LCA Evaluations

B. Ruggeri, S. Sanfilippo and T. Tommasi

Abstract The contents of this Chapter focus on the theoretical sustainable energy approach and its application to hydrogen and methane production, on the basis of results obtained from experimental tests on the Anaerobic Digestion (AD) technology. The evaluation of sustainability is pursued through the life cycle assessment (LCA), energy return on investment (EROI) and energy payback time (EPT) approaches. The EROI and EPT parameters are defined and applied to score the sustainability of the H_2/CH_4 energy carrier. The evaluation of the indirect energy following a life cycle assessment is consistent for the sustainability analysis. The sustainability of AD technology strongly depends on the reactor diameter: for values lower than 1 m the technology is not able to sustain the well-being of the society; the effect of the insulating material as well as the labor could be very important, and in this respect, thus, a sensitivity analysis on the sustainability is reported.

1 Introduction

The present energy crisis together with environmental issues, such as global warming, has persuaded men to search new energy sources (Balat 2008). Different renewable sources are now being exploited. Energy crops, wind power, water

B. Ruggeri (✉) · S. Sanfilippo
Department of Applied Science and Technology, Politecnico di Torino, Corso Duca degli Abruzzi 24, 10129 Turin, Italy
e-mail: bernardo.ruggeri@polito.it

S. Sanfilippo
e-mail: sara.sanfilippo@polito.it

T. Tommasi
Center for Space Human Robotics, Istituto Italiano di Tecnologia, Corso Trento 21, 10129 Turin, Italy
e-mail: tonia.tommasi@iit.it

A. Singh et al. (eds.), *Life Cycle Assessment of Renewable Energy Sources*, Green Energy and Technology, DOI: 10.1007/978-1-4471-5364-1_8,

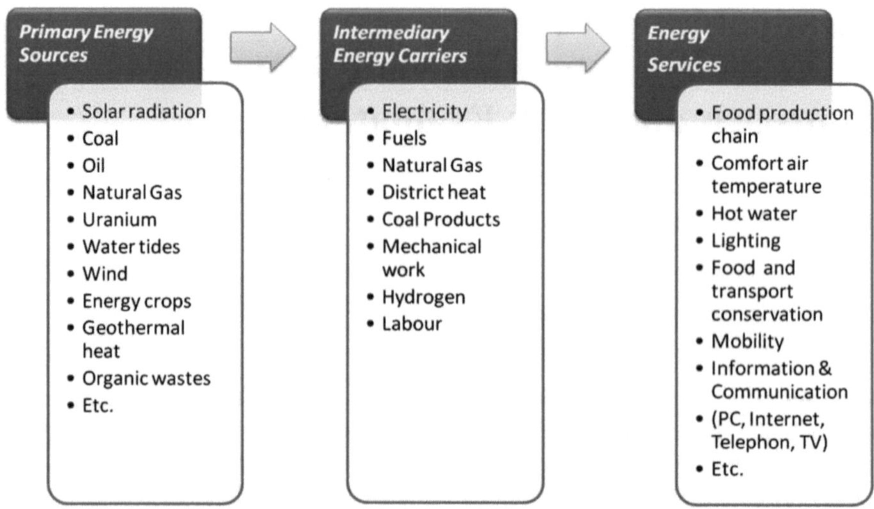

Fig. 1 General layout of energy flow

power, solar energy, and organic refuse from the food chain could offer possible solutions (Angenent et al. 2004).

However, we believe that it is also important to introduce the concept of *energy service*, here intended as the amount of energy required by the end-user as *useful energy*, i.e., the energy necessary to support human life, as outlined in Fig. 1. *Surplus energy* flowing from each block in Fig. 1 depends from the technology used, and it is of primary importance for society. Wealth, survival, art, army, and even civilization itself is a product of surplus energy. The interplay of *how much*, *what kind* (quality), and at *what rate* the energy is delivered determines the useful energy. It gives the ability to the society to divert attention from life-sustaining needs toward luxuries, such as art and scholarships including research and innovation for the exploitation of different energy sources.

Among the primary energy sources, organic waste material (Evans 2001) is approximately 60 % of daily refuse production. The technology pallet to use organic waste ranges from biological processes (Pfeffer and Lieman 1976) to thermal methods, such as gasification, pyrolysis, and incineration (Guéhenneux et al. 2005) including the direct conversion of organic matter into electrical energy through the use of Microbial Fuel Cell (Tommasi et al. 2012; Logan 2008; Aelterman et al. 2006) as reported in Fig. 2.

Taking in mind Fig. 2, in order to select the most appropriate technology, it is necessary to establish which criteria should be used to valorize the sources (Sentimenti and Biorgi 2006). In this context, economic criteria on their own appear to be inappropriate, because data can easily be manipulated according to the working hypothesis and the conclusions might not be completely reliable (Cleveland et al. 1984). Economists argue that the price of a technology or a fuel

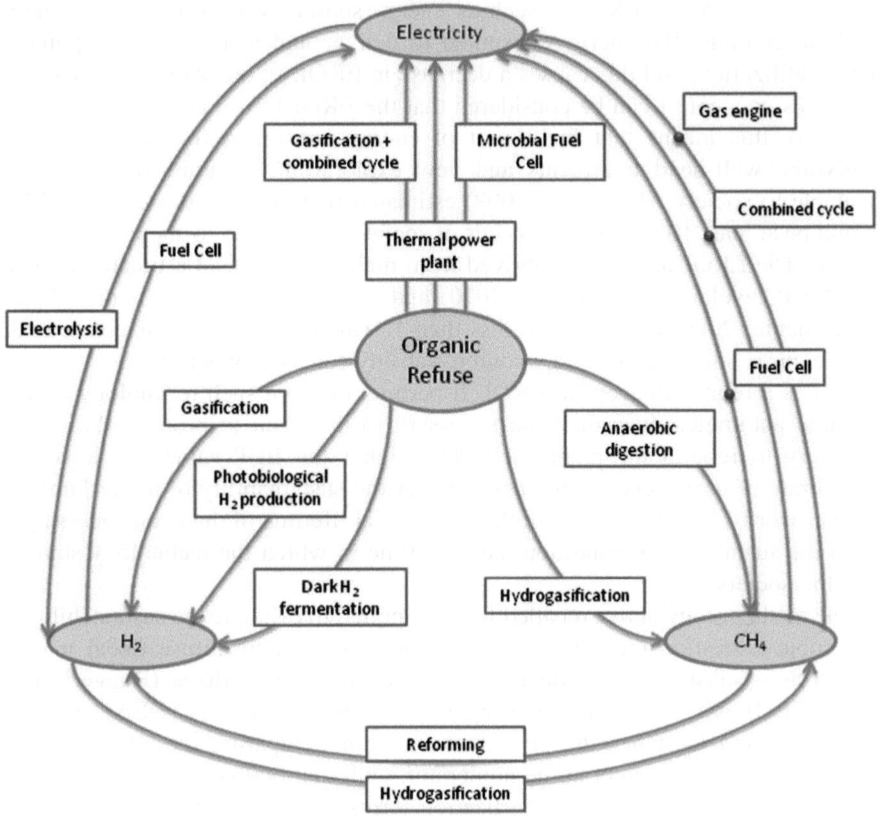

Fig. 2 Technologies able to produce energy using Organic Wastes

automatically captures all the relevant features, but in a finite resource scenario, this at least appears to be questionable. The life cycle assessment (LCA) (SETAC 1993), which takes into account all the aspects of such a technology (e.g., environmental impact, safety, toxicity, energy use, and social issues together with economics), is an alternative to a conventional economic analysis. One of the difficulties of selecting a technology is the need to measure the sustainability level of it.

To this aim, several approaches, ranging from a thermodynamic one (de Swan et al. 2004) to a more industrial-oriented alternative (Apazagic and Perdam 2000; De Simone and Popoff 1997), have been put forward in recent years to evaluate the sustainability (Azapagic 1999; Laws et al. 1984). Hall et al. (2009) with reference to energy sustainability, proposed that the most appropriate way to judge the relative merits of different energy sources is to evaluate the ratio between the amount of energy produced and the energy needed to produce it, known as the energy return on investment (EROI). EROI, in its simplest form, measures the output energy at the point of production or "mine mouth" (Murphy et al. 2011).

The evaluation of the EROI of such an energy source away from "mine mouth" needs to compute the energy consumed to deliver and to use it at the point of energy utilization, and this causes a decrease in EROI. In order to have some idea about this concept, it can be considered that the EROI for oil at "mine mouth" is about 20: this means that for 1 unit of energy consumed for extraction from reservoirs, well-head treatments and new exploration, 20 units of energy are available to society. Hall et al. (2009) estimated that at the end-user level EROI should be at least 10 to cover the needs of society/civilization to support an energy service. The EROI for ethanol derived from maize was instead estimated to be at best 1.3 (Cleveland and Costanza 2010) and according to some authors (Patzek and Pimentel 2006; Patzek 2004) less than 1. This implies that maize-based ethanol requires some other energy source, subsidy for its production.

EPT is a related concept to EROI. It permits to score such technology against the time parameter. It is the time necessary to the plant to produce the energy necessary to rebuild the plant itself. The higher the EPT value, the lower the annual rate of useful energy and hence lower the sustainability of the technology. In other words, EPT is the time of the operational lifetime of the plant necessary to reach the sustainability condition, i.e., the time in which the technology starts to feed the society.

The methodology above recalled will be applied to evaluate the sustainability of Anaerobic Digestion (AD) technology using food organic refuse (local marked refuse) as a substrate to produce biohydrogen plus biomethane (Ruggeri et al. 2010). EROI will be evaluated using the net energy analysis (NEA) approach (Cleveland and Costanza 2010) with the aim of comparing the amount of net energy delivered to society at the numerator, with the total energy required to run the plant as the indirect and amortization terms at the denominator. The terms *useful* for the energy delivered to society and *net* for the energy produced minus the direct energy spent to run the plant are used in this chapter.

A comparison with other energy technologies is shown. In order to inquire the influence of some choices, different insulation materials are also compared, and their impact on EROI and EPT is evaluated in this chapter.

2 Methodology

2.1 General Framework

According to the concept introduced by Röegen (1976), in order to have energy sustainability of such an energy technology, it is necessary that the technology must be vital. Like a biological system, an energy technology must be able to produce at least a quantity of useful energy that is able to sustain itself in order to sustain "others" energy service. It necessarily needs to use only a part of the energy source for its operational necessities and reproduction, and the remaining part will be used to feed civilization in an appropriate form. In other words, a technology is sustainable if produces a surplus energy as useful energy. Figure 3 reports a general picture of the energy terms involved in such technology in order to extract a useful energy from an energy source.

The evaluation of useful energy offers several advantages over the standard economic analysis (Röegen 1999): primarily, because it assesses the change in the physical scarcity of energy resources, then because it is a measure of the potential of such a technology to work in a sustainable way, and finally, because it is possible to rank alternative energy supply technologies according to their capacity to produce useful energy. In order to perform a useful energy analysis of such a technology, it is necessary to evaluate the direct energy and, moreover, the indirect energy required to produce it. Direct energy is the fuel or the electricity used directly to run a plant while indirect energy refers to the energy used to produce materials, to assemble parts of the plant (such as pumps, pipes, valves, etc.), to produce chemicals and all the other consumables, plus the energy consumed to produce fuels and electricity (Hammerschlag 2006). Amortization energy is the energy necessary to rebuild the plant taking into account the recycling or reuse options. Finally, it is important to take into account the energy used to sustain the labor to operate the plant. In this context, it is important to state that the energy terms need to be measured in a physical energetic unit; in some cases, the sum of direct and indirect energy is named embodied energy (Cleveland and O'Connor 2011).

The sustainability of $H_2 + CH_4$ produced by AD is investigated using a scale-up procedure (Najafpour 2007) along the diameter. The energy produced as H_2 and CH_4 is referred to their combustion enthalpies. The sustainability of the AD technology is estimated by evaluating the EROI and EPT parameters.

2.2 Indices to Evaluate Sustainability of ($H_2 + CH_4$) by AD

The energy assessment of a process by LCA approach involves the entire life cycle of the process, including raw material extraction and processing, manufacturing of the plant and its assembly, transportation, energy use for the operation of the plant, such as electrical energy and heat, multiple use of the plant and recycling and/or final disposal in the so-called decommissioning phase. Attention is here focused on the technology term in order to evaluate the most suitable technology to convert a source into an energy service in the most satisfactory sustainable way. This means that the same energy service could be furnished by using different sources and/or different technologies. Referring to Fig. 3, all included terms in the indirect energy need to be considered. The indirect energy will be considered by using LCA approach evaluating the global energy requirement (GER). The evaluation of the maintenance, amortization, and decommission energy terms is specific for each technology; in particular, the amortization term depends on lifetime of the technology.

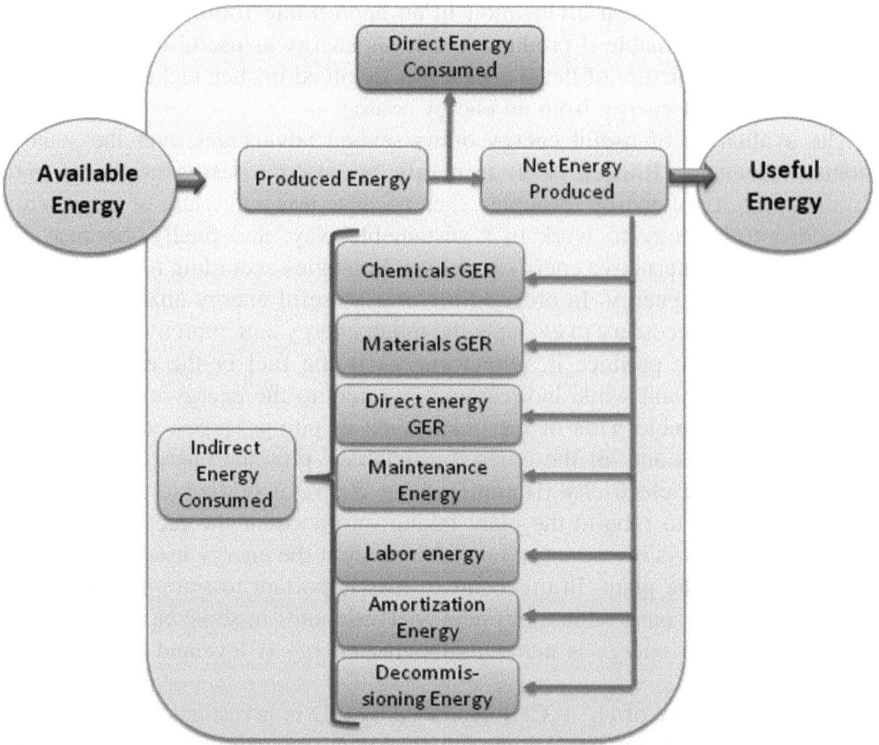

Fig. 3 Energy terms involved in an energy technology

Unlike other researchers in which EROI was used to evaluate the net energy of such energy sources (Cleveland and O'Connor 2011; Guilford et al. 2011; Brand 2009), we used EROI and EPT to evaluate the sustainability of $H_2 + CH_4$ produced by AD technology. The approach is quite similar, but some differences exist linked to the useful energy. Numerically speaking, only an EROI > 1 indicates a sustainable process. It is important to consider also the EPT parameter: if a technology has an EPT $= 7$ years and its operation lifetime is around 20 years, with a construction time of 2 years, the first energy unit able to support the energy service of the society will be available after 9 years, even if the technology has an EROI > 1.

2.3 EROI and EPT Tools

EROI is the ratio between the total amount of net energy delivered to society by a technology during its working lifetime and the amount of total indirectly energy in such process to produce energy (Murphy et al. 2011). It is a ratio between two energy quantities and is therefore dimensionless. In mathematical terms, EROI is:

$$\mathrm{EROI} = \mathrm{TNEP}/\mathrm{TIES} \tag{1}$$

TNEP is an acronym for *Total Net Energy Produced*: it represents the energy generated minus the direct energy necessary to run the plant itself. Direct energy, in general terms, is the electrical energy which should be produced in loco or taken from the grid and fuel (solid, liquid, or gas) to produce heat. According to Murphy et al. (2011), TIES is the *Total Indirect Energy Spent* elsewhere in the economy for the construction of the plant and for its operation. It includes the following energies: to produce the plant sections (vessel, pumps, valves, etc.), to produce the consumables, to prepare the site, to assemble the plant, to replace parts or to upgrade and, finally, to spend the energy for decommissioning. In addition, as indirect energy, we have to take into account the energy used to support the labor force in charge to the plant and the amortization energy as reported in Fig. 3.

It is important to point out that EROI should not to be confused with energy efficiency conversion, which is well depicted by First and Second Laws of classical thermodynamics, i.e., going from one form of energy to another one, such as upgrading oil in a refinery or converting diesel oil to electricity. EROI is only loosely related, at least in the short term, to the concept of return of monetary investment, but this aspect has not been considered in the present chapter. In the present case, the "energy source" refers to the organic refuse produced along the alimentary chain which is considered renewable and present in the world as long as the Humanity will exist.

A mathematical formula for EPT is:

$$\mathrm{EPT} = \mathrm{TIES}/(\mathrm{TNEP}/t_d) \tag{2}$$

TNEP and TIES have the same meaning as that of EROI; t_d is the facility's operation time.

Straight lines are usually used in the a priori estimation of EPT. For the evaluation of EPT, we have considered all the indirect energy including the amortization term, as spent during the construction time. Different assumptions can be made depending on the technology under study.

2.4 Spatial Boundaries

The selection of appropriate boundary conditions is a fundamental point in a sustainability analysis. The use of different boundaries in fact means using different inputs or outputs of the system under study and incomparable results could be generated (Murphy et al. 2011). In the present analysis, we adopted the LCA methodology to define the boundaries of the system. This methodology includes all the operations necessary to run a plant, from the introduction of the refuse till to

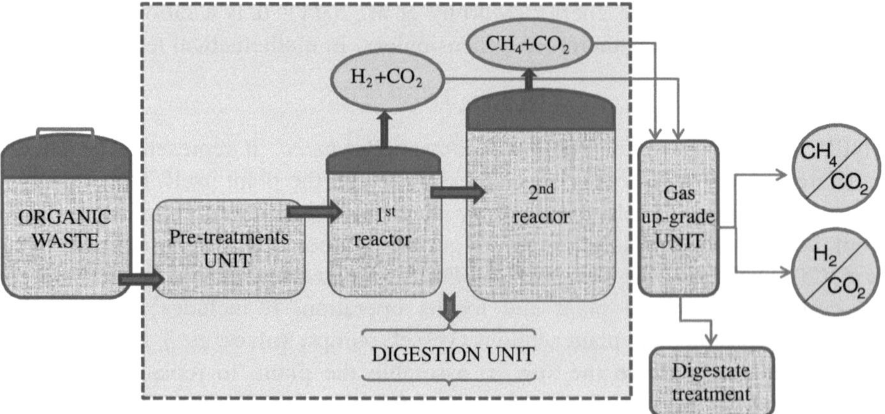

Fig. 4 Schematic view of the process involved in AD technology. The AD stages taken into consideration for the EROI Analysis are shown inside the block: pretreatment unit and digestion unit (two stages in series for the production of biohydrogen and biomethane, respectively)

the production of the biogas, as reported in the dashed area of Fig. 4. We have excluded the organic waste stock tank from the analysis, because it is independent of the technology used to valorize the refuse. The gas up-grading unit has also been excluded because it depends on the use of the gas and could be different (Ryckebosch et al. 2011); the biogas could be used to produce heat, electricity or as fuel for transportation purposes.

Referring to the layout of energy flow reported in Fig. 1, the present analysis concerns the sustainability evaluation of the AD technology as producer of an intermediary energy carrier such as ($H_2 + CH_4$) biogas as one of several candidate technologies as shown in Fig. 2. In EROI terms, the AD technology is equivalent to "mine mouth" for fossil sources. Bottom-up approach was used for the energy estimation of the technology along with a detailed process analysis to evaluate the input–output quantities. We decided to go ahead in this manner, which is different from an up-down approach (Carniege-Mellon 2009), because the scope was to analyze and compare different technologies in order to select the most energy sustainable one, and these technologies in many cases are still at an "infancy" or developing stage, and it is therefore difficult to find overall data on the technology.

2.5 Time Boundaries

Regarding the time boundary of the technology, we selected 15 years, with 300 working days per year as operational time of the plant. During this period, the generated energy as $H_2 + CH_4$ and all the spent energy were computed. We considered that 4 months were necessary for the site preparation and plant installation and 2 months for the decommissioning. We adopted the GER (Franzese et al. 2009), considering all the energy necessary to produce the materials, including the cost for the production of energy necessary to transform the raw material into goods (a pump for example), following the well consolidated methodology used in LCA. The organic refuse as an energy source was considered at its "natural state," which means that it only contains the low heating value (LHV) as the energy value, without any energy expenditure, for transport to the plant. This means that the plant is considered to be located in proximity of the point of generation of the refuse, as occurs in practice. As far as the direct energy necessary to run the plant as electricity and heat were concerned as internal (to the dashed boundaries of Fig. 4) by burning a produced quote of biogas. The indirect energy cost to produce these quantities was computed considering two terms: (1) the efficiency of an internal combustion engine and that of a heat exchanger, because we considered at the end of each fermentation cycle the recovery of the 50 % of the enthalpy contained in the broth; (2) the GER of the materials to construct the internal engine and the heat exchanger via computing their weight by commercial catalogs. As far as the energy spent for the maintenance of the plant, we have considered an energy cost of 15 % of the energy spent to build the plant. This means that some of the pumps, valves, piping, sensors, etc. were substituted during the operational time (Energy Business Report 2008). The energy amortization of the plant was evaluated as the initial GER of the materials and the energy cost for the installation of the plant, this value was distributed over the operational time of the facility. We have considered the consumption of sodium hydroxide and water as chemicals. The conversion into energy unit, in this case too, was made using GER.

2.6 Labor

The labor energy consumption deserves particular attention. It can be separated into three components: (1) the caloric value of food for the biological support of life; (2) both direct and indirect energy consumption necessary to produce, transport, conserve, and prepare food; (3) all the other direct and indirect forms of energy consumption linked to daily activities (clothing, appliances, fuel for transportation from the house to the factory, etc.). The energy spent on labor is intrinsically difficult to evaluate, in particular as far as the last contribution is concerned (Brown and Herendeen 1996; Cleveland and Costanza 2010). Generally, the labor contribution is often disregarded, but it could be of utmost importance when comparing different labor versus capital intensive technologies, for example, gasification versus energy crop cultivation. Considering the third term, some errors are introduced and it could be evaluated as pro-capita energy consumption of the Nation. Using the pro-capita energy, a false energy charge is calculated, either in the case the nation produces the plant or imports it: higher in the first case and lower in the second one, respectively. The pro-capita energy consumption depends to a great extend on the salary of the workers operating in the plant, and this can introduce a false energy charge on the technology under study. For all these reasons, only the energy spent to sustain labor as food was considered, via a dietary evaluation by LCA approach (Sanfilippo and Ruggeri 2009). The AD is not a labor-intensive technology, and hence, only one worker was considered necessary to run the plant, regardless of the plant dimensions. Finally, all the energy fluxes crossing the boundaries of the system were considered diameter dependent, and it has been used as scale-up parameter in scaling procedure.

3 Sustainability of Anaerobic Digestion

3.1 Introduction to the Anaerobic Digestion Process

AD is a naturally occurring decomposition process, by which organic matter is broken down to its simplest chemical components under anaerobic conditions. This process can be very useful to treat organic waste such as: sewage sludge, organic farm wastes, municipal liquid/solid wastes, green/botanical wastes as well as organic industrial and food commercial wastes.

The overall of anaerobic digestion process can be schematically divided into 4 sections, as shown in Fig. 4: pretreatment, digestion, gas upgrading, and digestate treatments. The key point is the digestion unit, which can work at different conditions, e.g., pH, temperature and hydraulic retention time and one or more stages.

Before being digested, the feedstock has to be pretreated. Various types of pretreatment can be adopted depending on the feedstock; the addition of water or on towing away undesirable materials such as large items and inert materials (e.g., plastic, glass, and metals) to allow a better digestate quality are generally applied. A more efficient digestion and higher energy production are obtained by means of acid or base as well as thermal pretreatments (Kim et al. 2009; Wang and Wan 2008a; Yang et al. 2007; Mu et al. 2006; Chen et al. 2002). The digestion process itself takes place in a digester, which can be classified in relation to the temperature, the water content of the feedstock, the number of stages (single or multi-stages and the type of biogas produced, that is methane or hydrogen (Ueno et al. 2007; Kraemer and Bagley 2005). During the natural anaerobic digestion process, some bacteria convert the organic material present in the digester into hydrogen, carbon dioxide, and water-soluble metabolites, such as ethanol, acetic, butyric, and propionic acids (Tommasi 2011). These bacteria usually live in close proximity to other bacteria that consume these metabolites, including hydrogen, and produce final products such as methane and CO_2. If the differences between *hydrogen forming bacteria HFB* which produce H_2 and *hydrogen-consuming bacteria HCB* are known, it is possible to design a 2-stage operation condition process (Lakaniemi et al. 2011). The combination of multistage processes with the production of two high-value gases, such as hydrogen and methane, is a solution which leads to several energy and environmental advantages: two separate fluxes of high-energy value gas (H_2 and CH_4), optimization of the AD process for the treatment of refuse and its control (Monnet 2003). The produced biogas (CH_4 and/ or H_2) can be used to create a source of income: biogas can be upgraded removing carbon dioxide and water vapor, and then, for example, used in a cogeneration unit as combined heat and power (CHP) to produce electricity and heat. The digestate either liquid or solid can instead be used as a fertilizer, or further processed into compost or high-value products, as bioproducts, e.g., acetic and butyric acids (Angenent et al. 2004).

3.1.1 Hydrogen and Methane Production in Two-Steps AD

Anaerobic digestion, from a biological point of view, is a multistep process that involves the action of various microbial species (Lyberatos and Skiadas 1999). Usually, such a process contains a particular step, the so-called rate-limiting step, which, being the slowest, limits the rate of the overall process (Hill 1977). However, the limiting step is not always the same over a wide range of operating conditions. It depends on the waste characteristics, hydraulic retention time, temperature, and many others (Speece 1983). The two-steps AD process is a process in which hydrogen and methane are produced in two separate bioreactors through the separation of hydrogen forming bacteria from methane forming bacteria (Tommasi 2011; Gómez et al. 2011) working in different conditions such as pH and hydraulic retention time. This partition, optimizing the fermentation process, permits the production of two high-value gases by splitting acetogenesis from methanogenesis and increases the overall energy production (89 %) compared with one-step processes (only hydrogen production ~33 %, only methane production ~84 %) as can be seen in Table 1.

Table 1 Stoichiometric energy efficiency of the reaction involved in H_2 and CH_4 production from AD with respect to the energy contained in 1 mol of glucose

Theoretical reaction involved in two-stage AD process	Energy yield (kJ/mol glucose)			
	H_2	CH_4	Total	Comparison (%)
Energy content in glucose	–	–	2,872	100
Theoretical maximum H_2 yield $C_6H_{12}O_6 + 6H_20 \rightarrow 12H_2 + 6CO_2$	2,870.4	–	2,870.4	99.9
Maximum H_2 yield from acidogenesis (1st step) $C_6H_{12}O_6 + 2H_2O \rightarrow 4H_2 + 2CO_2 + 2CH_3COOH$	956.8	–	965.8	33.3
Maximum CH_4 yield from standard AD $C_6H_{12}O_6 \rightarrow 3CH_4 + 3CO_2$	–	2,400	2,400	83.3
Maximum yield from two-steps ($H_2 + CH_4$) $C_6H_{12}O_6 + 2H_2O \rightarrow 4H_2 + 2CH_4 + 4CO_2$	956.8	1,600	2,556.8	89

For a sustainable energy point of view, it is necessary to energetically valorize the volatile fatty acids (VFAs) and other residue compounds present at the end of the first anaerobic step, which produces H_2 and VFAs as acetogenic fermentation. This valorization also permits the waste materials to be degraded as much as possible; the most adequate way is to use VFAs as a substrate for metanogenes to produce methane.

The energy analysis can be applied for only the H_2 or CH_4 production or for both AD processes in series to produce H_2 and CH_4, respectively.

The results of these analyses show that the net energy balance of a bioreactor producing H_2 in almost all conditions is never in the positive range (Ruggeri et al. 2010). On the contrary, two-steps (H_2 plus CH_4) in series show an increase in the produced energy and, consequently, the net energy balance becomes positive. In fact, from a thermodynamic point of view, during H_2-fermentation from glucose, only one-third of the energy available is converted to H_2, the other two-third remains occluded in the form of fatty acids. Therefore, one can obtain a positive net energy balance from an energy valorization of the end-liquid metabolites that accompany the H_2 production, due to the increment in the energy production. Temperature and pH play an important role on fermentative hydrogen production. Many studies (Akutzu et al. 2009; Wang and Wan 2008b; Mu et al. 2006; Zhang and Shen 2006; Nath and Das 2004; Hawkes et al. 2002; Lee et al. 2006) have shown that, in an appropriate range, increasing the temperature can increase the ability of hydrogen forming bacteria and archea bacteria to produce hydrogen and methane, during fermentation. Temperature is the most important parameter, from an energetic point of view, because it influences not only the energy produced, but also the energy necessary to run the bioreactor. Therefore, the temperature is the key parameter in the net energy balance of the technology. The present sustainability analysis of the AD of organic refuse is based on results experimentally evaluated by conducting test runs with market refuse pretreated with 2N NaOH at pH = 12 for 24 h (Bettoli 2010).

3.2 Net Energy Production in Anaerobic Digestion Process

The net energy produced in an anaerobic digestion process is the difference between the energy produced in the form of biofuels (H_2 and/or CH_4) and the direct energy used to run and maintain the system. The production of renewable energy (e.g., biofuels) in fact requires energy expenditure, as any other process. Pretreatment units also represent an expensive energy cost. To perform the energy balance in the present case, all the energy quantities have been evaluated in energy units per unit volume of bioreactor (MJ/L). Many factors can influence the net energy balance of anaerobic digestion such as the type of feedstock, environmental, geographical, and operational conditions.

In order to calculate the net energy, it is necessary to consider the energy balance of the anaerobic bioreactor, including the thermal and the electrical energy necessary to run the bioreactor. A detailed analysis of the net energy production of two-steps AD can be found in (Ruggeri et al. 2010) and only a brief introduction is given here. The net energy production E_{net} may be calculated as:

$$E_{\text{net}} = E_{\text{H2}} + \text{CH}_4 - \left(E_h + E_{\text{hp}} + E_l + E_m + E_p\right) \quad (3)$$

where

$E_{\text{H2+CH4}}$	is the energy produced (MJ/L)
E_h	is the heating energy necessary to reach the working temperature (MJ/L)
E_{hp}	is the heating energy necessary to reach the pretreatment temperature (MJ/L) if a thermal pretreatment is present
E_l	is the thermal energy loss, which depends on the outdoor ambient temperature and the duration of the fermentation (MJ/L)
E_m	is the electrical energy consumed for mixing (MJ/L) if a mixing system is present
E_p	is the electrical energy consumed for pumping (MJ/L)

The calculation of the net energy production requires the evaluation of the heat necessary to pretreat the organic refuse and the heat needed to keep the system at the working temperature. The heat required to keep the fermenting broth at the working temperature (T_w) is the sum of the heat necessary to warm up the feeding biomass from the ambient outdoor temperature (T_a) to T_w and the heat lost from the digester walls, which depends on the geography of the plant location, seasonal variations and obviously on the night/day oscillations. Figure 5 offers an overall view on the energies involved in the balance of an AD reactor, which is valid either in the case of producing H_2 or CH_4.

The energy balance of full-scale AD should be conducted in order to evaluate the quantity of net energy produced from a carbonaceous substrate as a function of two parameters, namely working temperature and the diameter of bioreactor. In the following sections, each term of Eq. (3) will be explained.

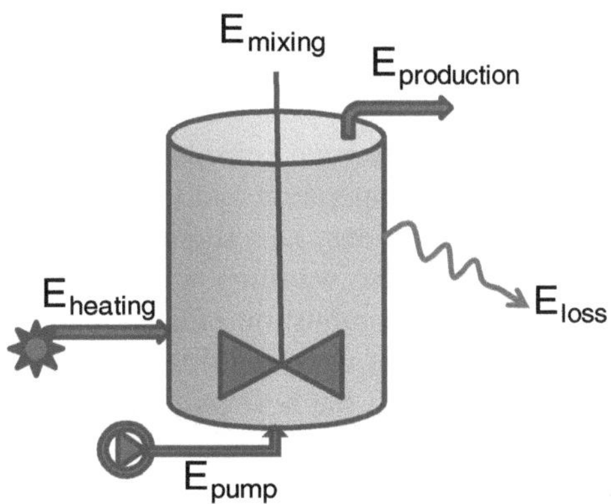

Fig. 5 Global view on the energies involved in the balance of an AD reactor

3.2.1 Energy Production

The produced energy is the total energy embedded in the produced gas, i.e., the energy contained in the amount of hydrogen and/or methane retrieved from a single batch run, with reference to the reactor volume and it can be calculated as:

$$E_{\text{produced}} = F \, * \, (P_{H_2}(T_w) \, * \, H_{H_2} + P_{CH_4}(T_w) \, * \, H_{CH_4}) \tag{4}$$

where

P_{H2} (T_w) and P_{CH4} (T_w)	are the specific productions of H_2 and CH_4, respectively, and refer to the amount of gas produced during a single-batch run. They are expressed as Nm^3 of H_2/CH_4 per unit of fermenting broth, which depends strongly on the working temperature.
H_{H2} and H_{CH4}	are the LHV (10.8 and 36.18 MJ/Nm^3, respectively)
F	is the filling coefficient of the reactor, which is usually equal to 90 % of the available volume.

3.2.2 Heating Energy

The energy required to warm up the fermenting broth mainly depends on its specific heat, the difference between the outdoor ambient and the working temperature of the bioreactor, and the efficiency of the heating system. The heating energy per unit volume of bioreactor can be calculated as follows:

$$E_h = (\rho * c_p * \Delta T \, * \, F) \, / \eta \tag{5}$$

where

ρ	is the biomass density (kg/m^3)
c_p	is the specific value of fermenting broth heat (kcal kg^{-1} $°C^{-1}$)
$\Delta T = T_w - T_a$	according to the season (°C)
η	is the global efficiency of the system to furnish heat taking into account η_{comb} and $\eta_{\text{heat exc}}$

ρ and c_p have been considered equal to those of water. The difference between the working temperature and the outdoor ambient ΔT was considered for different seasonal conditions, i.e., summer and winter conditions. To calculate the global efficiency of the warming system, a global combustion boiler efficiency was considered η: combustion efficiency ($\eta_{\text{comb}} \approx 0.8$) and heat exchanger efficiency ($\eta_{\text{heat exc}} \approx 0.6$) were multiplied to obtain the global efficiency ($\eta \approx 0.48$) necessary to furnish the heat required.

3.2.3 Thermal Energy Loss

The difference between the working temperature of the digester T_w and the pervading outdoor ambient temperature T_a is responsible for the heat loss from the fermenting broth. The amount of energy lost should be supplied from such a temperature control system and it depends on the insulation of the fermenting broth, the surface area exposed to the ambient and the duration of the batch run. The energy loss per unit volume of reactor can be calculated as follows:

$$E_l = (4.5 * k/s * \Delta t(T_w) * \Delta T/D)/\eta \quad (6)$$

where

k (Kcal h^{-1} m^{-1} °C^{-1})	is the thermal conductivity of the digester walls (e.g., material such as concrete or steel, coupled with an insulator, polystyrene foam, is an example)
s	is the thickness of the reactor/insulating walls
Δt (T_w)	is the total duration of fermentation (h)
$\Delta T = T_w - T_a$	according to the season (°C)
D	is the diameter of reactor

The resistance to heat transport is here only considered for the insulating material (*k*/*s*). This assumption leads to an overestimation of the insulator thickness for the same energy loss. Some explanations are given here about the above assumption. The heat flux from the bioreactor crosses three heat resistances in series. Therefore, the global thermal resistance U^{-1} is:

$$U^{-1} = 1/h_i + s/k + 1/h_e \quad (7)$$

where h_i and h_e are the internal and external convective heat transfers. A very thick insulator leads a higher resistance, due to a series of phenomena (Rohsenaw and Hartnett 1973) and both the convective coefficients, h_i and h_e can be disregarded; the situation graphically reported in Fig. 6 occurs, and hence, the controlling resistance will be that of the insulator, in terms of thickness s and heat conductivity *k*.

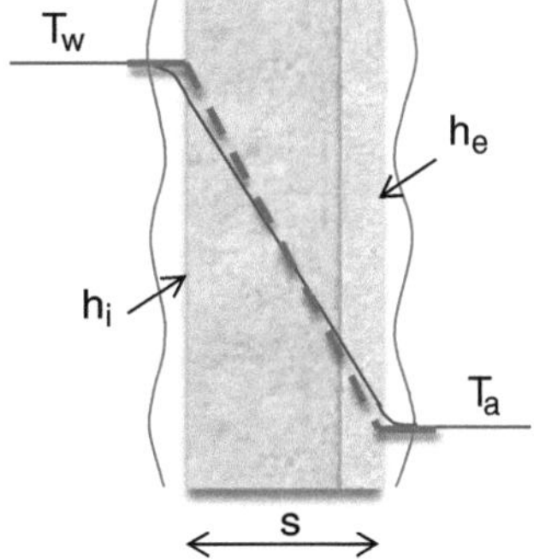

Fig. 6 Assumption used for the evaluation of heat loss across bioreactor wall

The insulating material is responsible for the main influence in the energy balance of AD for several reasons. It plays a particular role in limiting the heat loss and, at the same time, it contributes as GER to the total indirect energy consumed for construction materials. Insulating materials are solid and usually nonhomogeneous materials, characterized by a very low thermal conductivity value *k*, mainly due to the air enclosed in the pores of the material itself. The value of the coefficient of conductivity *k* [*W*/(*m***K*)] indicates the degree of ease with which a material allows the transport of heat, through collisions at the molecular level (Cocchi 1993). Thermophysical properties of some insulating materials are reported in Table 7.

3.2.4 Electrical Energy

Apart from the minor energy necessary to control the whole system, the larger quantity of electrical energy to run a bioreactor is consumed for mixing E_m and pumping E_p. A small energy input is necessary for E_p for pH control. Working in batch mode, the electrical energy is spent for filling and empting the bioreactor by a pump. In this case, it is possible to consider, as a first approximation, $E_p \sim 0$, compared with the electrical energy consumed for the agitation, considering the duration of the process of the order of weeks or months. However, E_p depends on the electric power of the pump:

$$E_p = P_{wpump} * \Delta t \tag{8}$$

The evaluation of the energy necessary to mix the fermenting broth could be computed by applying a turbulence scale-up criteria, taking into account a constant Reynolds number versus diameter. If one considers the constant ratio between the diameter of the bioreactor and that of the impeller as a geometrical scale-up, the following relations can be used to estimate the electrical power necessary to mix the broth (Nagata 1975):

$$Re \approx N_1 D_1^2 = N_D D_D^2 \tag{9}$$

$$P_W = (P_n * \rho) / (8 * g * \pi) * N_1^3 * D_1^6 * D_D^{-4} \tag{10}$$

where

1 and *D* are the bench scale and actual bioreactor, respectively

P_n is the power number and it can be evaluated by applying the procedure reported in Bailey and Ollis (1986) considering the *Re* of the bench scale bioreactor

P_w is the power required to have a defined *Re* in order to evaluate the energy consumed for mixing

It is necessary to take into account the running time, which depends on T_w:

$$E_m = P_W * \Delta t(T_w) / \eta_{el} \tag{11}$$

An efficiency factor of electrical energy conversion into mixing energy equal to 0.75 was considered. All the above-mentioned Equations could be implemented in an Excel sheet to perform the energy balance for each situation.

3.3 Indirect Energy in Anaerobic Digestion Process

When performing a sustainability analysis of a technology, great care should be taken in the evaluation of the energy and materials flows. The net energy and useful energy differ from each other because of the contribution of the total indirect energy (refers to Fig. 3). In mathematical terms, the useful energy can be evaluated from the difference between the net energy and the indirect energy. Equation (12) expresses, in mathematical terms, each contribution that should be considered to evaluate the total indirect energy having taken into account the GER, i.e., the sum of all the contributions of the energy life cycle (direct, indirect, capital, and feedstock energy):

$$E_{ind} = E_{chem} + E_{mat} + E_{diren} + E_{constr} + E_{main} + E_{decomm} + E_{amort} \tag{12}$$

where

E_{chem} is the GER of chemicals (MJ)
E_{mat} is the GER of construction materials (MJ)
E_{diren} is the GER of direct energy (MJ)
E_{constr} is the energy for plant building (MJ)
E_{main} is the energy for maintenance (MJ)
E_{lab} is the energy for labor (MJ)
E_{decomm} is the energy for decommissioning (MJ)
E_{amort} is the energy for amortization (MJ)

As previously stated, both direct and indirect energy need to be measured in a physical energy unit; hence, it is necessary to convert all the material flows into energy units. In the process, materials that were produced elsewhere are usually used. This leads to a higher consumption of energy, but without it, the process cannot take place. The GER allows one to convert and evaluate the energy content in each kilogram of material and is evaluated in energy units per unit mass of material. E_{chem} and E_{mat} were evaluated by utilizing the SimaPro 7.2.4 software (2010) and the Ecoinvent database (Ecoinvent 2007) (Table 2).

Table 2 GER values of the construction materials and chemicals

Steel	29,630 kJ/kg	(De Benedetti et al. 2007)
Polystyrene	105,800 kJ/kg	(Buwal 250 1996)
NaOH	6 kJ/kg	(Ecoinvent 2007)
Water	2 kJ/kg	(Ecoinvent 2007)

An analogous discussion should be made about direct energy: the scheme process should be followed for direct energy. The direct energy such as, e.g., electricity could be taken off from the grid; each unity of electricity has determined an energy expenditure that occurs elsewhere in order to be produced it (power plant construction, grid maintenance etc.). The indirect contribution of direct energy is calculated using a GER value expressed in an appropriate unit, e.g., kJ/kWh in the case of electricity. In the AD process here analyzed, the term E_{diren} is zero because the energy is produced in loco with a cogeneration plant using a quote of the biogas produced by the AD process itself. In the calculation of Eq. (12), E_{constr} is the energy consumed for the plant building, while E_{decomm} is the energy consumed for decommissioning: in the first case, it was considered that a workman works for 4 months to assemble the plant and 2 months to disassemble it. The energy consumed for maintenance operation (E_{main}) is evaluated as 15 % of the total energy expenditure for construction materials.

It is intrinsically difficult to evaluate the energy consumed for labor $E_{lab,}$ and generally, it is often disregarded; in this case, the scoring procedure is only valid for comparisons of technologies in the same category, i.e., capital-intensive or labor-intensive. In the present evaluation, it is calculated according to the statement of Sect. 2.6 considering the GER of a typical meal in the industrial world a worker needs to eat twice a day each workday in order to have the power necessary to work on the AD process, this was considered for 365 days per year.

Finally, E_{amort} is the energy necessary for amortization, and it is evaluated as the energy necessary to reconstruct the plant equal to E_{mat}.

4 Sustainability Evaluation

In this section, the result of the evaluation of the net and useful energy of the two-stage anaerobic fermentation process producing $H_2 + CH_4$ is reported by evaluating the above energy terms. All the energy terms are expressed as energy unit per volume evaluated over the operation time.

4.1 Net Energy and Useful Energy Production

An example of the evaluation of the net energy is reported in Table 3. It shows the following case study: AD of organic refuse taken from a local market, operated at 35 °C as the working temperature in batch mode producing H_2 and CH_4 in a two-steps digester. The substrate was pretreated with 2N NaOH at pH = 12 for 24 h. Regarding the volume of bioreactors, we considered the volume of bioreactor where hydrogen was produced as 1/20 of the volume of CH_4 for all the situations; results are reported to the diameter of methane bioreactor. Table 3 shows the net energy in different situations for the AD system under consideration.

Table 3 Net energy evaluation in the case of 4 and 10 m diameter, respectively

	$D = 4$ m	$D = 10$ m
E_{H2+CH4} (kJ/L)	1,123	1,123
E_{dir} (kJ/L)	171	161
E_{net} (kJ/L)	952	962

The useful energy production E_u can be calculated as follows:

$$E_u = E_{net} - E_{ind} \tag{13}$$

where E_{ind} is the total indirect energy consumed.

An example of the evaluation of the useful energy for the same situation is reported in Table 4.

From Table 4, one can see that the increase in the useful energy E_u along the diameter from $D = 4$ m to $D = 10$ m is only of 7 %.

In order to show the linkage between all the energy terms as contributions to the sustainability of AD, an analogical model of the process is presented in Fig. 7. This figure highlights the linkage between: (1) the energy production due to the knowledge of the technology; (2) the direct energy consumption necessary to run the technology; (3) the indirect energy; (4) the useful energy, i.e., the energy that the technology gives to society in a sustainable way.

It is interesting to conduct a detailed examination of the percent values: the theoretical available energy evaluated as the LHV of organic waste is 100 %. The percent value drops to 48 % as produced energy. This depends on the present knowhow on the fundamentals of AD technology or, in other words, the current knowledge on biochemistry and microbiology does not permit better results to be obtained. The percent value further decreases to 41 % as net energy, considering that the present technology of heat exchanger and electricity production technology have lead to an optimization of the system, the 7 % is consumed as direct energy. In the classical energy analysis approach, 41 % of energy is delivered to society and no other aspects need to be considered.

Table 4 Useful energy evaluation in the case of 4 and 10 m diameter, respectively

	$D = 4$ m	$D = 10$ m
E_{net} (kJ/L)	952	962
E_{chem} (kJ/L)	120.5	120.5
E_{mat} (kJ/L)	90	36
E_{diren} (kJ/L)	0	0
E_{main} (kJ/L)	18	7
E_{amort} (kJ/L)	210.5	156.5
E_{lab} (kJ/L)	29	1.8
$E_{constr} + E_{decomm}$ (kJ/L)	0.3	0.02
E_{ind} (kJ/L)	468	322
E_u (kJ/L)	484	640

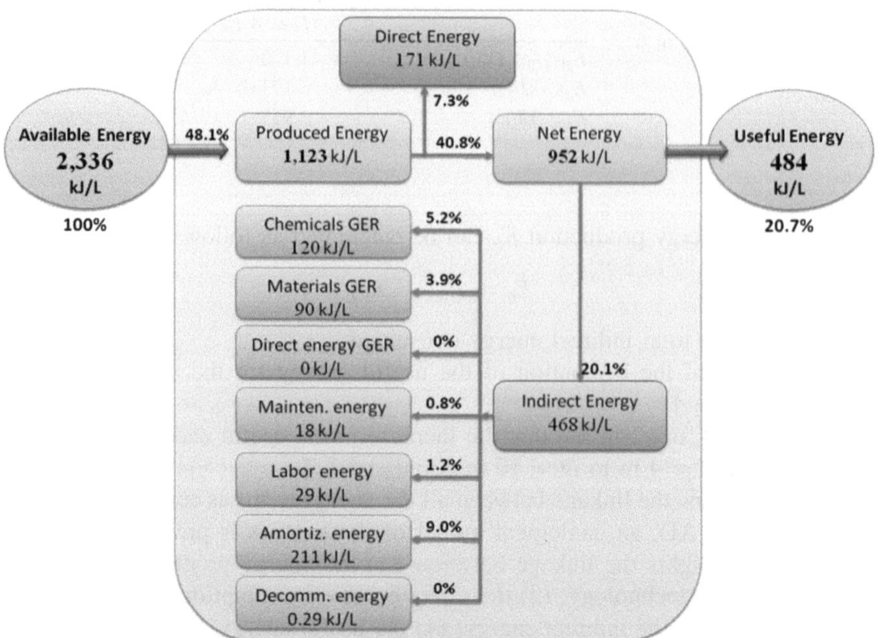

Fig. 7 Analogical model of the H_2 and CH_4 technology process at $T = 35$ °C with heat recovery and a diameter of the bioreactor equal to 4 m

From a global point of view, in terms of energy sustainability, it is also necessary to take into account the energy expenditure necessary for the production of materials and the energy flows in different part of the World. In this context, the useful energy effectively available from society adopting the AD technology is 21 %: at this point, it is clear that the complete ignorance of indirect energy is not justified.

One consideration should be made regarding the modality of supplying direct energy to run the plant: we have considered that the energy is produced through the cogeneration of a part of the produced methane. This energy can be furnished from different sources, for example, from a renewable one, such as solar energy or wind power. In this case, the degree of sustainability would not change because the use of renewable sources to furnish direct energy is removed from a different energy service in society. In other words, the quality of the source to produce direct energy in such a technology has no influence in the energy sustainability. In fact, with the present approach, the useful energy remains constant. However, using the exergy approach to score the sustainability of the technology, the use of renewable energy sources rather than no-renewable ones does make the difference, but in our opinion, this only introduces a *false-perception* of sustainability; obviously other impacts, for example, Global Warming, could be different.

4.2 The Evaluation of EROI and EPT

In this section, the sustainability of AD is evaluated by EROI and EPT. In this manner, it is possible to score the $H_2 + CH_4$ produced by the AD against other energy technologies; the AD is equivalent to "mine mouth" for fossil energy sources.

EROI and EPT are evaluated using the Eqs. (14) and (15). The results are shown in Figs. 8 and 9, considering two cases: with or without the contribution of labor in the Total Indirect Energy. The EROI and EPT values versus the reactor diameter are reported. The sustainability of the technology increases with the dimensions: without the labor contribution, EROI is higher than 1 for a diameter >1 m; with the labor contribution, the technology is sustainable for higher diameters than 2 m.

$$\mathrm{EROI} = E_{\mathrm{net}}/\,E_{\mathrm{ind}} \tag{14}$$

$$\mathrm{EPT} = t_d/\,\mathrm{EROI} \tag{15}$$

In Table 5, the values of EROI and EPT are reported for the two diameters considered in previous paragraph for the evaluation of useful energy.

In spite of the fact that the energy labor as only food contribution in this analysis, a great effect of labor on the EROI for low diameters of the bioreactor was found.

A comparison of the evaluated EROI and EPT of $H_2 + CH_4$ values with other energy technologies ranks the AD technology in a good position among renewable and fossil energy sources (see Table 6). The sensitivity of EROI and EPT to the indirect energy of materials has been investigated: the main case previously described, with polystyrene foam as the insulator, is compared with other cases, in order to permit a comparison between different insulating materials. Physical and thermophysical properties are reported in Table 7. In the sensitivity evaluation, net energy value has been kept constant, i.e., k/s = cost., which means that the thickness of the insulating walls varies according to thermal conductivity value.

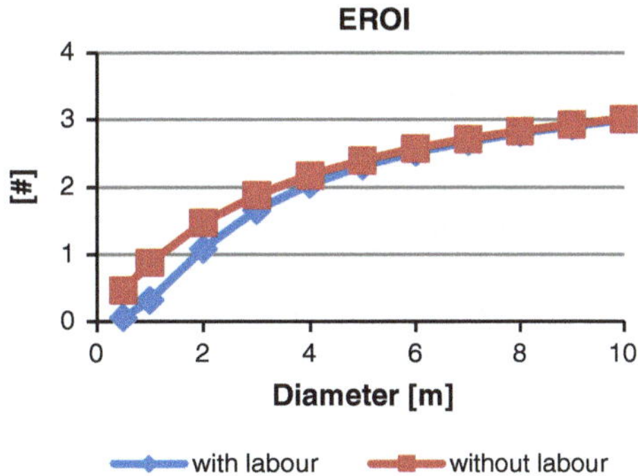

Fig. 8 EROI of an AD process with and without the labor contribution

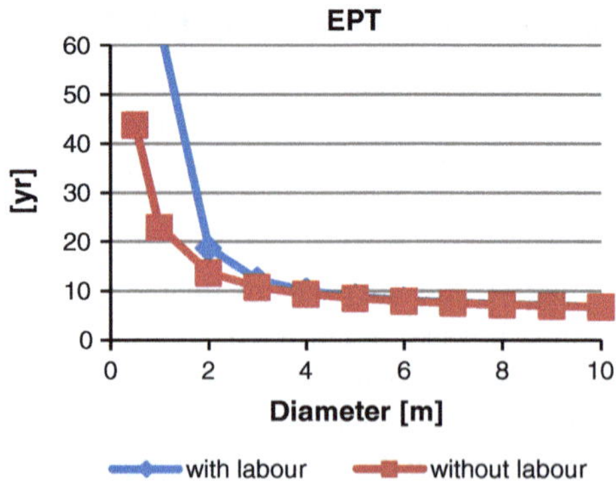

Fig. 9 EPT of an AD process with and without the labor contribution

Table 5 EROI and EPT evaluation for different diameters considering labor contribution

Diameter (m)	EROI (#)	EPT (year)
4	2.03	9.84
10	2.99	6.69

Figures 10 and 11 show the EROI and EPT for different insulating materials: polyurethane, cork, and sheep wool permit similar performances to those of polystyrene foam, while better results could be achieved using lime foam, straw, and raw clay; the use of recycled paper as an insulating material has the worst impact on both the EROI and EPT.

Table 6 Values of EROI of several energy technologies

Technology	EROI (Elliot)	EROI (Hore-Lacy)
Hydroelectric[a]	50–250	50–200
Mini hydroelectric[a]	30–270	
Oil XIX century[a]	50–100	5–15
Oil today[a]		
Wind turbine [a]	5–80	20
Nuclear power[a]	5–100	10–60
Photovoltaic Si[a]	3–9	4–9
Photovoltaic film[a]		25–80
Natural gas[a]		5–6
Anaerobic digestion (present estimation)	0–3	

[Data with [a] are derived from Sentimenti and Biorgi (2006)]

Table 7 Physical and thermophysical properties of different materials

Insulation materials	k [W/(m*K)]	ρ (kg/m^3)
Polyurethane	0.03	35
Polystyrene foam	0.035	25
Cork	0.04	100
Sheep wool	0.04	28
Lime foam	0.045	100
Straw	0.058	175
Recycled paper	0.07	400
Raw clay	0.132	700

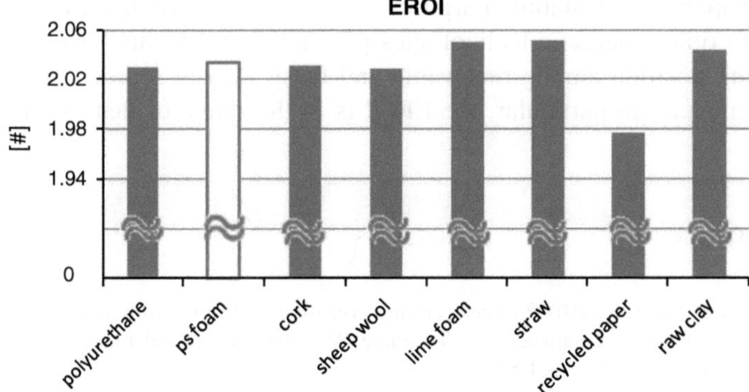

Fig. 10 EROI evaluation of different insulation materials considering labor for D = 4 m

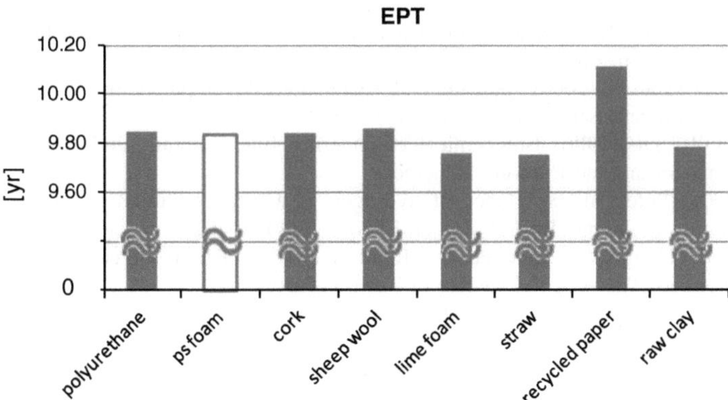

Fig. 11 EPT evaluation of different insulation materials considering labor for D = 4 m

5 Conclusion

The degree of sustainability of H_2/CH_4 energy carrier via the Anaerobic Digestion technology has been studied through an evaluation of the useful energy and a determination of the EROI and EPT parameters. The technology resulted to be sustainable for all the diameters higher than 2 m; an EROI > 10 is never obtained. The use of an analogical model to evaluate the useful energy of the studied technology has shown that more than 20 % of the available energy present in the organic refuse can be furnished to society as useful energy. This value depends to a great extent on the material that is used to insulate the plant. The best case was obtained considering straw, while the worst case was referred to the use of recycled paper for insulating purposes. A comparison of the evaluated EROI values with other energetic technologies places the AD technology in an acceptable ranking position among renewable and fossil energy sources for higher bioreactor diameters; in particular, the EROI is in the range of that of natural gas.

References

SimaPro 7.2.4 Software (2010) Product ecology consultants, Pré Consultants, Amersfoort

Aelterman P, Rabaey K, Clauwaert P, Verstraete W (2006) Microbial fuel cell for wastewater treatment. Water Sci Technol 54:9–15

Akutzu Y, Li YY, Harada H, Yu HQ (2009) Effect of temperature and substrate concentration on biological hydrogen production from starch. Int J Hydrogen Energy 34:2558–2566

Angenent LT, Karim K, Al-Dahhan MH, Domiguez-Espinosa R (2004) Production of bioenergy and biochemicals from industrial and agricultures waste water. Trends Biotechnol 22(9):477–485

Azapagic A (1999) Life cycle assessment and its application to process selection, design and optimisation. Chem Eng J 73:1–21

Azapagic A, Perdan S (2000) Indicators of sustainable development for industry: a general framework. Process Saf Environ Prot 78(4):243–261

Bailey JE, Ollis DF (1986) Biochemical engineering fundamentals, 2nd edn. McGraw Hill Education International Editions, Singapore

Balat M (2008) Potential importance of hydrogen as a future solution to environmental and transportation problems. Int J Hydrogen Energy 33:4013–4029

Bettoli S (2010) Indagine Sperimentale con Reattore Batch su Rifiuti Mercatali per la Produzione di Bioidrogeno. Thesis at Politecnico di Torino, Turin

Brandt AR (2009) Converting oil shale to liquid fuels with the Alberta Taciuk processor: energy input and greenhouse gas emission. Energy Fuels 23:6253–6258

Brown MT, Herendeen RA (1996) Embodied energy analysis and EMERGY analysis: a comparative view. Ecol Econ 19:219–235

Buwal 250 (1996) Life cycle inventories for packaging. Swiss Federal Office of the Environment, Forests and Landscape. Switzerland

Carniege-Mellon (2009) Economic input–output life cycle Assessment. Pittsburg Pennsylvania available on line at: http://www.eiolca.net/. Cited on 5 Oct 2011

Chen CC, Lin CY, Lin MC (2002) Acid-base enrichment enhances anaerobic hydrogen production process. Appl Microbiol Biotechnol 58:224–228

Cleveland CJ, Costanza R (2010) Net energy analysis. In: Encyclopedia of Earth. National Council for Science and Environment. Available via http://www.eoearth.org/article/Net_energy_analysis. Cited 22 Jan 2011

Cleveland CJ, O'Connor P (2011) Energy return of investment (EROI) of oil shale. Sustainability 3:2307–2322

Cleveland CJ, Costanza R, Hall CAS, Kaufmann R (1984) Comparing different energy processes from energy and the U.S. economy: a biophysical perspective. Science 225:890–897

Cocchi A (1993) Elementi di termofisica Generale e Applicata. Progetto Leonardo Press, Bologna

De Simone LD, Popoff F (1997) Eco-efficiency: the business link to sustainable development. MIT, Cambridge

de Swaan AJ, van der Kooi H, Sankaranarayanan K (2004) Efficiency and sustainability in the energy and chemical industries. Marcel Dekker Inc, New York

Debenedetti B, Maffia L, Rossi S (2007) From materials to eco-materials: life-cycle environmental approach for insulation products in building applications. Proceedings of the 8th international conference of eco-materials, Brunel University, UK

Ecoinvent (2007) Ecoinvent data v2.0., Final reports Ecoinvent 2000 N.o 1–25. Swiss Centre for Life Cycle Inventories, Dubendorf

Energy Business Report (2008) Landfill gas as an energy source technology and opportunities 166 case studies. http://energybusinessreport.com

Evans G (2001) Biowaste and biological waste treatment. James and James Science Publishers, London

Franzese PP, Rydberg T, Russo GF, Ulgiati S (2009) Sustainable biomass production: a comparison between gross energy requirement and energy synthesis methods. Ecol Indic 9:959–970

Gómez X, Fernández C, Fierro J, Sánchez ME, Escapa A, Morán A (2011) Hydrogen production: two stage processes for waste degradation. Biores Technol 102(18):8621–8627

Guéhenneux G, Baussand P, Brothier M, Poletiko C, Boissonnet G (2005) Energy production from biomass pyrolysis: a new coefficient of pyrolytic valorisation. Fuel 84(6):733–739

Guilford MC, Hall CAS, O'Connor P, Cleveland CJ (2011) A new long term assesment of energy return on investment (EROI) for U.S. oil and gas discovery and production. Sustainability 3:1866–1887

Hall CAS, Balogh S, Murphy DJR (2009) What is the minimum EROI that a sustainable society must have? Energies 2:25–47

Hammerschlag R (2006) Ethanol's energy return on investment: a survey of the literature 1990-present. Environ Sci Technol 40(6):1744–1750

Hawkes FR, Dinsdale R, Hawkes DL, Hussy I (2002) Sustainable fermentative hydrogen production: challenges for process optimization. Int J Hydrogen Energy 27:1339–1347

Hill CG (1977) An introduction to chemical engineering kinetics and reactor design. Wiley, New York

Kim J, Park C, Kim TH, Lee M, Kim S, Kim SW et al (2009) Effects of various pretreatments for enhanced anaerobic digestion with waste activated sludge. J Biosci Bioeng 95(3):271–275

Kraemer JT, Bagley DM (2005) Continuous fermentative hydrogen production using a two-phase reactor system with recycle. Environ Sci Technol 39(10):3819–3825

Lakaniemi AM, Koskinen PEP, Nevatalo LM, Kaksonen AH, Puhakka JA (2011) Biogenic hydrogen and methane production from reed canary grass. Biomass Bioen 35:773–780

Laws D, Scholz RW, Shiroyama H, Susskind L, Suzuki T, Weber O (1984) Expert views on sustainability and technology implementation. Int J Sustain Dev World Ecol 11:247–261

Lee KS, Lin PJ, Chang JS (2006) Temperature effects on biohydrogen production in a granular sludge bed induced by activated carbon carriers. Int J Hydrogen Energy 31:465–472

Logan L (2008) Microbial fuel cells. Wiley, New Jersey

Lyberatos G, Skiadas IV (1999) Modeling of anaerobic digestion—A review. Global Nest Int J 1(2):63–76

Monnet F (2003) An introduction to anaerobic digestion of organic waste, Technical Reports, Scotland. Available on http://www.biogasmax.co.uk. Cited on 10 Sept 2011

Mu Y, Zheng XJ, Yu HQ, Zhu RF (2006) Biological hydrogen production by anaerobic sludge at various temperatures. Int J Hydrogen Energy 31:780–785

Murphy DJ, Hall CAS, Dale M, Cleveland C (2011) Order for chaos: a preliminary protocol for determining the EROI of fuels. Sustainability 3:1888–1907

Nagata S (1975) Mixing: principles and applications. Wiley, New York

Najafpour GD (2007) Bioprocess scale-up. In: Najafpour GD (ed) Biochemical engineering and biotechnology, cap, vol 13, 1st edn. Elsevier, Amsterdam

Nath K, Das D (2004) Improvement of fermentative hydrogen production: various approaches. Appl Microbiol Biotechnol 65:520–529

Patzek TW (2004) Thermodynamics of the corn-ethanol biofuel cycle. Critical Rev Plant Sci 23(6):519–567

Patzek TW, Pimentel D (2006) Thermodynamics of energy production from biomass. Critical Rev Plant Sci 24(5–6):329–364

Pfeffer JT, Liebman JC (1976) Energy from refuse by bioconversion, fermentation and residue disposal processes. Res Rec Conserv 1(3):295–313

Röegen NG (1976) Dynamic models and economic growth. In: Energy and the economic myths. Pergamon Press, New York

Röegen NG (1999) The entropy law and economic process. Harvard University Press, Cambridge

Rohsenaw WM, Hartnett JP (1973) Handbook of heat transfer. McGraw-Hill Press, New York

Ruggeri B, Tommasi T, Sassi G (2010) Energy balance of dark anaerobic fermentation as a tool for sustainability analysis. Int J Hydrogen Energy 35:10202–10211

Ryckebosch E, Drouillon M, Vervaeren H (2011) Techniques for transformation of biogas to biomethane. Bio Bioenergy 35:1633–1645

Sanfilippo S, Ruggeri B (2009) LCA Alimentazione: stima del consumo energetico per la produzione, il trasporto e la preparazione del cibo in Italia. La Rivista di Scienza dell'Alimentazione 38(4):1–16

Sentimenti E, Biorgi U (2006) Energia ed economia. Tpoint Eni Tecnologie 8(2):22–26

SETAC (1993) Guidelines for life-cycle assessment: a code of practice. Brussels, Belgium

Speece R (1983) Anaerobic biotechnology for industrial wastewater treatment. Environ Sci Technol 17:416–426

Tommasi T (2011) Anaerobic biohydrogen production: experimental evaluation of design bioreactor parameters for dark biohydrogen production using organic wastes. LAP LAMBERT Academic Publishing, Saarbrücken

Tommasi T, Ruggeri B, Sanfilippo S (2012) Energy valorisation of residues of dark anaerobic production of Hydrogen. J Cleaner Prod 34:91–97

Ueno Y, Fukui H, Goto M (2007) Operation of a two-stage fermentation process producing hydrogen and methane from organic waste. Environ Sci Technol 41(4):1413–1419

Wang J, Wan W (2008a) Comparison of different pretreatment methods for enriching hydrogen-producing bacteria from digested sludge. Int J Hydrogen Energy 33:2934–2941

Wang J, Wan W (2008b) Effect of temperature on fermentative hydrogen production by mixed cultures. Int J Hydrogen Energy 33:5392–5397

Yang M, Yu HQ, Wang G (2007) Evaluation of three methods for enriching H_2-producing cultures from anaerobic sludge. Enzyme Microb Technol 40:947–953

Zhang Y, Shen J (2006) Effect of temperature and iron concentration on the growth and hydrogen production of mixed bacteria. Int J Hydrogen Energy 31:441–446

Life-Cycle Assessment of Wind Energy

E. Martínez Cámara, E. Jiménez Macías and J. Blanco Fernández

Abstract This chapter looks at wind power from the viewpoint of life-cycle assessment (LCA). Such analyses have, of course, been conducted at various times throughout the development of wind power, and their results have varied as the designs and main characteristics of wind turbines have evolved. For that reason, modern double-fed induction generator (DFIG) multimegawatt turbines are considered here, as this is the most frequently used type on wind farms. On that basis, a comprehensive LCA is conducted on a wind turbine, covering all phases from its manufacture to its decommissioning and the processing of waste at the end of its useful lifetime.

1 Introduction

The idea of using wind power to produce electricity dates back to the nineteenth century. The field is generally considered to have been pioneered by Charles F. Brush, an outstanding inventor, thinker, and entrepreneur of the time, known mainly for his dynamo and his arc lights. He registered over 50 patents and founded the Brush Electric Company, which after subsequent sales and mergers eventually became General Electric, now one of the world's biggest power companies. Brush was the first man to install and operate a wind turbine to generate electricity (in 1888, in order to charge the batteries at his home in Cleveland, Ohio). It stood 60 ft (18.29 m) high, measured 56 ft (17.07 m) in diameter and

E. Martínez Cámara (✉) · J. Blanco Fernández
Department of Mechanical Engineering, University of La Rioja, Logroño, La Rioja, Spain
e-mail: eduardo.martinezc@unirioja.es

E. Jiménez Macías
Department of Electrical Engineering, University of La Rioja, Logroño, La Rioja, Spain

A. Singh et al. (eds.), *Life Cycle Assessment of Renewable Energy Sources*, Green Energy and Technology, DOI: 10.1007/978-1-4471-5364-1_9,

was connected to a 12 kW dynamo. It charged the batteries at the house for 20 years. His turbines had 144 blades, whereas modern turbines have just 3 (Rivkin et al. 2012).

Renewable energy sources, and particularly wind power, have undergone considerable development in recent years. This new boom in wind power dates back to the early 1990s and is due mainly to the need to find viable alternatives to fossil fuels, reserves of which are finite and will in the long term be incapable of sustaining current levels and trends in consumption around the world.

The wind turbines now in use are models developed in recent years. There are numerous types of turbine capable of generating electricity, but most of those currently installed use horizontal-axis technology. Their rated power outputs range from 500 kW to 5 MW. They are used basically in direct connections to the electricity grid and are grouped into wind farms to take advantage of economies of scale and facilitate monitoring and maintenance tasks.

Figure 1 shows the world-wide trend over time in total generating capacity from wind power. There has been a clear, sustained year-by-year increase in capacity in recent years, to almost 200 GW in 2010. However, a breakdown of these data by regions (Fig. 2) clearly shows the influence of the worldwide recession, with less capacity being installed in 2010 than in 2009, especially in Europe and North America.

The basic idea underlying the generating of electricity via modern, horizontal-axis wind turbines of the kind that can be found in most wind farms around the world is to convert wind energy into mechanical energy, by using the force of the aerodynamic thrust of the blades to generate torque on the main shaft. That mechanical energy is then turned into electricity by means of a generator. The disadvantage of this system compared with other conventional generators is that it can only generate electricity when there is enough wind. Moreover, it is not currently possible, from the economical point of view, to store the electricity generated in order to use or to transfer it to the grid at a later time, because this type of energy is noncontrollable and is subject to fluctuations depending on changes in wind speed. From the viewpoint of electricity grids as a whole, the

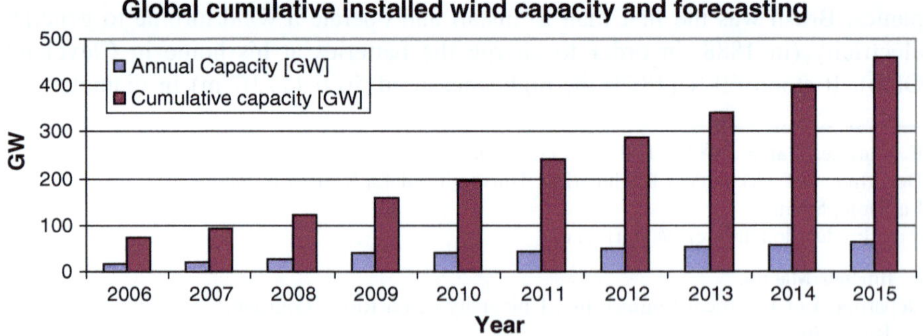

Fig. 1 Global cumulative installed wind capacity and forecasting

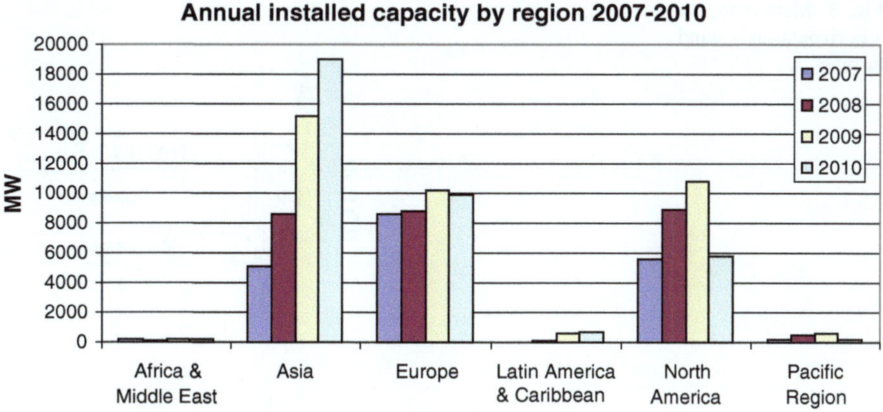

Fig. 2 Annual installed capacity by region 2007–2010

requirements for the connection of new wind farms are becoming increasingly stringent. Continual efforts are being made to improve the integration of this type of energy into the grid, with control systems being established to regulate the reactive energy produced by wind farms, to vary the voltage or frequency supplied by farms at the point of connection, or to prevent the complete disconnection of wind farms when there are small voltage gaps in the grid, etc. This partly offsets the negative effect of the noncontrollability of wind farms within the grid and at the same time increases the wind energy capacity installed in the system, acting as basic energy that enables the number of conventional power plants operating to be reduced, or at least reducing the amount of fossil fuel required for their operation.

2 Wind Turbines

As mentioned, most of the wind turbines in use today for large-scale energy production around the world are three-blade, horizontal-axis units. The basic components of such turbines are shown in Fig. 3:

- Foundations
- Tower
- Nacelle
- Rotor

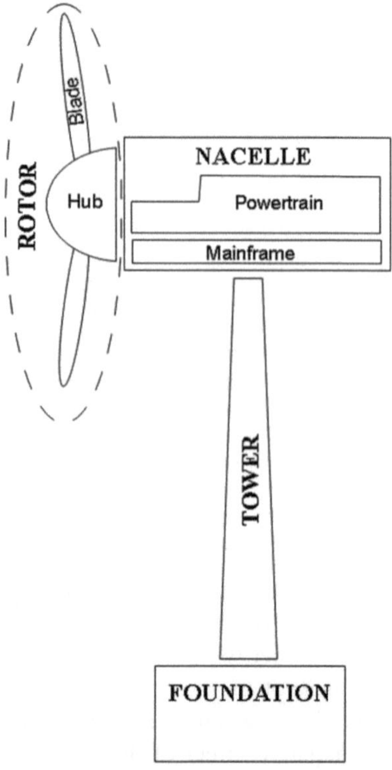

Fig. 3 Main components of a horizontal-axis wind turbine

2.1 Foundation

For a wind turbine, to be stable, its foundations must be capable of bearing the loads to which the turbine is subjected. The type of foundations built depends on the type and consistency of the soil where the unit is to be erected, and on which type of unit.

Obviously, the foundation requirements are considerably more complex for off-shore wind farms than for land-based farms. Various types of off-shore foundations may be used, depending on the depth of the sea at the site where the turbines are erected:

- Monopile (4–25 m): A hollow steel pile, plus grout injected between pile and transition piece.
- Surface-level foundations: Made of concrete and steel with a large base which sits on the terrain
- Jackets (30–35 m): Similar to lattice towers used in offshore oil and gas projects.
- Tripods: Monopile adaptation, broadening the footprint by adding three piled connections close to the seabed.

- Multipile (up to 40 m): Another monopole adaptation. Several piles, connecting above the water via a grouted transition piece, driven into the seabed forming a larger foundation footprint than a typical monopile.

2.2 Tower

Towers must be built not just to bear the weight of the nacelle and rotor but also to absorb the loads caused by variations in wind speed.

Various types of tower can be used:

- Tubular steel towers: Most large wind turbines have tubular steel towers built in 20–30 m sections with flanges at each end, which are bolted together on site. They are tapered, i.e., their diameter decreases toward the top, to increase their strength and save on materials.
- Concrete towers built on site: These towers are subject to height limitations. They have the advantage that there is no need to transport the tower sections to the point of installation of the turbine.
- Prefabricated concrete towers: These are made of segments that are placed one on top of another.
- Lattice towers: These are made of steel sections. Their main advantage is their cost, as they require only half as much material as a tubular tower with no additional supports and provide the same degree of stiffness. Their main disadvantage is their appearance. For esthetic reasons, lattice towers are hardly ever used in large, modern wind turbines. They are common, however, in India and can also be found elsewhere, e.g., in the USA and in Germany.
- Hybrid towers: Towers can also be built using combinations of the above methods. The most common variant is to build the bottom part of cement and the top of tubular steel. Indeed, this seems likely to be the tendency for the large, multimegawatt turbines of the future.

The height of the towers used for modern wind turbines varies as shown in Table 1.

Table 1 Tower height depending on rated power and rotor diameter

Tower height (m)	Rated power (kW)	Rotor diameter (m)
65	600–1,000	40–60
65–115	1,500–2,000	70–80
120–130	4,500–6,000	112–126

2.3 Rotor

The rotor converts wind energy into mechanical rotational movement. It comprises the turbine blades and the hub (which joins the blades to the main shaft). The hub is the center of the rotor. It is made of cast steel or iron. Depending on the type of wind turbine, hubs may be connected to the low-speed shaft of the gearbox or directly to the generator if the turbine has no gearbox. A high percentage of the turbines installed today are still fitted with gearbox to raise the rotor revolutions to the rated speed of the generator. However, gearboxes can be problematic, so there is a tendency for modern multimegawatt turbines to be built without them in order to reduce maintenance requirements.

Currently, most rotors have three blades and a horizontal shaft and are between 40 and 90 m in diameter. Conventional windmill rotors for pumping water use 16 or more blades and are made of metal, but three-blade rotors are more efficient for generating power in large turbines; they also make for better weight distribution, which allows more stable rotation.

Rotor blades are made mainly of fiberglass or carbon fiber reinforced with plastic. In profile, they are similar to aircraft wings, and they work on the same principles of thrust. The wind passing over the bottom of the blade generates high pressure, while low pressure is generated at the top. This force, plus the force of resistance, causes the rotor to rotate. Each manufacturer makes blades according to its own concepts and characteristics.

2.4 Nacelle

The nacelle encases and supports all the machinery of the wind turbine and must be able to turn in order to face the wind direction. It is therefore attached to the tower by bearings. Nacelle designs vary from one manufacturer to another in line with the designs of the turbines and the location of the powertrain components (main shaft, gearbox, generator, coupling, and brake).

The gearbox converts the 18–50 rpm rotation of the rotor to the rated rotation speed of the generator, i.e., approximately 1,750 rpm. The generator rotation speed depends on the frequency of the electrical current and on how many pairs of poles the machine has. Apart from enabling low rotor rotation speeds to be coupled to high generator rotation speeds, the gearbox also enables the unit to withstand a wide range of variations in wind speed.

The generator is what turns mechanical energy into electricity. Double-fed asynchronous generators are the most widespread type in use on today's high-power wind turbines.

Along with the main components mentioned previously, it is also necessary to install a coupling between the generator and the gearbox. Flexible couplings are usually utilized for this purpose. Mechanical brakes are also fitted to the powertrain.

The standards usually applied in wind turbine design require turbines to be equipped with two independent braking systems, so both aerodynamic and mechanical brakes are usually fitted: the aerodynamic system is located at the tip of the blades or over the whole rotor blades to change the pitch angle. On most turbines, the second system is usually a mechanical disk brake, which is intended mainly to be used should the aerodynamic system fail, and when the turbine is stopped for repair or maintenance work.

3 Wind Turbine LCA

3.1 State of the Art

The relevant scientific literature contains various life-cycle analyses of wind power and wind turbines (Ardente et al. 2008; Ben et al. 2008; Góralczyk 2003; Tryfonidou and Wagner 2004; Wiese and Kaltschmitt 1996; Gürzenich et al. 1999; Uchiyama 1995; Nadal 1998; Haack 1981; Krohn 1997; Uchiyama 1996; Schleisner 2000; Lenzen and Munksgaard 2002; Lenzen and Wachsmann 2004; Crawford 2009; Weinzettel et al. 2009; Martínez et al. 2009). Most of them focus on energy analysis and CO_2 emissions, and their results vary considerably. This should come as no surprise in view of the significant developments in the field of wind energy in recent years: in a very short time, wind turbines have been developed with widely differing electricity generating capacities, ranging from rated power outputs of just a few kW to the latest large multimegawatt turbines (7.5 MW). Moreover, issues such as estimated useful lifetime and equivalent hours of annual production significantly affect the final results of any study conducted. In spite of the variability observed, a clear tendency is found in which CO_2 emissions per kWh produced decrease as the rated power of the turbine increases. The results also show levels of emissions and environmental impact that are far lower than those of other conventional sources of electricity generation.

With this in mind, the present study focuses on current, multimegawatt wind turbines, seeking to obtain LCA data for wind power at the present time.

3.2 LCA of a 2 MW Wind Turbine

This wind power LCA focuses on the type of wind turbine most widely used all over the world at this time: the multimegawatt double-fed induction generator (DFIG).

3.2.1 System Limits

Cutoff criteria for the system must be defined if the results of the analysis are to be understood and interpreted correctly. The criteria applied in this case include the following:

- The construction phase of the main components of the wind turbine.
- Transportation of the wind turbine to its specific site in the wind farm.
- Direction, installation, and commissioning of the turbine at its final operating site.
- Decommissioning of the turbine and subsequent treatment of the waste produced.

For a diagram showing these system cutoff criteria, see Fig 4.

All the electrical switchgear for the electricity generated by the turbine, which is used to distribute that energy to the grid, is excluded from the system. This includes the following:

- Medium voltage wiring
- The substation transformer
- The electrical distribution and transmission network

3.2.2 Functional Unit

The functional unit defined here for the LCA of the electricity generated by a wind turbine is the production of 1 kWh of electricity. Choosing a functional unit that is appropriate for the task being performed is of course an important factor in ensuring that the results obtained are meaningful. What is being studied here is the

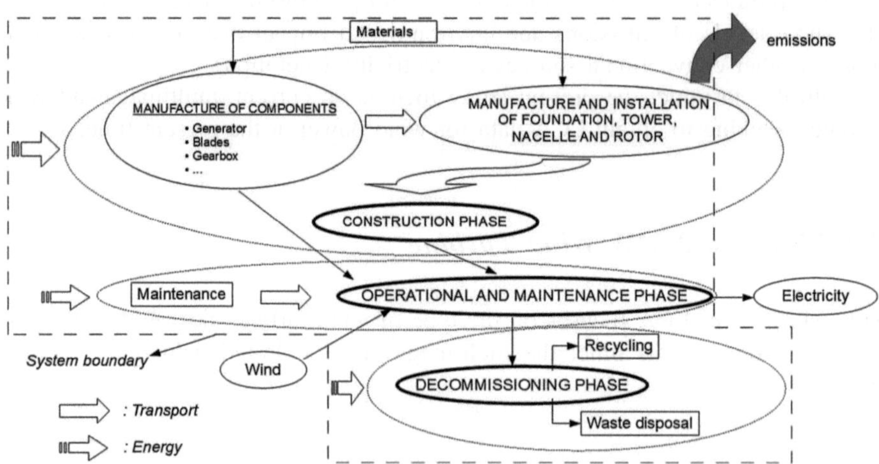

Fig. 4 LCA system limits (Martínez et al. 2009)

environmental impact of wind energy in the context of other types of renewable and nonrenewable energy sources, so selecting 1 kWh as the functional unit gives a clearer outlook on results, and facilitates comparisons, if desired, with the environmental impacts resulting from the generation of 1 kWh with such other sources.

3.2.3 Data Used

When an LCA is conducted on a wind power system, data need to be collected on each of the main components of the wind turbine and on the various subcomponents that make up those main components. These components are highly varied in their nature and characteristics and may include various types of mechanical, electrical, and electronic parts. This makes it difficult to obtain all the information needed from the various suppliers in order to perform an LCA on each and every part used in the turbine. It is therefore necessary to draw up a suitable life-cycle inventory including the most important and significant components of the unit, such as the foundations, the tower, the nacelle, and the rotor. Most of the inventory data have been obtained directly from the companies that produce the wind turbines (generator, gearbox, blades, etc.). The exception corresponds to the dry-type transfomer, where the information was obtained from a third company that manufactures dry transformers similar in volume, power, and tension. In individual cases, where it was not possible to obtain reliable and verified data, commercial databaseEcoinvent (Boustead and Hancock 2003; Frischknecht and Rebitzer 2005; Frischknecht et al. 2005) was utilized.

Data on energy expenditure and materials used in manufacturing the main components of a wind turbine were supplied by Gamesa, a major worldwide manufacturer. Table 2 gives a general summary of the materials used in the main components of turbines and the energy used in the manufacturing processes associated with those main components.

Along with materials and energy consumption, the transportation of components from their places of manufacture to the final location of the turbine on the wind farm must also be taken into account. Once the wind turbine has been erected on site and commissioned, it will require preventive and corrective maintenance to ensure that it remains in working order for most of its useful lifetime. Logically, all these operations must also be taken into account in the LCA conducted, and information must be compiled on how much oil and grease is used, on how many filters are replaced, on the transportation of materials and workers, etc.

3.2.4 Assumptions

When conducting an LCA on a system as complex as a wind energy generation system, limitations must be set on the level of detail applied in the compilation of data. Only thus can the LCA be completed within reasonable limits in terms of

Table 2 Summary of the materials and manufacturing energy used in the main components (Martínez et al. 2009)

Component	Materials			Energy	
Foundation	700	Concrete	T	5.12	MWh
	25	Iron	T		
	15	Steel	T		
Tower	143	Steel	T	47.2	MWh
Nacelle	18.5	Iron	T	287.11	MWh
	21.69	Steel	T		
	0.35	Silica	T		
	3.5	Copper	T		
	0.8	Fiberglass	T		
	1.2	Resin	T		
Rotor	11.89	Resin	T	33.1	MWh
	7.92	Fiberglass	T		
	14	Cast iron	T		

effort, resource requirements, and time with no loss of quality in the final results obtained.

As mentioned above, the first step is to establish cutoff criteria for the system to be studied. The study conducted here includes all the elements of the foundations on which the wind turbine is supported, plus the other main components (tower, nacelle, and rotor). It excludes all the elements that comprise the system of connection with the electricity distribution and transmission network (medium voltage lines and transformer substation).

The main cutoff criterion used is the weight of each component in relation to the total weight of the system under study. Despite the use of this cutoff criterion, which would be not valid in general cases, when applying it we have taken into account the necessity of not to exclude the components or materials which, in spite of its relatively low weight in the system, may present a significant environmental impact. This considerably reduces the number of small components that need to be analyzed individually and for which specific data must be compiled, but does not appreciably affect the final results. It also makes the LCA more flexible, enabling major changes to be incorporated into the system more quickly and easily.

The data used to define and characterize each of the main components are taken from a database supplied by the component manufacturer. The main data collected are the following: raw materials required for manufacture, energy consumed in the associated manufacturing processes, and details of the transportation of materials and components.

The basic assumptions made in the wind power system LCA are the following:

- The weight of the main components is used as the cutoff criterion to determine whether or not they are included in the life-cycle inventory. The sum total of the components considered in the analysis accounts for 95 % of the weight of the foundations, 95 % of that of the tower, 85 % of that of the nacelle, and 85 % of that of the rotor.

- All data on the emissions from and characteristics of the energy used in the manufacturing processes are taken from the Ecoinvent database and refer to Spain, because that is where the turbine manufacturer's plants are located.
- Modern, multimegawatt wind turbines are assumed to have a useful lifetime of 20 years.
- The estimates made in regard to the decommissioning of the turbine at the end of its useful lifetime and the subsequent processing of waste products are based on decommissioning projects prepared by the company that holds the operating rights to the turbine. Basically, it is considered that 90–95 % of the metal (iron, steel, and copper) is recycled, PVC plastics, fiberglass and concrete are landfilled, and oils and other types of plastic are incinerated.
- The annual output from the wind turbine is worked out on the basis of an appropriate figure in equivalent hours of production to ensure the economic viability of a wind farm, i.e., 2,000 equivalent hours of production per year. For a turbine with a rated power of 2 MW, this is equivalent to an annual output of 4 GWh.
- In the context of corrective maintenance work on the turbine, in the course of its useful lifetime, it is estimated that one replacement generator will need to be installed due to malfunction (the complete new generator is considered, including manufacturing and assembly).

Finally, allocation as per the recommendations of standard ISO 14044 (which sets out the characteristics to be met by life-cycle analyses) is not used (Guinée et al. 2001; ISO ISO 2006a, b). The only function considered in the system analyzed and in all its components is that of generating electricity, so all the environmental impacts associated with the system are allocated to electricity generation. It is considered that the material recycled at the end of the useful life replaces virgin material used in the manufacturing stage of the turbine.

3.2.5 Methodology and Impact Categories Analyzed

When conducting an LCA, an environmental impact assessment method must be selected, and it must be decided which impact categories are to be analyzed. In this case, the assessment method chosen is CML Leiden 2000 (Guinée et al. 2001), and the categories are the following:

- Abiotic depletion: this category is linked to the extraction of minerals and fossil fuels associated with the inputs of the system under analysis and their effects on human health and the ecosystem.
- Acidification: this category is linked to the effect of various acidifying substances on the soil, groundwater, surface water, organisms, ecosystems, etc.
- Eutrophication: this category is linked to excess micronutrients in the environment as a result of emissions of nutrients into the air, soil, and water.

- Freshwater aquatic ecotoxicity: this category is linked to the effects of emissions into the air, soil, and water on freshwater ecosystems.
- Global warming (GWP100): this category is linked to the effects of greenhouse gas emissions on human health and the environment.
- Human toxicity: this category is linked to exposure to toxic substances and their effects on human health.
- Marine aquatic ecotoxicity: this category is linked to the effects of emissions on marine ecosystems.
- Ozone layer depletion (ODP): this category is linked to the proportion of UV-B radiation that reaches the surface of the Earth.
- Photochemical oxidation: this category is linked to reagents, mainly ozone, whose appearance in the environment may have effects on human health and ecosystems and may damage crops.
- Terrestrial ecotoxicity: this category is linked to the effects of emissions on land-based ecosystems.

4 Analysis of Results and Future Trends

The results for the production of 1 kWh of electricity via wind power are shown in Table 3, broken down into the various impact categories studied in the LCA and at the characterization stage. The intention is to prevent the potential subjectivity associated with other phases of the LCA and to facilitate the comparison of results with other LCA in the fields of renewable and conventional energy sources.

For example, a comparison with the environmental impact associated with electricity generation according to the Spanish energy mix for the year 2000 shows that the figures obtained here are between 89 and 99 % lower, according to the impact category considered. In the specific case of CO_2 emissions associated directly with climate change (see the Global Warming category), the figure for 1 kWh of electricity generated via wind power is 98 % lower than the figure for the Spanish energy mix.

Another interesting result can be provided by an examination of how long wind turbines need to operate to offset their environmental impact from the time of their manufacture through their operational lifetime and maintenance to their decommissioning, and the processing of the resulting waste. Taking as the starting point for this analysis the fact that the energy generated by a wind turbine avoids the need for an equivalent amount of energy to be produced via conventional sources, another comparison with the Spanish energy mix can be made. The Spanish energy mix is used for the purposes of comparison because the turbine studied is located in Spain and the manufacturer is Spanish, but the results can be extrapolated to any other country or location. Under this premise, the environmental impact caused by the turbine is offset in between 53 and 784 days, depending on the impact category considered.

Table 3 Results of the LCA by impact categories (Martínez, et al., 2009)

Impact category	Total	Maintenance	Tower	Foundation	Rotor	Nacelle
Abiotic depletion [kg Sb eq]	3.75E −05	2.78E−06	7.28E−066	4.39E−06	1.88E−05	4.33E −06
Acidification [kg SO_2 eq]	5.43E −05	3.51E−04	1.35E−03	1.56E−03	2.61E−03	6.96E−04
Eutrophication [kg PO_4^- eq]	5.68E −06	4.98E−11	1.41E−10	8.69E−11	1.83E−10	6.11E−11
Freshwater aquatic ecotoxicity [kg 1,4-DB eq]	2.81E −03	6.48E−03	1.40E−03	3.63E−04	4.36E−04	6.84E−03
Global warming (GWP100) [kg CO_2 eq]	6.58E −03	8.19E−05	1.65E−03	4.00E−04	2.43E−04	4.43E−04
Human toxicity [kg 1,4-DB eq]	1.55E −02	3.25E−01	1.69E+00	4.48E−01	1.04E+00	9.26E−01
Marine aquatic ecotoxicity [kg 1,4-DB eq]	4.41E+00	2.78E−05	4.89E−05	1.48E−05	1.55E−05	4.99E−05
Ozone layer depletion (ODP) [kg CFC-11 eq]	5.21E −10	5.10E−07	1.84E−07	1.06E−07	6.75E−07	6.51E−07
Photochemical oxidation [kg C_2H_4]	2.13E −06	7.64E−06	5.34E−06	3.53E−06	1.94E−05	1.84E−05
Terrestrial ecotoxicity [kg 1,4-DB eq]	1.56E −04	3.24E−07	1.71E−06	8.,25E−07	1.91E−06	8.98E−07

A more detailed examination of the results obtained reveals that the main components with the greatest impact on the environment are the rotor, the tower, and the nacelle. The impact of the rotor is derived mainly from the amount of fiberglass used in the blades and in the nose cone that covers the hub. The impact associated with this material is increased by the fact that it is not recycled at the end of the turbine's useful lifetime. In this area, there is a clear potential for reducing environmental impact if the possibility of recycling fiberglass is considered in the future, even if it is only as a replacement for other types of plastic in applications other than wind turbine blades.

In the case of the nacelle, one of the elements that have the most environmental impact is the copper used in the wiring, and others include the electronic components and the fiberglass used in the casings that cover and protect the components that make up the powertrain and the associated systems on the turbine.

The main component of the tower is the steel from which it is manufactured. Much of the impact associated with it is offset by the fact that this steel is recycled at the end of the turbine's useful lifetime.

All these results refer to state-of-the-art wind turbines currently available on the market, but the trend in wind power is toward even larger turbines with greater rotor diameters and higher power ratings. It is therefore reasonable to assume that the increased scale of future wind turbines will make for even greater reductions in environmental impact in electricity generated in this way.

Moreover, wind power technology is gradually maturing, and more and more efforts are being devoted into improving the operating and maintenance conditions of wind turbines. For example, more and more commercial wind turbines without gearboxes are being installed because gearboxes are among the most failure prone components in turbines. This is done in an attempt not only to reduce maintenance and breakdown costs but also to increase the effective production time of turbines over the course of their useful lifetimes.

5 Conclusions

This study looks at wind power from the viewpoint of life-cycle assessment. Such analyses have, of course, been conducted at various times throughout the development of wind power, and their results have varied as the designs and main characteristics of wind turbines have evolved. For that reason, the latest DFIG multimegawatt turbines are considered here, as theirs is the most numerous type currently in use on wind farms. On that basis, a comprehensive LCA is conducted on a wind turbine, covering all phases from its manufacture to its decommissioning and the processing of waste at the end of its useful lifetime.

The results clearly show how low the environmental impact of wind power is in the various impact categories studied, especially when compared to the figures for other, conventional sources of electricity generation. This confirms the positive nature of wind power in all environmental and climate change-related aspects, but does not take into account other essential elements such as the financial and technical viability of installing such systems within a specific electrical grid. LCA can thus be confirmed as a potentially important tool in the field of energy provided that it is used as just one of the means of support for decision making by the relevant authorities and other players in the field of energy system development.

References

Ardente F, Beccali M, Cellura M, Lo Brano V (2008) Energy performances and life cycle assessment of an Italian wind farm. Renew Sustain Energy Rev 12(1):200–217

Ben Amar F, Elamouri M, Dhifaoui R (2008) Energy assessment of the first wind farm section of Sidi Daoud, Tunisia. Renew Energy 33(10):2311–2321

Boustead I, Hancock GF (2003) Implementation of life cycle assessment methods. Ecoinvent report no. 3. In: Frischknecht R, Jungbluth N (eds) Handbook of industrial energy analysis. Swiss Centre for Life Cycle Inventories, Dübendorf, pp 22–28

Crawford RH (2009) Life cycle energy and greenhouse emissions analysis of wind tur-bines and the effect of size on energy yield. Renew Sustain Energy Rev 13(9):2653–2660
Frischknecht R, Rebitzer G (2005) The ecoinvent database system: a comprehensive web-based LCA database. J Cleaner Prod 13(13–14):1337–1343
Frischknecht R, Jungbluth N, Althaus H-J, Doka G, Dones R, Heck T et al (2005) The ecoinvent database: overview and methodological framework. Int J Life Cycle Assess 10(1):3–9
Góralczyk M (2003) Life-cycle assessment in the renewable energy sector. Appl Energy 75(3–4):205–211
Guinée JB, Gorree M, Heijungs R, Huppes G, Kleijn R, van Oers L, Wegener Sleeswijk A, Suh S, Udo de Haes HA, de Bruijn JA, van Duin R, Huijbregts MAJ (2001) Life cycle assessment: an operational guide to the ISO standards. Kluwer, Amsterdam
Gürzenich D, Methur J, Bansal NK, Wagner H-J (1999) Cumulative energy demand for selected renewable energy technologies. Int J Life Cycle Assess 4(3):143–149
Haack BN (1981) Net energy analysis of small wind energy conversion systems. Appl Energy 9:193–200
ISO (2006a) ISO 14040: environmental management—life cycle assessment—principles and framework. International Standard Organization, Geneva
ISO (2006b) ISO 14044: environmental management—life cycle assessment—requirements and guidelines. International Standard Organization, Geneva
Krohn S (1997) The energy balance of modem wind turbines. Wind Power Note 16:1–16
Lenzen M, Munksgaard J (2002) Energy and CO_2 life-cycle analyses of wind turbines-review and applications. Renew Energy 26(3):339–362
Lenzen M, Wachsmann U (2004) Wind turbines in Brazil and Germany: an example of geographical variability in life-cycle assessment. Appl Energy 77(2):119–130
Martínez E, Sanz F, Pellegrini S, Jiménez E, Blanco J (2009) Life-cycle assessment of a 2-MW rated power wind turbine: CML method. Int J Life Cycle Assess 14(1):52–63
Nadal G (1998) Life cycle direct and indirect pollution associated with PV and wind energy systems. SC de Bariloche, Argentina: Fundación Bariloche, Av 12 de Octobre 1915, CC 138
Rivkin DA, Toomey K, Silk VL (2012) Wind turbine technology and design. The art and science of windpower. Jones & Bartlett Publishers, Burlington
Schleisner L (2000) Life cycle assessment of a wind farm and related externalities. Renew Energy 20:279–288
Tryfonidou R, Wagner H-J (2004) Multi-megawatt wind turbines for offshore use: aspects of life cycle assessment. Int J Global Energy Issues 21(3):255–262
Uchiyama Y (1995) Life cycle analysis of electricity generation and supply systems. In: Electricity, health and the environment: comparative assessment in support of decision making, IAEA-SM-338/33, Vienna, Austria
Uchiyama Y (1996) Life cycle analysis of photovoltaic cell and wind power plants. In: Assessment of greenhouse gas emissions from the full energy chain of solar and wind power and other energy sources, IAEA Working material, Vienna (Austria)
Weinzettel J, Reenaas M, Solli C, Hertwich EG (2009) Life cycle assessment of a floating offshore wind turbine. Renew Energy 34(3):742–747
Wiese A, Kaltschmitt M (1996) Comparison of wind energy technologies with other electricity generation systems—a life-cycle analysis. In: European Union wind energy conference. Stephens & Associates, Bedford (UK)

Comparing Various Indicators for the LCA of Residential Photovoltaic Systems

Ruben Laleman, Johan Albrecht and Jo Dewulf

Abstract This chapter presents a broad environmental evaluation of residential Photovoltaic (PV) systems. It focuses mainly on how variations in irradiation levels, assessment methodology, and the lifetime of the solar panel influence the perception regarding its sustainability. Data from the Ecoinvent Life-Cycle Assessment (LCA) database and the literature were used and various Life-Cycle Impact Assessment (LCIA) methods were considered for six different PV types. The results indicate that variations in irradiation levels, methodology and lifetime can significantly influence the final results and conclusions of a LCA. By carefully selecting assumptions and methodology, one can clearly influence the perceived environmental impact of PV systems. In addition, we state that multidimensional indicators should be used along with the one-dimensional ones. Also, the choice of the perspective (Hierarchist, Egalitarian or Individualist) has a major impact on the final results. In our opinion, the current focus on Greenhouse Gas (GHG)-emissions and energy efficiency ignores important environmental impact dimensions such as resource depletion.

R. Laleman (✉) · J. Albrecht
Faculty of Economics and Business Administration, Ghent University, Tweekerkenstraat 2, 9000 Ghent, Belgium
e-mail: Ruben.laleman@ugent.be

J. Albrecht
e-mail: Johan.albrecht@ugent.be

J. Dewulf
Research Group ENVOC, Ghent University, Coupure Links 653, 9000 Ghent, Belgium
e-mail: Jo.dewulf@ugent.be

A. Singh et al. (eds.), *Life Cycle Assessment of Renewable Energy Sources*, Green Energy and Technology, DOI: 10.1007/978-1-4471-5364-1_10,

1 Introduction

The importance of photovoltaic technology as a low-carbon alternative for fossil-driven electricity production has increased markedly in recent years. Photovoltaic systems (PV systems) have evolved from a small niche player into an international market of several billions. The average annual growth rate of globally installed capacity between 2005 and 2010 exceeded 49 % (Fig. 1). This growth can be attributed to the combination of a steep decline in production costs and continued government support, the latter mainly in Europe and specifically Germany (Frondel et al. 2008). By 2010, the global capacity was estimated to be 40 GW, 80 % of which is installed in Europe (REN21 2011).

Governments around the globe promote the diffusion of PV technology because it is deemed to be a renewable, green and clean technology. However, no energy technology is 100 % sustainable. In this chapter, we will aim to evaluate the environmental impact of PV technology and compare this with alternative sources for electricity production. This will be done using various Life-Cycle Impact Assessment (LCIA) methodologies.

The whole life-cycle (from cradle to grave) of a residential roof-top PV system will be taken into account. Such an installation consists of many parts: the PV panels themselves, a support system to fix the panels on the roof, electric wiring and an inverter to convert the direct current (produced by the PV system) into alternating current that can be consumed by the household, or injected into the grid (Fig. 2). The LCA data of all these parts are included in the LCA database of Ecoinvent (v2.0) and will be included in this chapter's discussion.

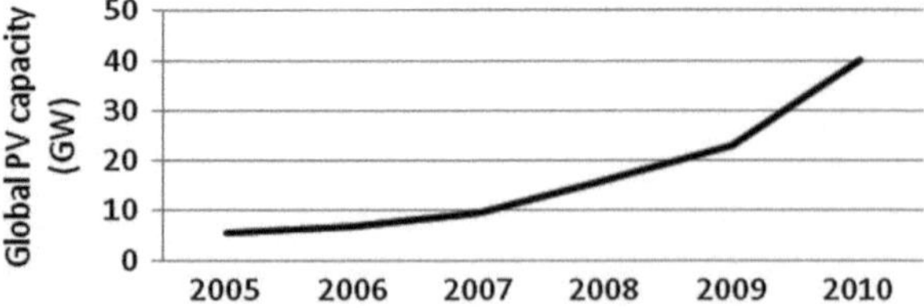

Fig. 1 Global PV capacity increased significantly in the past 5 years (*source* figure based on data found in REN21, 2011; original data from EPIA and PV news)

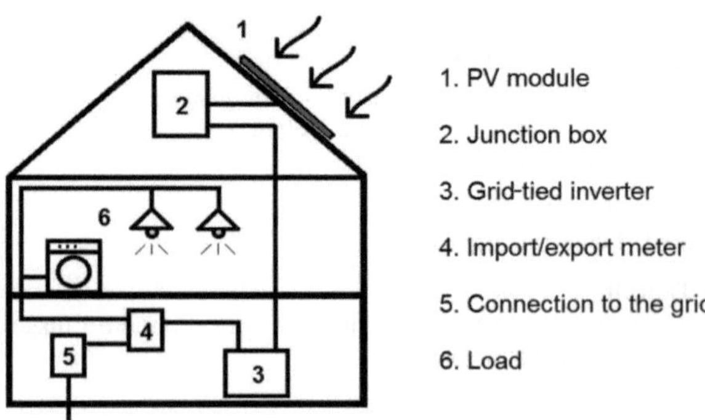

Fig. 2 Overview of a typical residential PV system (Greenpeace International, EPIA 2008)

The environmental impact of the PV system will be quantified by using various environmental indicators. We will mention some one-dimensional indicators: Global Warming Potential (GWP), Cumulative Energy Demand (CED), Energy Payback Time (EPT), Net Energy Ratio (NER), and Fossil Energy Requirement (FER). Also, one multidimensional indicator will be discussed here, namely the Eco-Indicator 99 with its three perspectives: Hierarchist (H,A), Egalitarian (E,E) and Individualist (I,I). Many other methods can be found in the literature; a nonexhaustive overview can be found in the Ecoinvent report n° 3 (Frischkneicht and Jungbluth 2007). The main difference between a one-dimensional indicator and a multidimensional is that the latter combines multiple environmental effects into one number, and thus aims to provide a more nuanced picture of the environmental impact of a good or service.

Most of the LCA studies, however, focus only on one-dimensional indicators, mainly related to the impact on climate change, (fossil) energy use and energy efficiency. For example, two recent reviews by Varun and his colleagues focus almost entirely on GHG emissions (Varun et al. 2009a, b). Others mainly consider the energy payback time (Ito et al. 2003; Kato et al. 1997; Mason et al. 2006). This one-dimensional view has the advantage of being relatively simple to interpret. On the other hand, it has the obvious disadvantage of leaving many—possibly important—parameters unexplored. The Eco-Indicator 99 method can be a tool to overcome this issue and obtain a broader perspective.

Comparing results from the literature is often not straightforward since a multitude of assumptions regarding the lifetime of the panel and the solar irradiation is used. These two parameters are hugely important to estimate the impact of PV electricity. As Table 1 shows, the irradiation in a sunny region can be twice as high as a sun-deprived region. This variation in irradiation will obviously affect the amount of electricity produced by a given PV system during its lifetime.

In this chapter, we will compare the results from the one-dimensional indicators with the Eco-Indicator 99 (EI 99) and we will show in detail how different methods and assumptions shape the final results of an environmental impact assessment. All the results are evaluated in detail and compared with data found in the literature. The goal is to obtain a very broad, nuanced, and clear picture of the environmental impact of a residential PV system.

Table 1 Horizontal irradiation

Low irradiation (kWh/m^2/year)		Moderate irradiation (kWh/m^2/year)		High irradiation (kWh/m^2/year)	
Brussels	960	Istanbul	1,320	Seville	1,700
Cologne	960	Bordeaux	1,300	Cyprus	1,750
London	980	Turin	1,340	Malta	1,770
Stockholm	940	Minneapolis	1,430	San Francisco	1,715
Vancouver	1,100	Seattle	1,200	Los Angeles	1,788

Sources EU (Suri et al. 2007), *US* National Renewable Energy Laboratory, 2010, *Canada*: National Resources Canada, 2010

Six different types of PV systems will be evaluated: Cadmium Telluride (CdTe), $CuInSe_2$ (CIS), ribbon Si, multicrystalline Si (multic-Si or poly c-Si), monocrystalline Si (mono c-Si), and amorphous (a-Si).

In the next section, we present a brief overview of the LCIA methods. In Sect. 3, the environmental impact of various PV types is evaluated using the Ecoinvent database, which is compared with results from the literature in Sect. 4. In Sect. 5, the impact of PV electricity is calculated and compared with other technologies' impact. Section 6 contains the discussion and conclusions. Throughout the whole chapter, we focus on the impact of assumed lifetime and irradiation, and the methodology used, on the results and the perceived sustainability of PV-produced electricity.

2 LCIA methods

2.1 Eco-Indicator'99

The Eco-Indicator assessment method (EI 99) was developed by PRé consultants in 1999 and offers a broad perspective on the environmental impact of a good or service (PRé consultants 2001). For this reason, many authors have used it to analyze the environmental impact of a wide variation of products, ranging from red clay (Bovea et al. 2007), beer (Cordella et al. 2008), water-based UV-lacquers (Dreyer et al. 2003), desktop PC's (Duan et al. 2009) and wind turbines (Lenzen and Munksgaard 2002; Martinez et al. 2009a, b). Two papers, by the same author, were found that applied the EI 99 assessment method to PV systems (Jungbluth 2005; Jungbluth et al. 2008a, b).

The EI 99 method has the advantage that the different aspects of environmental impact can easily be visualized and summed up in one final result, namely the "eco-point". PRé consultants, however, advises against the use of the score in an absolute way by stating that "The absolute value of the points is not very relevant as the main purpose is to compare relative differences between products or components" (Goedkoop et al. 2000)[1]

The environmental impact of a good or service is quantified using three main dimensions, namely human health (HH), ecosystem quality (EQ), and the depletion of nonrenewable resources (*R*). The first step in the calculation of the overall environmental impact score is the quantification of the impact for these three dimensions. The unweighted results obtained in this first step are referred to as the "Characterization Results" and have different units. To obtain a single score (with a single unit namely impact points or eco points), these results are normalized and weighted (Table 2). To cope with the issue of subjectivity in the weighting step,

[1] Quote from Préconsultancy report, manual for designers, page 4, http://teclim.ufba.br/jsf/ecodesign/dsgn0212.pdf.

Table 2 Normalization and weighting factors for the three perspectives

	Normalization			Weight (%)		
	Hierarchist (H,A)	Egalitarian (E,E)	Individualist (I,I)	(H,A)	(E,E)	(I,I)
HH	0.0154 DALY's[a]	0.0155 DALY's	0.00825 DALY's	40	30	55
EQ	5, 130 PDF[b]$_*$m$^2_*$y	5,130 PDF$_*$m^2$_*$y	4,510 PDF$_*$m^2$_*$y	40	50	25
R	8,410 MJ surplus[c]	5,940 MJ surplus	150 MJ surplus	20	20	20

(*Source* Ecoinvent report n°3 (Frischkneicht and Jungbluth 2007); (PRé consultants 2001))

[a] Disability Adjusted Life Years: Years of life lost trough disability or early death

[b] Potentially Disappeared Fraction: % of species that disappear due to environmental load

[c] Mega Joules surplus (increase in energy needed for resource extraction)

three different perspectives were developed: Hierarchist, Individualist and Egalitarian. Each perspective is based on a different ranking of preferences, values, and attitudes (Table 3).

The weighting and normalization factors for the three perspectives (Table 2) used in the Ecoinvent database (Ecoinvent report n°3) are taken directly from the original report by PRé consultants (2001). Notice the big difference between the normalization factor for resources used in the Individualist perspective (150 MJ) compared to the Hierarchist (8,410 MJ) and Egalitarian perspective (5,940 MJ). This difference in normalization will have a big impact on the results.

The *Individualist* does not consider the risk of a near fossil fuels depletion as credible. To the Individualist, the only resource depletion that matters is mineral extraction. With a share of 20 % in total impact, the amount of mineral extraction will have a big influence on the total score, especially for the production of PV systems, which is quite mineral-intensive.

The *Hierarchist* perspective is considered to represent the view of the "average scientist" and is used as the default setting. The Hierarchist, therefore, follows the IPPC assessment reports to consider the effects of climate change ((PRé consultants 2001), methodology report, p 18).

The *Egalitarian* view pays more attention to future generations and is considered as rather risk averse. The Egalitarian looks at the very long term and puts a high value on ecosystem quality. However, this can result in overestimating risk. This short discussion already illustrates that the outcome of an environmental analysis should always be evaluated with care. PRé consultants, the developers of the EI 99 method, stated that researchers should use the 3 different perspectives and carefully

Table 3 General properties of the different EI 99 perspectives

Perspective	Timeframe	Manageability	Evidence
Hierarchist (H,A)	Short and long term are balanced	Proper policy can avoid many problems	Based on consensus
Egalitarian (E,E)	Very long term	Problems can lead to catastrophe	All possible effects
Individualist (I,I)	Short term	Technology can avoid many problems	Only proven effects

(*Source* (PRé consultants 2001))

compare the results. Sustainability and intergenerational equity are complex concepts and, by consequence, one should accept that sustainability assessments imply complex trade-offs. It is therefore crucial to have a good understanding of the assumptions and weighting methods incorporated in the EI 99 method.

2.2 Global Warming Potential

The Global Warming Potential (GWP) assessment method, developed by the Intergovernmental Panel on Climate Change (IPCC 1997; IPCC 2001), is frequently used in energy research to investigate the impact of a product or a service on global warming (Bravi et al. 2007; Heller et al. 2004; Lechon et al. 2008; Mohr et al. 2009). Three GWP methods have been developed, each for a different time span (20, 100 and 500 y). In this study, the 100 y method was used. Using the 20y or 500y time span has no significant impact on the overall results.

2.3 Cumulative Energy Demand

The Cumulative Energy Demand (CED) is a very popular LCIA method, especially in renewable energy technology research (Huijbregts et al. 2006; Jungbluth et al. 2007a, b; Jungbluth et al. 2008a, b; Alsema 1998; Alsema 2000; Alsema and Nieuwlaar 2000; Alsema and de Wild-Scholten 2005; De Wild-Scholten and Alsema 2006). The CED aims to quantify all the energy that is consumed (or wasted) during the life cycle of a product. The CED is usually expressed in terms of primary energy (MJ_{prim}). In Ecoinvent, a different unit is used, namely energy equivalents (MJ-eq).

2.4 Energy Pay-back Time

The Energy Pay-back Time (EPT) is a frequently used parameter because of its input–output format and its ease to interpret. The EPT is, however, not straightforward to calculate. The formulas used to calculate the EPT are briefly summarized below (based on (Alsema 2000; Frischkneicht and Jungbluth 2007; Jungbluth et al. 2007a, b; Pacca et al. 2007). The first step is to calculate the Yearly Energy Output (YEO [kWh/year]) of the energy technology. There are two ways to do so. One starting from the Output Ratio (OR);

$$\text{YEO} = \text{OR} \cdot \text{Power} \tag{1}$$

With OR = Output Ratio [kWh/kWp/year]

Power = Total installed power, determined at STC[2] [kWp]

The other is based on irradiation (*R*), efficiency (θ) and performance ratio (*p*);

$$\text{YEO} = R \cdot A \cdot \theta \cdot p \tag{2}$$

with R = Yearly irradiation [kWh/m^2/year]
A = Active Surface of the PV module [m^2]
θ = Conversion Efficiency [%]
p = Performance Ratio [%]

The EPT [years] can now be calculated by dividing the CED by the YEO, on the condition that both components are expressed in identical units (kWh *or* MJ of *primary energy*);

$$\text{EPT} = (\text{CED} / \text{YEO}) \cdot C \tag{3}$$

with CED=Cumulative Energy Demand [MJ_{prim}]
YEO = Yearly Energy Output [MJ_{el}/year]
C = Conversion coefficient [MJ_{el}/MJ_{prim}]

To convert the YEO from electrical energy to primary energy, one has to include the efficiency of the electricity supply in the region of interest. The Conversion coefficient (*C*) indicates how efficient the generation of electricity is, on average, for the generations assets in a particular region (Gürzenich and Wagner 2004; Alsema and Nieuwlaar 2000). In this chapter, a Conversion coefficient of 0.35 [MJ_{el}/MJ_{prim}] is used. It is remarkable that some authors do not incorporate *C* (Koroneos et al. 2006; Pacca et al. 2007), a practice that results in EPT's that are approximately 3 times longer.

Keep in mind that conversion coefficient (*C*) has nothing to do with conversion efficiency (θ). The former is 30–40 % and refers to the grid electricity production; the latter is 10–20 % and refers to the efficiency of the solar PV system.

2.5 Net Energy Ratio

Another indicator for energy efficiency which is commonly used is the Net Energy Ratio. This indicator takes into account the total lifetime of a PV system. It can be interpreted as: the amount of energy that a technology can produce relative to the total amount of energy that was consumed, over the total life cycle.

$$\text{NER} = \textit{lifetime}/\text{EPT} \tag{4}$$

The NER is therefore an indication of the "life-cycle energy efficiency" of the technology. If it is lower than 1, the technology is by definition not renewable, since more energy was consumed than produced. In other words, this would entail a

[2] STC = standard test conditions (25°C; 1,000 W/m; AM 1,5).

net-energy loss when considering all the steps in the energy production chain. The difference between NER and EPT is that, for a given technology, the NER score will be more favorable if its lifetime is longer, while the EPT will remain the same.

2.6 Fossil Energy Requirement

The Fossil Energy Requirement provides a score for the "renewability" of a kWh of produced electricity. Unlike the NER, it only contains the nonrenewable part of the electricity. It is thus not equal to the inverse of the NER. The FER can be calculated by dividing the total amount of nonrenewable energy required for production (i.e., the nonrenewable part of the CED as defined in Sect. 2.3) by the total lifetime energy production (LEO) of the PV system (Cherubini et al. 2009). A low FER indicates that the electricity produced has a high "renewability"[3]

$$\mathrm{FER} = \mathrm{CED}_{\mathrm{non-ren}} / \ \mathrm{LEO} \left[\mathrm{kWh}_{\mathrm{prim}} / \mathrm{kWh}_{\mathrm{el}}\right] \tag{5}$$

Cherubini et al. (2009) also mentions the Cumulative Energy Requirement (CER), which contains both the renewable and the nonrenewable part;

$$\mathrm{CER} = \mathrm{FER} + \mathrm{RER} \left[\mathrm{kWh}_{\mathrm{prim}} / \mathrm{kWh}_{\mathrm{el}}\right] \tag{6}$$

with CER = Cumulative Energy Requirement
FER = Fossil Energy Requirement
RER = Renewable Energy Requirement

The Renewable Energy Requirement (RER) is, according to Cherubini et al., 2009, equal to 1 for most of the renewable energy sources, except for biomass. Biomass electricity production has a high RER since a lot of (renewable) energy is needed to produce biomass. By consequence, the CER for electricity from biomass is very high, even higher than for fossil-fueled technologies. These high scores could be misleading. We therefore chose to only mention the FER's here.

3 LCA of a 3 kWp PV system

3.1 Introduction

Table 4 shows that the efficiency of a PV module varies substantially according to the type of solar cell used, with the lowest efficiencies for amorf-Silicon-type cells (a-Si; about 7 %) and the highest for monocrystalline Si solar cells (mono c-Si; about 14 %). The column titled "active surface" shows how much m^2 is needed to

---

[3] Cherubini et al. (2009) uses MJ/MJe, however, we prefer to use $\mathrm{kWh}_{\mathrm{prim}}/\mathrm{kWh}_{\mathrm{el}}$.

Table 4 Properties of the PV systems that will be investigated in this chapter

Cell type	Cell eff[a] (%)	Module eff[a] (%)	Module eff[b] (%)	Active surface[a] (m^2/kWp)	Active surface[b] (m^2/kWp)	Weight[a] (kg/m^2)
Mono c-Si	15.3	14.0	14	7.1	7	14.6
Multi c-Si	14.4	13.2	13	7.6	8	14.6
Ribbon Si	13.1	12.0	11	8.3	9	14.6
a-Si	6.5	6.5	7	15.4	14	8.2
CIS	10.7	10.7	10	9.4	10	17.6
CdTe	7.6	7.1	10	14.1	10	19.0

[a] Ecoinvent report n°6; [b] Raugei and Frankl (2009)

obtain a PV system of one kWp. It is not surprising that higher efficiencies result in a lower active surface (m^2/kWp). This is important because households are generally restrained by the surface of their roofs to install a residential PV system.

The data found in the Ecoinvent (v 2.0) report about PV systems (report n°6) are very similar to more recently published data by Raugei and Frankl (2009), except for CdTe type systems, which have a considerably lower efficiency according to the Ecoinvent report. In this paper, the Ecoinvent "module efficiency" will be used. Up until now, thin film type solar systems such as CdTe, a-Si and CIS modules have been more commonly used for large scale PV systems. Thin film PV systems only had a market share of 10–15 % in 2008 (International Energy Agency 2010), they are thus not the main focus of this chapter. However, in the future, these thin film types are likely to become more important and gain market share as their per kWp costs decrease (EPIA 2010)

To conduct our analysis, we made the following assumptions: the residential PV systems are not integrated but installed on top of slanted roofs; a standard 3 kWp installation is considered; the conversion coefficient (*C*) is $0.35 MJ_{el}/MJ_{prim}$ and the output ratio remains constant (module efficiency loss over time is not incorporated). In the literature review, however, also large scale installations are evaluated.[4]

3.2 Energy Indicators

3.2.1 Cumulative Energy Demand

The results of a CED analysis based on the Ecoinvent database for the different types of slanted roof, nonintegrated, residential PV systems are presented in Fig. 3. The CED is normally presented as a ratio: CED/kWp (MJ-eq/kWp) or CED/m^2

[4] The environmental impact per kWp for large scale systems is generally slightly lower compared to small scale systems due to economies of scale.

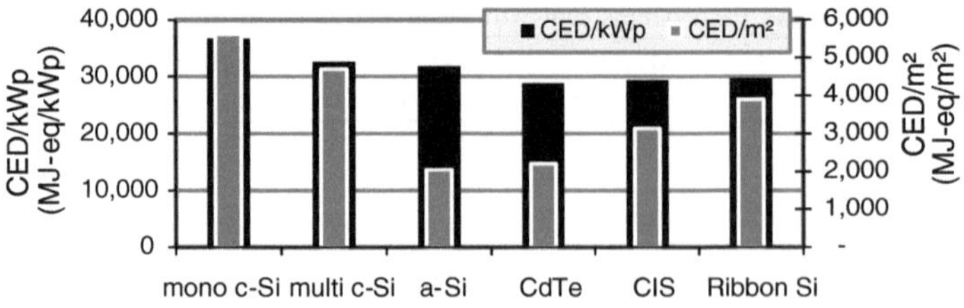

Fig. 3 CED/kWp and CED/m^2 for residential 3 kWp PV systems (Ecoinvent v2.0)

(MJ-eq/m^2). As the graph shows, the results differ strongly according to the selected ratio. From an energy efficiency perspective, it is advisable to use the CED/kWp ratio, because it incorporates the differences in conversion efficiency.

According to the Ecoinvent database, the CED/kWp for CdTe, CIS and ribbonSi PV systems is less than 30.000 MJ-eq/kWp. The recent technologies appear to be more energy efficient than the "old" crystalline Si-based technologies. As these new technologies have a steep learning curve, further significant energy savings can be expected. Mono c-Si, the oldest technology, has a high CED/kWp because of the energy intensive process that is required to produce the mono Si crystals (Alsema and Nieuwlaar 2000). As a result, the mono c-Si type is less attractive from an energy efficiency point of view.

3.2.2 Energy Payback Time

The EPT is calculated as explained in Sect. 2.4 ($C = 0.35$ [MJ_{el}/MJ_{prim}]). YEO and C are equal for all PV systems for a given location, which, according to the Eq. 3, results in EPT and CED/kWp being perfectly proportional. In other words, a PV type that has a low CED/kWp ratio will automatically have a relatively low payback time. Figure 4 shows this clearly, with mono c-Si having the highest CED/kWp and, by consequence, the highest EPT for a given yearly energy output (YEO) and conversion efficiency (*C*).

The grey bars in Fig. 4 show the EPT for 3 European countries. We selected Belgium (BE) as a "low-irradiation" region ($R = 946$kWh/m^2/y; OR = 725kWh/kWp), Switzerland as a "medium irradiation" region (CH, $R = 1{,}117$kWh/m^2/y; OR = 848 kWh/kWp) and Spain as a "high irradiation" region (ES, 1,660kWh/m^2/y; OR = 1,282 kWh/kWp) (Jungbluth et al. 2007a, b). The graph shows how the irradiation level significantly influences the energy payback time, with EPT's in Belgium being almost twice as long as the EPT's in Spain.

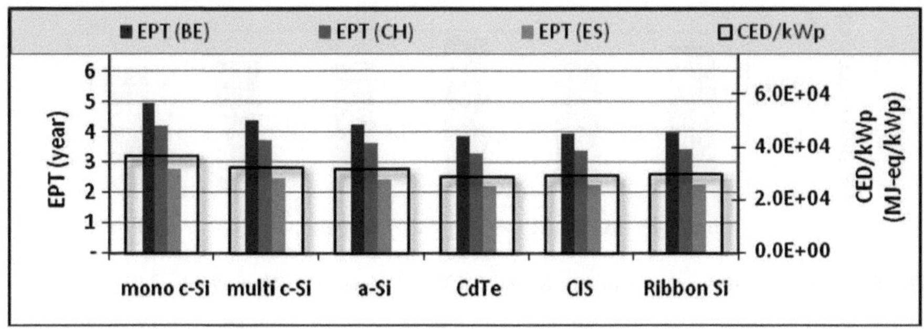

Fig. 4 CED/kWp and EPT for various types of residential 3 kWp PV systems in different regions (Ecoinvent v2.0)

3.2.3 Net Energy Ratio

Most authors mention a life expectancy of 25–30 years for well-maintained PV systems (International Energy Agency 2010; Fthenakis et al. 2008; IPCC 2011). As a result, the Net Energy Ratio of PV systems installed in regions with a low solar irradiation is about 5 (year/year), in sunny regions like Spain, PV systems can even have NER's up to 12 (lifetime/EPT = 30y/2.5y = 12) or more. In other words, a residential PV system in Spain can produce at least 12 times more energy than it consumed during its life cycle.

3.3 Global Warming Potential

The GWP gives an indication of the amount of greenhouse gasses (GHG's) emitted during the life cycle of the PV system. The results are shown in Fig. 5. There are some differences between the GWP (kgCO_2-eq) and EPT (or CED/kWp) results, indicating that higher energy use apparently does not automatically lead to higher GHG emissions—even though the two are clearly related to each other. Consider, for example, the a-Si type PV system; it has an EPT that is about the same as a multi c-Si PV system, the GWP, on the other hand, is relatively high ($\pm$ 6,000 kgCO_2-eq). Overall, the "new" technologies, such as CdTe, CIS, and ribbon Si, have a relatively lower impact on global warming ($\pm$ 5,000 kgCO_2-eq for a 3 kWp rooftop installation).

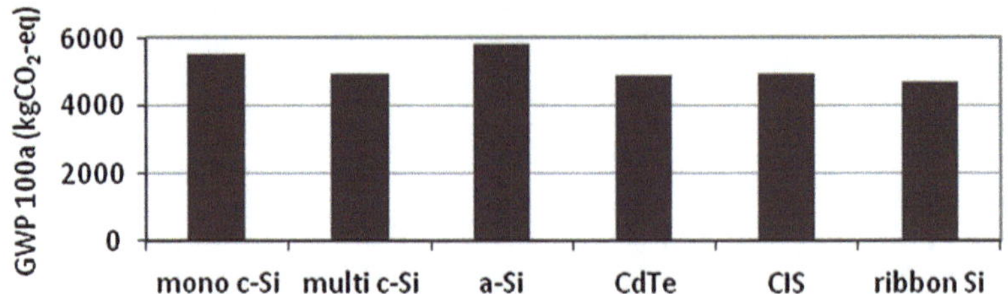

Fig. 5 GWP for various types of residential 3 kWp PV systems (Ecoinvent v2.0)

3.4 Eco-Indicator 99

3.4.1 Eco-Indicator 99 (H, A)

From Fig. 6, we can see that the CdTe PV system has the highest impact score (450 points) according to the EI 99 (H,A). This is surprising since the technology performed very well according to the GWP and EPT indicators. The other thin film technologies, CIS and ribbon Si, have a much lower impact (317 and 353 points, respectively). These findings are consistent with Jungbluth et al. (2008a, b).

Figure 6 also shows that "depletion of fossil fuels" and "respiratory effects" both have a significant share in the overall Eco-Indicator impact score. This is probably linked to the energy use during the PV-panel production and transport. Research has shown that in many cases fossil fuel extraction, respiratory effects, climate change, acidification, and carcinogenics are closely related to each other,

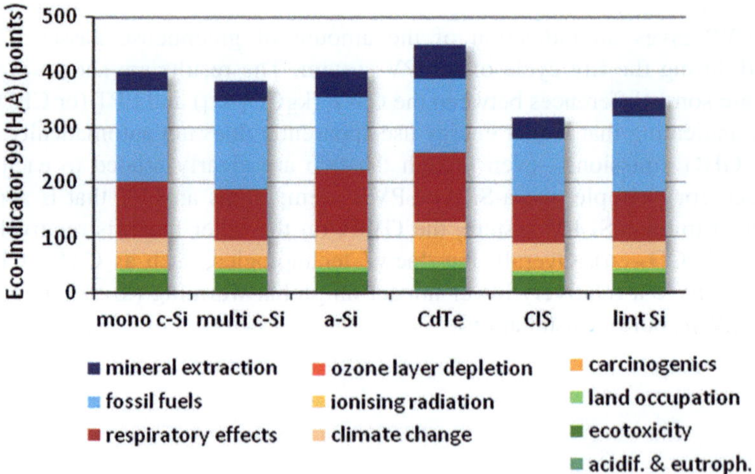

Fig. 6 Eco-Indicator '99 (H,A) results for various types of residential 3 kWp PV systems (Ecoinvent v2.0)

and to the use of fossil energy (Huijbregts et al. 2006). Indeed, the burning of fossil fuels has many adverse impacts, on climate, health, and the ecosystem. Therefore, decreasing fossil energy use during production of PV systems will decrease the impact of all of these factors (Mohr et al. 2009).

3.4.2 Influence of the Used EI 99 Perspective

When comparing the three different EI 99 perspectives, the results for the Individualist perspective stand out (Fig. 7). According to the Individualist perspective, the high amount of mineral extraction necessary for the production of PV systems is so important that other impact factors become negligible. The Individualist assigns high scores to CdTe (3,764 points) and a-Si (3,020 points).

The very high scores attributed to mineral extraction by the Individualist are due to the different ways by which resource depletion is normalized in the EI 99 method. The Individualist does not take into account the increasing energy intensity of fossil fuel depletion. This has as a consequence that all resource depletion impact is caused by mineral extraction. The view of the individualist seems rather extreme. It is hard to believe that one would find the depletion of fossil fuels to be of no importance. More and more evidence is emerging that the energy needed to extract fossil fuels is increasing. The increasing interest in oil from tar sands and shale gas is an indicator of this fact.

In the Hierarchist and Egalitarian perspectives, on the other hand, resource depletion includes mineral extraction *and* fossil fuel use. As Fig. 7 shows, this difference in opinion about fossil energy has a big effect on the normalization of resource depletion as a whole.[5]

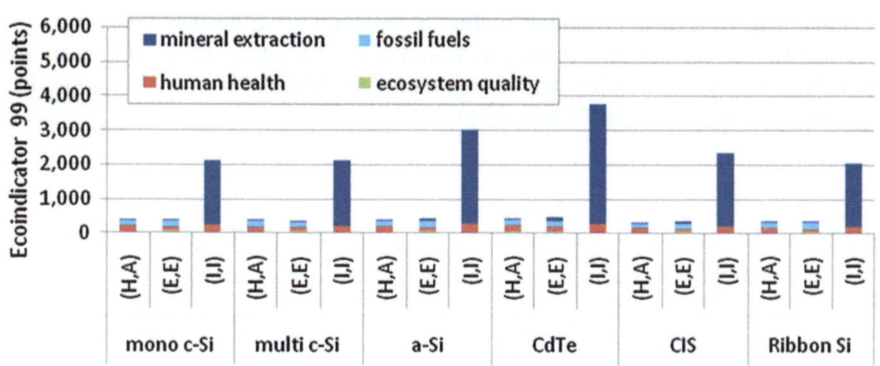

Fig. 7 Comparison of the EI 99 perspectives for 3 kWp PV systems (Ecoinvent v2.0)

[5] More details on how these high figures for the EI 99 individualist perspective come about can be found in the literature Laleman et al. (2011).

An important conclusion than can be drawn from this is that using a weighted multidimensional method, such as the Eco-Indicator, can lead to surprising results. The advantage of providing an overall score that takes into account many types of environmental impact and provides an elegant overview is unfortunately related to the disadvantage of the subjectivity of the weighting step. One cannot have one (a single overall score) without the other (some kind of weighting).

3.4.3 Mineral Extraction

Given the large impact that the individualist places on mineral extraction, it can be interesting to have a closer look at this particular aspect of "sustainability" which is rarely considered in the literature (Kato et al. 1997; Ito et al. 2003; Pehnt 2006; Raugei et al. 2007a, b). Table 5 shows that the production of PV systems requires a vast amount of minerals, even if most of the metals are assumed to be recycled (ecoinvent report n°6). However, when compared to the weight of 3 a kWp PV system, the results are not that surprising (Table 4). Removing the aluminum frames can significantly reduce the CED—and by consequence the environmental impact—of a solar panel (Alsema and Nieuwlaar 2000) but this is rarely done in practice.

4 Literature Review

4.1 Energy Payback Time Review

The EPT has been a popular measure to estimate the environmental impact of PV systems. The overview presented in Fig. 8 indicates that the data from the Eco-invent database are in line with results found in the literature. Generally speaking, the EPT is 4–5 years in a low-irradiation case (950 kWh/m^2/y) and 2–3 years in a high irradiation case (1,700 kWh/m^2/y).

Table 5 Life-cycle crude ore extraction of iron, aluminum, and copper for a 3 kWp PV installation (Ecoinvent v2.0) and total weight of a 3 kWp PV system (calculations based on Ecoinvent report n°6)

Mineral Ore	Mono c-Si	Multi c-Si	a-Si	CdTe	CIS	Ribbon Si
Fe (kg)	103.23	106.50	189.13	112.94	77.43	77.22
Al (kg)	75.95	80.54	145.04	135.17	89.66	87.95
Cu[d] (kg)	25.30	25.41	26.37	39.44	24.14	25.56
Weight of 3 kWp PV system (PV modules only)						
Weight (kg)	311	333	379	804	496	364

[d] Is the sum of all the Copper ore types available in Ecoinvent (v2.0)

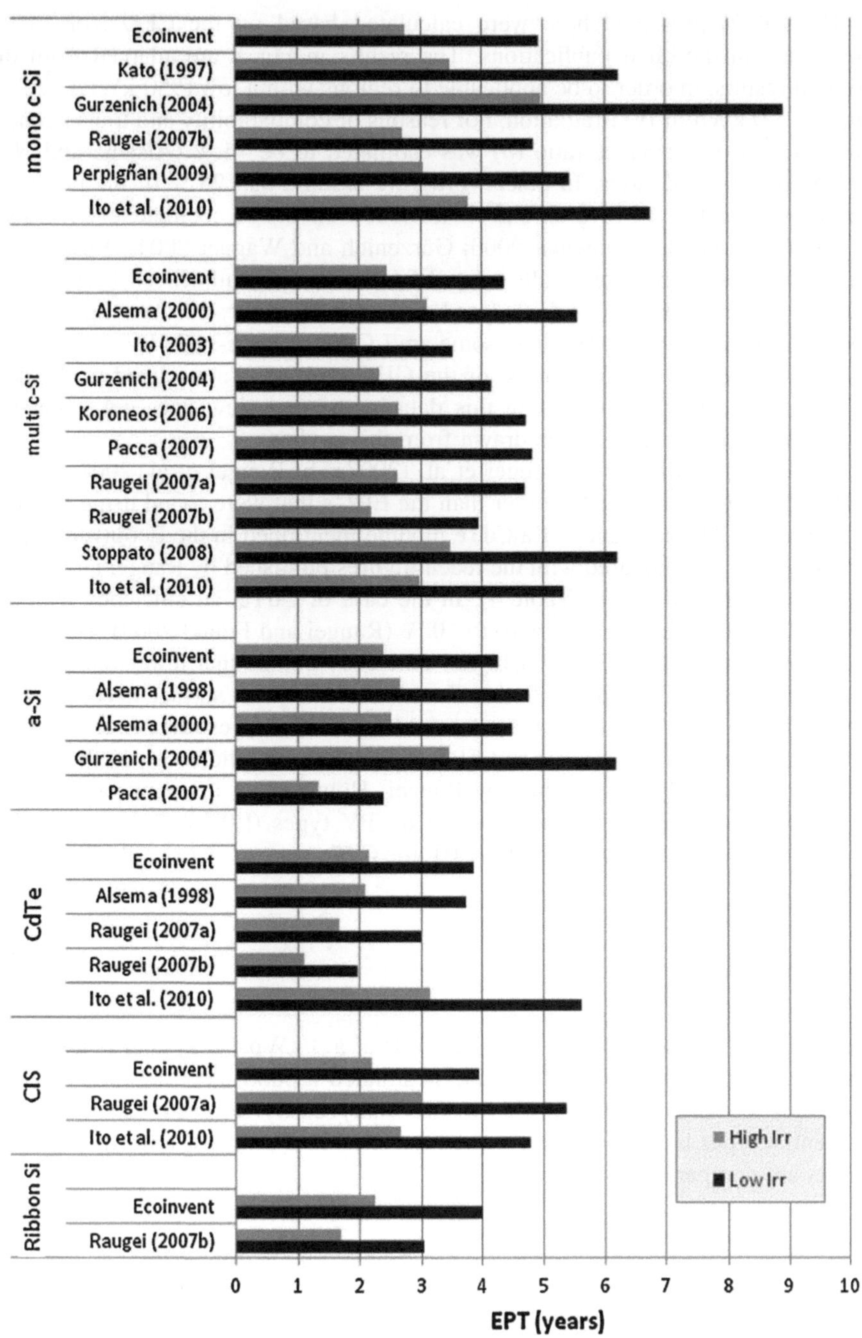

Fig. 8 Review of EPT's from previous publications and those calculated with Ecoinvent,adapted for a low and a high solar irradiation

The EPT's presented here were calculated based on the CED's or EPT's mentioned in the cited publications. The results in Fig. 8 are adapted from the original results, in order to be applicable to regions with a low (950 kWh/m^2/y) or high (1,700 kWh/m^2/y) irradiation. For reasons of comparability and transparency, the system's performance ratio (p) was estimated to be 75 % (Alsema and Nieuwlaar 2000) in all cases. In other words, we assume that 25 % of the produced electricity is lost in the inverter and cables. The Conversion coefficient (C) was set at 0.35 (Alsema and Nieuwlaar 2000; Gürzenich and Wagner 2004). Keeping all these parameters constant results in an EPT that is only influenced by the Conversion efficiency (θ) and the CED (see Eq. 2). Not all of these data are specifically applied to a household PV system, some are LCA's of large-scale PV systems (Ito et al. 2010). However, since most of the GHG emissions are related to the production of the module, including this data in the overview does not affect the overall conclusions that can be drawn from this review.

The EPT's published by Raugei et al. (2007a, b; Raugei et al. 2007a, b) for CdTe PV systems are much shorter than the EPT's that were found using the data in Ecoinvent. The efficiency of a CdTe module mentioned in the Ecoinvent report n°6 is rather low compared with the recent figures published by Raugei and Frankl (Raugei and Frankl 2009) (Table 4). In the case of CdTe, an efficiency increase from 7 % (according to Ecoinvent) to 10 % (Raugei and Frankl 2009) results in a decrease in the EPT by 30 %. A higher efficiency entails that fewer modules are needed to obtain a 3 kWp installation. In this case, the total surface needed for a 3 kWp PV installation would decrease from 43 m^2 to 30 m^2, resulting in less support structure and thus a lower CED and EPT. An increase in efficiency can thus partly explain the low EPT's published by Raugei. However, in a recent paper by Ito et al. (2010), the EPT is quite equal for all PV types (EPT = 2–3 y with high irradiation), not indicating a shorter EPT for CdTe systems (Ito et al. 2010).

4.2 GWP Review

From Fig. 9, we can conclude that the GWP of a 3 kWp PV system is about the same for all types of PV modules and is situated around 6,000 kg CO_2-eq. It is important to mention that the PV systems analyzed by Ito et al. (2010) are not residential types but large-scale systems installed in the desert, nevertheless their results are comparable with ours.

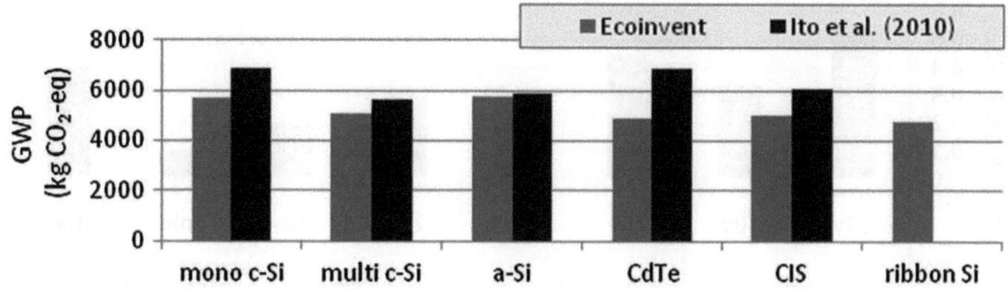

Fig. 9 GWP of a 3 kWp PV system (Ito et al. 2010 and Ecoinvent)

5 LCA of 1 kWh of Electricity

5.1 Introduction

To obtain the impact of 1 kWh of electricity, we can simply divide the impact of a 3 kWp PV system by the Lifetime Energy Output, which is calculated as follows:

$$LEO = YEO \cdot \text{lifetime}[\text{kWh}] \tag{7}$$

with LEO = Lifetime Energy Output [kWh]
YEO = Yearly Energy Output [kWh/year]
Lifetime = Expected lifetime of the PV system [years]

In the Ecoinvent report n°6, a lifetime of 30 years is assumed. An IEA-PVPS report (International Energy Agency 2009) also suggests to use a lifetime of 30 years. Some authors, however, suggest lifetimes of 20 or 25 years (Varun et al. 2009a). Therefore, two lifetimes will be evaluated, a pessimistic estimate (20 y) and a realistic estimate (30 y). Also the impact of the irradiation is important, since higher irradiation results in more electricity production (a higher YEO) and thus a lower per kWh impact (see Eqs. 2, 3, and 4).

As the focus of this chapter is on residential systems, which is dominated by multi c-Si PV systems (Raugei and Frankl 2009), we will from here on mention only the results for multi c-Si systems. This does not have any major implication for the overall conclusions since the environmental impact of the production of a PV system is relatively similar for all types of solar PV systems. More important here—considering the impact of *1 kWh of PV-produced electricity*—is the influence of the irradiation and assumed lifetime, as will be discussed below.

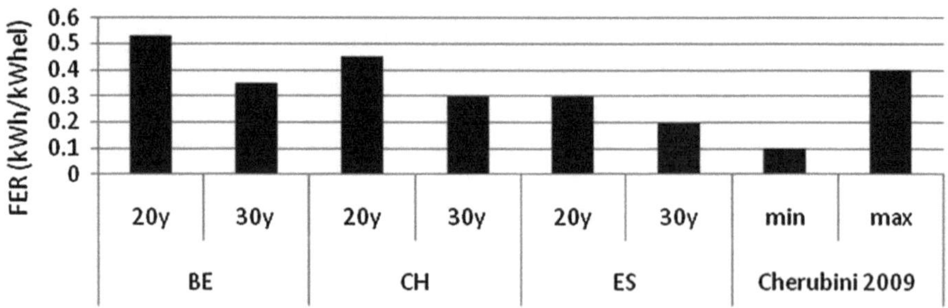

Fig. 10 The impact of lifetime and irradiation on the FER a 3 kWp multi c-Si PV systems (results based on calculations from Ecoinvent v2.0 compared with Cherubini et al. 2009)

5.2 Energy Efficiency of Electricity Production

5.2.1 Fossil Energy Requirement

The FER of 1 kWh of PV-produced electricity is lower than 1 in all the scenarios (Fig. 10), which indicates that PV electricity is renewable. However, the results are highly dependent on the assumed lifetime and irradiation. The FER in the most pessimistic case, (20 years lifetime, installed in Belgium) is higher than 0.5 kWh/kWhel. In the most optimistic case (30 years lifetime, installed in Spain), the FER is about 0.2 kWh/kWhel. These results are on average slightly higher than those found in Cherubini et al. (2009) but nevertheless comparable.

Note that the renewability of the electricity production of the country where the PV cells and modules are produced also has a big impact on the FER since only nonrenewable energy is incorporated. The FER would be very low, for example, if the cells and modules were produced in Norway or high when produced in China. The results from Ecoinvent represent a European case, with about 18 % of the CED of a PV system being renewable. Recently, most PV panels sold in Europe are produced in China, which would normally result in higher FER's. It is, however, unlikely that the FER for a kWh of PV-produced electricity would be higher than 1, even in a pessimistic scenario.

Table 6 shows that, compared with wind and hydro, PV electricity production is relatively energy intense. Biomass, on the other hand, has a FER similar to that of PV. The fossil-fueled technologies are of course not renewable and thus have a high FER (FER > 1 for nonrenewables).

The FER's for fossil technologies like coal and gas should be slightly higher than the inverse of the efficiency of the power plant. For example, if a coal plant has an efficiency of 35 % than the FER must be higher than 2.8 (2.8 = 1/0.35).[6] The FER will be slightly higher than this figure since other life-cycle energy

[6] The efficiency of a plant indicates how much electricity (MJ_{el} or kWh_{el}) is produced from a MJ of primary energy, the energy contained in the fuel. The efficiency is thus a percentage with

Table 6 Fossil energy requirement for various technologies

FER (kWh_{prim}/kWh_{el})			Min	Average	Max
PV			0.10	0.32	0.63
Wind	Onshore		0.01	0.05	0.13
	Offshore		0.03	0.03	0.03
Biomass	Pellets/Waste	Dedicated	0.06	0.19	0.40
		Co-Firing	0.05	0.12	0.20
Geo.			0.01	0.06	0.15
Hydro	Small		0.01	0.02	0.04
	Big		0.03	0.03	0.03
Nuclear			2.80	3.05	3.30
Gas			1.70	2.35	3.00
Coal			2.14	2.84	4.20

Sources calculations based on data from (Pehnt 2006; Varun et al. 2009a; Viebahn et al. 2007; Lenzen and Munksgaard 2002; Manish et al. 2006; Cherubini et al. 2009; Djomo et al. 2011)

costs—such as the construction of the power plant and the transport of the coal—are not incorporated into the plant efficiency, but should be incorporated into the life-cycle energy use. In other words, the inverse of the efficiency can be seen as an indicator of maximal energy efficiency. The data in Table 6 seem to confirm this statement. It can be seen that the lowest FER's for fossil technologies are related to relatively high efficiencies of 40–45 % (1/0.45 = 2.2).[7]

5.2.2 Discussion

The FER is similar to the inverse of the NER, there are, however, some important differences. First, the FER only takes into account the nonrenewable part of the energy needed, whereas the NER includes the total amount of energy. Second, the unit of the FER is kWh/kWh_{el} and does not take into account the conversion efficiency. The unit of the NER is kWh_{prim}/kWh_{prim} and, since it is calculated from the EPT, does incorporate the conversion efficiency. It is important to be aware of these differences and avoid misinterpretations.

In a recent report, the IPCC published some data on the energy efficiency of various technologies that could lead to such misinterpretations (Special Report Renewable Energy Sources). In chapter 9 of their report, the sustainability of renewables is examined in detail, based on an extensive literature review (IPCC 2011). The IPCC mentions energy ratios (ER) [kWh_{el}/kWh_{prim}]—which are the inverse of the above-mentioned FER's—for fossil-fueled technologies that are

(Footnote 6 continued)
kWh_{prim}/kWh_{el} or MJ_{prim}/MJ_{el} as a unit. As shown, the FER has the inverse as unit (MJ_{el}/MJ_{prim}). This is why the inverse of the efficiency is a rough indicator of the FER.

[7] The very low minimal FER for gas was found in Cherubini et al. 2009, however, an explanation for this low value is not given in the paper.

between 2 and 20. This is equivalent to a FER of 0.5–0.05. These numbers indicate a technology that is renewable, since the electricity produced (kWh_{el}) exceeds the primary energy consumed (kWh_{prim}). The reason for these high-energy ratios is that the report does not include the energy content of the fuel, which is somewhat surprising, and results in the strange notion that the energy ratio of a fossil technology can be higher than one.

The above discussion illustrates the difficulties that arise when there is a lack of general methodology and definitions. The need, thus, for a more uniform approach seems clear. The approach followed by Cherubini et al. (2009) seems to be a good way forward, in the authors' opinion. By mentioning both the renewable (Renewable Energy Requirement) and nonrenewable (Fossil Energy Requirement) energy need, a more balanced view of the environmental impact of the technology is provided. In general, the definition of an energy indicator should always be stated clearly such that misinterpretations can be avoided; unfortunately, this is not always the case.

5.3 Global Warming Potential of 1 kWh of electricity

5.3.1 GWP of 1 kWh of PV electricity

Figure 11 shows that both the estimated lifetime and the irradiation have a strong impact on the global warming potential of a PV-produced kWh of electricity. A PV system installed in Belgium (BE) with an estimated lifetime of only 20 years produces electricity with a GWP of 116 gCO_2-eq per kWh. This is almost twice the impact of a kWh produced in Switzerland (CH) with a PV system that has an expected lifetime of 30 years (66 gCO_2-eq). Under optimal conditions—a PV system installed in Spain with a lifetime of 30 years—the GWP could be reduced to 44 gCO_2-eq.

The bottom part of Fig. 11 shows that results from Ecoinvent data are similar to the literature. However, one can still find authors claiming that GHG emissions could be as high as 160 gCO_2-eq per kWh. This seems to be very pessimistic and is not in line with Ecoinvent data, nor with most of the literature. Overall, we can safely conclude that the GHG emissions from a kWh of crystalline PV systems produced electricity are likely to be in the range of 50–100 gCO_2-eq/kWh, with the lower end being valid for sunny regions like Spain and Italy, and the higher value being valid for regions like Belgium, the United Kingdom, and Germany.

5.3.2 Global Warming Potential of 1 kWh of Electricity

In this section, the results for the GHG emissions of a residential PV system are compared with results from various renewable and nonrenewable electricity production technologies. The data in Table 7 show that, overall, the emissions from PV systems are much lower than emissions from fossil-fueled electricity

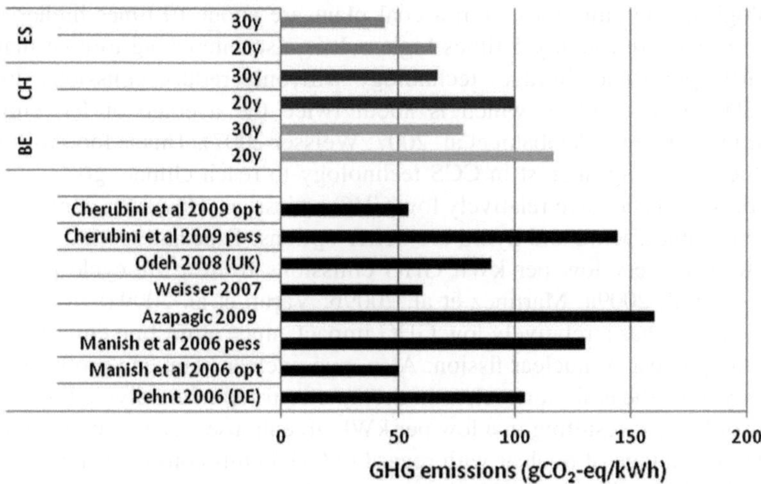

Fig. 11 The impact of lifetime on the global warming potential of 1 kWh of PV electricity in Belgium, Switzerland and Spain (*sources* Ecoinvent, and the literature (Cherubini et al. 2009; Odeh and Cockerill 2008; Weisser 2007; Azapagic 2009; Manish et al. 2006; Pehnt 2006). *UK* United Kingdom, *DE* Germany, *Pess* pessimistic estimate, *Opt* optimistic estimate

Table 7 Life-cycle GHG emissions for various technologies

GHG emissions (g CO_2-eq/kWh)			Min	Average	Max
PV			50	99	160
Wind	Onshore		4	17	40
	Offshore		9	13	17
Biomass	Pellet/Waste	Dedicated	2	66	122
		Co-Firing	7	55	150
Geothermal			7	39	90
Hydro	Small		2	16	43
	Big		10	14	18
Nuclear			8	34	108
Gas			360	474	720
	CCS		114	121	130
Coal			800	1,019	1800
	CCS		130	190	280
Lignite			900	1,000	1100
	CCS		130	165	200
Oil			662	815	1080

Sources (Cherubini et al. 2009; Odeh and Cockerill 2008; Weisser 2007; Azapagic 2009; Manish et al. 2006; Pehnt 2006; Viebahn et al. 2007; Djomo et al. 2011; Martinez et al. 2009a; The Environment Agency 2009; Varun et al. 2009a, b; Varun et al. 2009a; Lenzen 2010; Jaramillo et al. 2007)

technologies. The emissions from a coal plant are about 10 times higher. Emissions from gas are roughly 5 times higher. It is also interesting to note that CCS (Carbon Capture and Storage) technology will only reduce emissions down to about 200 gCO_2-eq/kWh, which is about twice the average of PV emissions (Jaramillo et al. 2007; Viebahn et al. 2007; Weisser 2007). This is food for thought given the increasing interest in CCS technology to reach climate goals.

PV thus seems to have relatively low GHG emissions. However, compared with other renewables, wind and hydro are better options. Especially, wind technology seems to have very low per kWh GHG emissions from a life-cycle perspective (Martinez et al. 2009a; Martinez et al. 2009b; Varun et al. 2009a).

Nuclear also has a relatively low GHG impact, since almost no emissions occur during the process of nuclear fission. Also, one nuclear plant can produce a lot of electricity, thus the emissions released during construction are divided over a huge number of kWh's, resulting in a low per kWh greenhouse gas impact. On the other hand, the advantage of nuclear with regard to GHG emissions is overshadowed by the big debate on its safety. Nuclear appeared to be an option to fight climate change a few years ago, however, since the Fukushima accident the political feasibility of nuclear had decreased dramatically.

The GHG emissions related to biomass electricity production are in the same range as PV, according to most of the literature. Biomass is, however, a special case, since a lot of the emissions can be attributed to the harvesting and transport phases. Lowering GHG emissions in these steps is thus crucial to obtain a low-carbon energy source. Many factors influence the overall GHG-balance of a kWh of biomass electricity. Land use change, for example, can significantly contribute to the overall emissions, be it in a positive (carbon capture and soil improvement) or very negative way (such as the burning of rainforest to replace it with biomass plantations) (IPCC 2011). The big influence of the assumed land-use impact partly explains the large variations in emissions that can be found in the literature. In the papers mentioned here, the emissions vary from 2 to 150 gCO_2-eq/kWh. The variation reported by the IPCC (IPCC 2011) is even larger, ranging from −600 gCO_2-eq/kWh up to +300 gCO_2-eq/kWh, depending on which assumptions and methods were applied. These figures suggest that the debate on the sustainability of biomass for electricity production is not likely to end soon.

5.4 Eco-Indicator 99 Analysis of 1 kWh of Electricity

5.4.1 Comparing Different Perspectives

The results that were found using the EI 99 method for a residential 3 kWp PV system in Sect. 3.4.2 varied greatly depending on the selected perspective. Figure 12 shows that this observation remains valid for the comparison of different energy technologies. In order to simplify the results, only the three main impact categories (resource depletion, human health, and ecosystem quality) are shown.

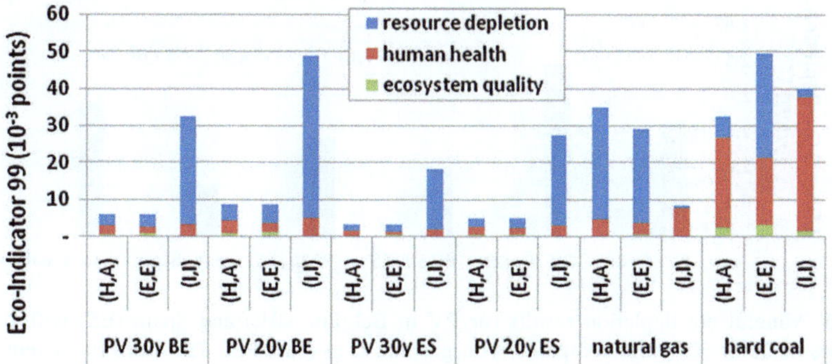

Fig. 12 EI 99 results for PV in Belgium (BE) and Spain (ES) with assumed lifetimes of 20 or 30 years, compared with gas and coal (based on data from Ecoinvent v2.0)

Generally speaking, resource depletion is the most important contributor to the overall environmental impact score, except for hard coal electricity generation, where impact on human health is more important.

The Individualist (I,I) prefers natural gas and puts a high weight on the resource depletion related to the production of PV systems. Not surprisingly, as the individualist is very much concerned with human health, the score related to human health impact for coal is very high since the burning coal is related to unhealthy emissions. The results in the Hierarchist (H,A) and Egalitarian (E,E) views are rather similar, with the exception of the impact score for hard coal. Coal will only become scarce in the very long term; this could explain the relatively high impact according to the Egalitarian view since it is more considered about long term effects (see Table 3). Notice that coal generally has the highest impact score, even higher than Individualists' estimate for PV systems, assuming the latter are installed in sunny regions.

The view of the Hierarchist and the Egalitarian is very clear; they prefer PV systems as electricity sources instead of fossil-based electricity production. According to these two perspectives, the overall score for PV is many times smaller compared with natural gas or coal. The score for gas is much higher here, since fossil fuel depletion is considered to have an unfavorable impact on future generations' ability to obtain sufficient amounts of natural gas with a reasonable effort.

In a recent article by Martínez et al. (2009a), an EI 99 score of about 0.001 points is assigned to 1 kWh of wind energy from a 2 MW turbine in Spain. This is about 3 times lower than the score we found for PV systems in the most optimistic case (0.003 eco points, high solar irradiation, and a 30 years life expectance) using the Hierarchist perspective. According to our results, PV systems have good properties, but according to Martínez et al. (2009a), the use of wind turbines has less environmental impact. Unfortunately, only one perspective was used by Martínez et al. (2009a), the impact of different perspectives was not analyzed.

Jungbluth (2005) has also compared the LCA results from PV systems with those of wind, hydro, biomass and gas plants (and others). In his paper, the EI 99

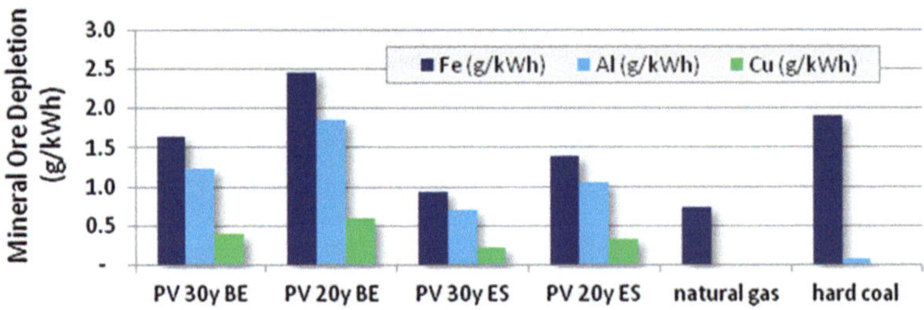

Fig. 13 Mineral ore depletion results for PV in Belgium (BE) and Spain (ES) with assumed lifetimes of 20 or 30 years, compared with gas and coal (based on data from Ecoinvent v2.0)

perspective is compared with other indicators, such as the GWP, the "ecological scarcity 97" Index and the CED_{nonren}. He finds that PV has a better profile compared with natural gas when using EI 99, GWP, and CED_{nonren} methods, the score for scarcity is, however, similar. When compared to the other renewables, PV had the worst score for all the selected LCIA methods.

5.4.2 Mineral Extraction for 1 kWh of Electricity

The total amount of minerals needed for the production of a 3 kWp PV system is relatively high (Table 5). If the data from table 5 is divided by the total amount of electricity produced, the "per kWh mineral use" is obtained. In Fig. 13, this is compared with the "per kWh mineral use" of other technologies. The results from the Ecoinvent database (v2.0) show that the amount of copper and aluminum ore extraction is much higher for PV-electricity, compared with the fossil-based technologies. This clearly indicates that PV has a big impact on mineral depletion, compared with the other technologies, especially for the more expensive metals like copper (Cu) and aluminum (Al). The results for iron depletion (Fe), on the other hand, are similar to the results for gas and coal.

When comparing Figs. 13 and 12, it is clear that, according to the Individualist, the energy technologies that are very mineral-intensive (such as PV) have a high environmental impact. Technologies that need very small amounts of minerals, such as gas, have the lowest score. Despite the fact that most of the metals are considered to be recyclable, the impact of PV systems on mineral depletion is not to be ignored. Our results indicate that about 2 grams of iron ore and 1.5 grams of aluminum ore (Bauxite) are needed for the production of 1 kWh of PV electricity. An article by Pehnt (Pehnt 2006) mentions comparable figures of 3.3 g of iron ore and 1.2 grams of Bauxite. If PV systems are to become a major contributor to the electricity supply, an efficient recycling program would be advisable. Fortunately, this issue is receiving increasing attention.[8]

[8] For more information see http://www.pvcycle.org/.

6 Conclusions and Discussion

In this chapter, the environmental impact of residential 3 kWp PV systems is evaluated using various life-cycle impact assessment methods (LCIA methods). In general, the results indicate that, relative to fossil-based energy sources, PV systems have a lower environmental impact, even in regions with a low solar irradiation, and even if only lifetimes of 20 years are assumed. However, when compared to other renewables, the results are not that promising. Wind technology, for example, seems to have an impact that is significantly lower. Biomass has an impact that is on average similar, although one must keep in mind that the sustainability of biomass is still a heavily debated topic.

The energy payback time (EPT) of residential PV systems is less than 5 years in regions with a low solar irradiation, such as Belgium and the United Kingdom and about 2–3 years in regions with a high irradiation (such as the South of Europe and central/south U.S.). As most authors consider lifetimes of PV systems to be at least 20 years, we can conclude that they are indeed a renewable source of electricity, since more energy is produced than consumed. The lifetime energy production is roughly 5 times higher than the lifetime energy consumption in region with little sun. This could rise up to 12 times in sunny regions. Other energy efficiency indicators such as the Fossil Energy Requirement (FER) confirm that PV technology is indeed a renewable technology, with FER's lower than 1 in even the most pessimistic case.

Since minimizing global warming is a frequently mentioned reason for governments to stimulate the use of PV systems, the Global Warming Potential (GWP) of PV was estimated and compared with other energy technologies. We found that the GWP of 1 kWh of PV electricity is on average much lower than that of fossil power plants, similar to biomass, but higher than wind or hydro plants.

The EI 99 method with three perspectives offers a much broader perspective when compared to one-dimensional indicators. However, the EI 99 results differ greatly depending on the used perspective. The Individualist view, that considers fossil fuels to be unlimited and only looks at the short term effects, puts a high weight on the amount of mineral extraction that comes with PV production. According to the Individualist, gas fired power plants are a better option for electricity production than PV systems. The Hierarchist and Egalitarian views, however, do consider the use of fossil fuels to have a negative impact on the well-being of future generations. According to these latter perspectives, PV systems are a better option than gas or coal fired plants.

It is important to stress that the results obtained by the EI'99 method can be misleading, since it involves a weighting step that can be considered quite arbitrary. One can argue that the very high weight that the Individualist gives to mineral extraction is far from what most people would consider logical or rational. It does, however, point out aspects that a one-dimensional indicator does not. The EI 99 method showed us that other aspects then GHG emissions and energy efficiency are also relevant. In the case of PV, the EI 99 method stressed that PV systems are quite mineral-intensive compared with fossil fuel alternatives.

A closer look at the aspect of mineral extraction revealed that about 1.6 grams of iron ore and 1.2 grams of aluminum ore are necessary for the production of 1 kWh of PV electricity, this relatively high amount of minerals that are needed for the production of a PV system should be acknowledged. This fact is not mentioned nor stressed in much of the literature, which mainly focuses on emissions.

This chapter has tried to show that one cannot overstress the fact that LCIA methods should always be evaluated with care. The chosen methodology and assumptions clearly have a major impact on the overall conclusions drawn from a LCA. Most reviews mention only energy efficiency and greenhouse gas parameters or only mention results for PV systems in regions with a high irradiation. Hopefully, this chapter will contribute to a more careful and complete approach toward life cycle analysis in the renewable energy literature.

References

Alsema EA (1998) Energy requirements of thin-film solar cell modules-a review. Renew Sustain Energy Rev 2(4):387–415

Alsema EA (2000) Energy pay-back time and CO2 emissions of PV systems. Prog Photovoltaics Res Appl 8(1):17–25

Alsema EA, de Wild - Scholten MJ (2005) The real environmental impacts of crystalline silicon PV modules: an analysis based on up-to-date manufacturers data. 20th European Photovoltaic Solar Energy Conference and Exhibition, Barcelona, Spain, ECN Solar Energy

Alsema EA, Nieuwlaar E (2000) Energy viability of photovoltaic systems. Energy Policy 28(14):999–1010

Azapagic A (2009) Life cycle assessment as a tool for sustainable management of ecosystem services (26th June 2009). London. http://www.bath.ac.uk/research/seminars/esrc/june09/downloads/azapagic.pdf

Bovea MD, Saura U, Ferrero JL, Giner J (2007) Cradle-to-Gate study of red clay for use in the ceramic industry. Int J Life Cycle Assess 12:439–447

Bravi M, Coppola F, Ciampalini F, Pulselli FM (2007) Comparing renewable energies: estimating area requirement for biodiesel and photovoltaic solar energy. Energy Sustain 105:187–196

Cherubini F, Bird N, Cowie A, Jungmeier G, Schlamadinger B, Woess-Gallasch S (2009) Energy- and greenhouse gas-based LCA of biofuel and bioenergy systems: key issues, ranges and recommendations. Resour Conserv Recycl 53(8):434–447

Cordella M, Tugnoli A, Spadoni G, Santarell F, Zangrando T (2008) LCA of an Italian lager beer. Int J Life Cycle Assess 13(2):133–139

De Wild-Scholten MJ, Alsema EA (2006) Environmental life cycle inventory of crystalline silicon photovoltaic module production. Life-Cycle Anal Tools Green Mater Process Sel 895:59–71

Djomo S, El Kasmioui O, Ceulemans R (2011) Energy and greenhouse gas balance of bioenergy production from poplar and willow: a review. Glob Change Biol Bioenergy 3(3):181–197

Dreyer LC, Niemann AL, Hauschild MZ (2003) Comparison of three different LCIA methods: EDIP97, CML2001 and Eco-indicator 99—Does it matter which one you choose? Int J Life Cycle Assess 8(4):191–200

Duan H, Eugsterb M, Hischierb R, Streicher-Porteb M, Lia J (2009) Life cycle assessment study of a Chinese desktop personal computer. Sci Total Environ 407(5):1755–1764
EPIA (2010) Global Market outlook for photovoltaics until 2014
Frischkneicht R, Jungbluth N (2007) Ecoinvent report n°3. Implementation of life cycle impact assessment methods
Frondel M, Ritter N, Schmidt CM (2008) Germany's solar cell promotion: dark clouds on the horizon. Energy Policy 36:4198–4204
Fthenakis VM, Kim HC, Alsema EA (2008) Emissions from photovoltaic life cycles. Environ Sci Technol 42(6):2168–2174
Goedkoop M, Effting S, Colligno M (2000) PRé consultants. http://teclim.ufba.br/jsf/ecodesign/dsgn0212.pdf
Greenpeace International, EPIA (2008) Solar Generation V. http://www.greenpeace.org/raw/content/international/press/reports/solar-generation-v-2008.pdf
Gürzenich D, Wagner HJ (2004) Cumulative energy demand and cumulative emissions of photovoltaics production in Europe. Energy 29(12–15):2297–2303
Heller M, Keoleian G, Mann M, Volk T (2004) Life cycle energy and environmental benefits of generating electricity from willow biomass. Renew Energy 29(7):1023–1042
Huijbregts MAJ, Rombouts LJA, Hellweg S, Frischknecht R, Hendriks AJ, Van de Meent D, Ragas AMJ, Reijnders L, Struijs J (2006) Is cumulative fossil energy demand a useful indicator for the environmental performance of products? Environ Sci Technol 40(3):641–648
International Energy Agency (2009) Methodology guidelines on life cycle assessment of photovoltaic electricity. http://www.iea-pvps.org/products/download/rep12_09.pdf
International Energy Agency (2010) Technology roadmap solar photovoltaic energy
IPCC (1997) Revised 1996 IPCC guidelines for national greenhouse gas inventories, Intergovernmental Panel on Climate Change, I-III
IPCC (2001) Climate change 2001: the scientific basis, Third assessment report of the IPCC. http://www.grida.no/climate/ipcc_tar/wg1/
IPCC (2011) Special Report renewable energy resources. http://srren.ipcc-wg3.de/report
Ito M, Kato K, Sugihara H, Kichimi T, Song J, Kurokawa K (2003) A preliminary study on potential for very large-scale photovoltaic power generation (VLS-PV) system in the Gobi desert from economic and environmental viewpoints. Sol Energy Mater Sol Cells 75(3–4):507–517
Ito M, Komoto K, Kurokawa K (2010) Life-cycle analyses of very-large scale PV systems using six types of PV modules. Curr Appl Phys 10:S271–S273
Jaramillo P, Griffin W, Matthews H (2007) Comparative life-cycle air emissions of coal, domestic natural gas, LNG, and SNG for electricity generation. Environ Sci Technol 41(17):6290–6296
Jungbluth N (2005) Life cycle assessment of crystalline photovoltaics in the swiss ecoinvent database. Prog Photovoltaics 13(5):429–446
Jungbluth N, Tuchschmid M, ESU-services Ltd., Uster (2007) Ecoinvent report n°6. Part XII: Photovoltaics
Jungbluth N, Dones R, Frischknecht R (2007) Life cycle assessment of photovoltaics: update of the ecoinvent database (trans: Fthenakis VDASN). Materials Research Society, Boston, MA, pp 33–42
Jungbluth N, Dones R, Frischknecht R (2008) Life cycle assessment of photovoltaics: update of the ecoinvent database. In: Fthenakis V, Dillon A, Savage N (eds) Life-cycle analysis for new energy conversion and storage systems. Materials Research Society, Warrendale, pp 33–42
Jungbluth N, Tuchschmid M, de Wild-Scholten M (2008) Life Cycle assessment of photovoltaics: update of ecoinvent data v2.0. http://www.esu-services.ch/cms/fileadmin/download/jungbluth-2008-LCA-PV-web.pdf
Kato K, Murata A, Sakuta K (1997) An evaluation on the life cycle of photovoltaic energy system considering production energy of off-grade silicon. Sol Energy Mater Sol Cells 47(1–4):95–100

Koroneos C, Stylos N, Moussiopoulos N (2006) LCA of multicrystalline silicon photovoltaic systems—Part 1: Present situation and future perspectives. Int J Life Cycle Assess 11(2):129–136

Laleman R, Albrecht J, Dewulf J (2011) Life Cycle Analysis to estimate the environmental impact of residential photovoltaic systems in regions with a low solar irradiation. Renew Sustain Energy Rev 15(1):267–281

Lechon Y, de la Rua C, Saez R (2008) Life cycle environmental impacts of electricity production by solarthermal power plants in Spain. J Solar Energy Eng-Trans ASME 130(2):7

Lenzen M (2010) Current state of development of electricity generating technologies: a literature review. Energies 3:462–591

Lenzen M, Munksgaard J (2002) Energy and CO2 life-cycle analyses of wind turbines—review and applications. Renew Energy 26(3):339–362

Manish S, Pillai IR, Banerjee R (2006) Sustainability analysis of renewables for climate change mitigation. Energy Sustain Dev 10(4):25–36

Martinez E, Sanz F, Pellegrini S, Jimenez E, Blanco J (2009a) Life cycle assessment of a multi-megawatt wind turbine. Renew Energy 34(3):667–673

Martinez E, Sanz F, Pellegrini S, Jimenez E, Blanco J (2009b) Life-cycle assessment of a 2-MW rated power wind turbine: CML method. Int J Life Cycle Assess 14(1):52–63

Mason JE, Fthenakis VM, Hansen T, Kim HC (2006) Energy payback and life-cycle CO2 emissions of the BOS in an optimized 3.5 MW PV installation. Prog Photovoltaics, 14(2):179–190

Mohr N, Meijer A, Huijbregts MAJ, Reijnders L (2009) Environmental impact of thin-film GaInP/GaAs and multicrystalline silicon solar modules produced with solar electricity. Int J Life Cycle Assess 14(3):225–235

Odeh NA, Cockerill TT (2008) Life cycle analysis of UK coal fired power plants. Energy Convers Manage 49(2):212–220

Pacca S, Sivaraman D, Keoleian GA (2007) Parameters affecting the life cycle performance of PV technologies and systems. Energy Policy 35(6):3316–3326

Pehnt M (2006) Dynamic life cycle assessment (LCA) of renewable energy technologies. Renew Energy 31(1):55–71

PRé consultants (2001) The eco-indicator 99 A damage oriented method for Life Cycle Impact Assessment. Methodology report

Raugei M, Frankl P (2009) Life cycle impacts and costs of photovoltaic systems: current state of the art and future outlooks. Energy 34(3):392–399

Raugei M, Bargigli S, Ulgiati S (2007a) Life cycle assessment and energy pay-back time of advanced photovoltaic modules: CdTe and CIS compared to poly-Si. Energy 32(8): 1310–1318

Raugei M, Frankl P, Alsema EA, de Wild-Scholten M, Fthenakis V, Kim HC (2007) Life cycle assessment of present and future photovoltaic systems. Presentation at AIST Symposium "expectations and advanced technologies in Renewable Energy", Chiba, Japan, 11 October 2007. http://www.clca.columbia.edu/papers/Raugei_AIST_FINAL.pdf

REN21 (2011) Renewables 2011 global status report. http://www.ren21.net/Portals/0/documents/Resources/GSR2011_FINAL.pdf

Suri M, Huld TA, Dunlop ED, Ossenbrink HA (2007) Potential of solar electricity generation in the European Union member states and candidate countries. Sol Energy 81(10):1295–1305

The Environment Agency (2009) Minimizing greenhouse gas emissions from biomass energy generation

Varun, Bhat IK, Prakash R (2009a) LCA of renewable energy for electricity generation systems-A review. Renew Sustain Energy Rev 13(5):1067–1073

Varun, Prakash R, Bhat IK (2009b) Energy, economics and environmental impacts of renewable energy systems. Renew Sustain Energy Rev 13(9):2716–2721

Viebahn P, Nitsch J, Fischedick M, Esken A, Schuwer D, Supersberger N, Zuberbuhler U, Edenhofer O (2007) Comparison of carbon capture and storage with renewable energy technologies regarding structural, economic, and ecological aspects in Germany. Int J Greenhouse Gas Control 1(1):121–133

Weisser D (2007) A guide to life-cycle greenhouse gas (GHG) emissions from electric supply technologies. Energy 32(9):1543–1559

Hydropower Life-Cycle Inventories: Methodological Considerations and Results Based on a Brazilian Experience

Gil Anderi da Silva, Flávio de Miranda Ribeiro and Luiz Alexandre Kulay

Abstract Hydropower studies are among the least discussed themes concerning energy systems life-cycle assessments (LCA). This scarcity may be related both to the relevance of this energy source in only a small group of countries and to the difficulty in obtaining the necessary data to conduct those evaluations. The present chapter gives some useful insights regarding specifically the construction of hydropower life-cycle inventories (LCI), aiming to help LCA practitioners involved with environmental evaluation of this source of electricity. Starting with a brief review of the available studies, methodological considerations are given, separated by each LCI methodological step. These recommendations are illustrated with a case study of Itaipu Hydroelectric Power Plant, the largest energy plant in operation worldwide, leading to conclusions and recommendations for further studies. With the present chapter, the authors hope not only to guide, but encourage the conduction of new hydropower plants LCIs, in order to improve the actual LCA databases used for renewable energy sources comparison.

G. A. da Silva (✉)
President of Brazilian Life-Cycle Association, Coordinator of Pollution Prevention Group—University of Sao Paulo (GP2-USP), Sao Paulo, Brazil
e-mail: ganderis@usp.br

F. de Miranda Ribeiro · L. A. Kulay
Researcher at Pollution Prevention Group—University of Sao Paulo (GP2-USP), Sao Paulo, Brazil
e-mail: flv.ribeiro@gmail.com

L. A. Kulay
e-mail: luiz.kulay@poli.usp.br

A. Singh et al. (eds.), *Life Cycle Assessment of Renewable Energy Sources*, Green Energy and Technology, DOI: 10.1007/978-1-4471-5364-1_11,

1 Introduction

With the growing interest on environmental issues, fostered by academic, governmental, and private initiatives, life-cycle assessment (LCA) has become an increasingly prominent tool in evaluating the environmental performance of products and services, presenting benefits in comparison with more conservatives and less holistic approaches.

Notwithstanding their diverse benefits, LCA's ability to support and influence decision-making processes in which the environmental dimension is taken into account is conditioned to some structural factors of the methodology, such as data quality and representativeness. Moreover, the interest of potential LCA users depends on the relationship of cost and duration of the study versus its precision and depth. Those aspects led LCA practitioners to make some investments on databases development, which are constructed, generally, by life-cycle inventories (LCI), which can be understood as a set of information regarding material and energetic flows consumed from the environment, or disposed in it, from an anthropic activity. Among the most requested LCIs are those dealing with energy sources. Every material or product needs energy in different stages of their life cycles. This energy can be obtained from various sources in different locations.

Since the beginning of LCA studies, during the 1970s, energetic LCIs have been developed for systems located in developed countries, in which the electricity generation is usually focused on thermal sources (by fossil fuels like coal, gas, and oil) and nuclear. On this context, very few countries use hydropower as a significant source of electricity, and consequently, LCIs for hydropower are quite scarce. Complementarily, the greatest parts of the existent studies focus on very few aspects—generally greenhouse gas (GHG) emissions from the reservoir flooding.

In terms of LCA, hydropower is not a traditional subject, because most of the total environmental load relies on civil construction, and not on a transformation process that could be described in terms of an input/output balance. Besides due to the intrinsic characteristics of hydropower generation, one study can hardly be representative of any other enterprise or plant in the same segment. Hydropower uses the local land relief to create the dam and associated reservoir. Each project has its own conception and methods, significantly altering the environmental burdens associated with the construction procedures and materials choices, deeply changing both the construction machines used and the materials consumption, modifying their associated life cycle.

This chapter aims to give a contribution for LCA practitioners presenting and discussing methodological considerations about LCI for hydropower in order to better understand the subject. It starts with a brief review of the most prominent studies on the area and some general considerations regarding each specific definition within the "scope definition" step of the LCA methodology, focusing on materials and energy flows (and not only on greenhouse gas emissions). Special attention is given for the requirements for product system modeling. In order to illustrate those arguments, a case study is also presented. It is based on a Brazilian LCI study of Itaipu Plant, which was the largest hydropower installation in the world (Ribeiro and Silva 2010) during the period in which the study was performed. In the topic regarding final recommendations, some suggestions about future and complementary developments related to the topic are made.

2 Review of Available Hydropower LCI's

A remarkable initiative on the area of LCA of energy systems was the workshop conducted by the Organization for Economic Co-operation and Development (OECD) on 1992 (OECD 1992). At that time, a very important discussion was conducted regarding the importance and the difficulties of this kind of study, including the necessity of local data collection to obtain representativeness of the results.

In 1994, a study of Oak Ridge National Laboratory (ORNL) was published (ORNL 1994). This study, contracted by the United States Department of Energy, first quantifies the environmental and social costs for six USA hydropower plants, adopting a life-cycle approach, in order to establish monetary value for them afterward.

Similar to this initiative, and using the ORNL methodology, the European Union launched an ambitious project called ExternE. In an effort involving more than 50 teams from 15 countries during 3 years, the project estimates the externalities of different energy options in Europe, considering their life-cycle. In 1995, the report of hydropower generations was launched (EUROPEAN COMISSION 1995), but it contained very little quantitative information, since the study intended to work with monetary values for the externalities.

Something different from these initiatives was the analysis conducted by Vattenfall AB, a Nordic energy company which decided to carry out an LCA study of its electricity generation options with the purpose to inform its customers regarding their environmental performance. The first study, published on 1996 (Brännstrom-Norberg et al. 1996), describes the product system of hydropower generation, presents elements of scope definitions and bring some inventory data.

A literature survey focusing environmental analysis of hydropower generation systems revealed two important studies: the first published in 1999 (Swedish Environmental Management Council 2002a), and the other on 2002 (Swedish Environmental Management Council 2002b). Both are considered by many experts the most complete LCA for hydropower published up to now. These studies contain definitions, inventory results, impact assessment and conclusions.

The International Energy Agency (IEA) has also two publications where LCA of energy generation is discussed. The first one (IEA 1998), dated from 1998, is a discussion about the environmental implication of renewable energy options. The second one (IEA 2000), from 2000, proposes mitigation measures for hydropower schemes. Nevertheless, these two publications do not perform the LCA in its methodological fullness. In any case, the considerations made by both studies lead to important conclusions for life-cycle impacts of hydropower and also give some typical range of values for environmental burdens.

A remarkable project was published on 2000 by the Scientific Certification System (SCS) of California to the Washington Public Utility District. In this study, a methodological approach based on LCA, called Life-Cycle Stressor Effect Assessment was proposed as an alternative for environmental certification of electricity enterprises (Carringnton 2000). The technique is basically composed by definition and inventory steps, followed by a peculiar impact assessment, which makes the difference from traditional LCA. In order to demonstrate the method, a case study was conducted on Lake Cheelan Hydropower Plant, using some of the definitions from the Vatenfall's case (Vattenfall 1999).

In 2001, some attention was given to the theme of energy systems LCI's, since its relevance as source of environmental burdens had risen up. To foment the discussion on this theme, the United States Environmental Protection Agency (USEPA) promoted a workshop (USEPA 2001), where various experts debated practical questions regarding electricity data for LCI. Based on different studies, the experts concluded that each case demands a proper evaluation. However, depending on the use to which the results of the study are provided, some generic considerations could be made.

One last remarkable reference was published on 2002 in which some researchers adopted LCA results obtained from literature, to make a comparison of various electricity generation options (Gagnon et al. 2002). For each of the electricity sources, the authors compiled emission values, land use indexes, and the energy payback ratios, presenting the results as typical values of environmental performance for hydroelectric projects.

3 Methodology Recommendations for Hydropower Life-Cycle Inventory Scope Definition Modeling

Even though each hydropower represents a different case, it is possible to consolidate some methodological considerations for hydropower LCI development, based on the experience from literature studies, the Itaipu LCI, and also from two conferences on energy LCI's: the first one held in 1992 at Paris, France (OECD 1992); and, the second in 2001 at Cincinnati, United States (USEPA 2001).

The results of these conferences—that are not exclusive to hydropower, but for energy LCI's in general terms—revealed that besides the inherent complexity of electricity generation systems, some specific factors bring additional difficulties to develop electricity LCI. They are related to geographical extension; the large seasonal variation of energy supply; significant differences of energy sources and plant configurations (even when the method of generation is the same); rapid technological evolution; and long-time scenarios (Krewitt 2001). These factors lead LCI experts to establish general conventions.

Although there are some studies using a life-cycle perspective to evaluate hydropower, most of them are limited to a few environmental impacts, mainly GHG emissions and the energy pay-back (ORNL 1994; IEA 2000; Carrignton 2000). Other studies are focused on establishing monetary values to environmental impacts (ORNL 1994; EUROPEAN COMMISSION 1995); and very few are designed to produce a complete LCI dedicated to hydropower (Brännstrom-Norberg et al. 1996; Vattenfall 1999). Evaluating those studies, the most relevant findings are presented in the following.

3.1 Product System Modeling

As already commented, many initiatives have been taken to define a "typical" process for hydropower construction and operation. However, local and technological differences between plants make the results of those attempts inaccurate.

It was verified that all the founded hydropower LCI studies first obtain the inventory of a single plant, generally the one with the higher installed power, and then, the whole system was approximated by an extrapolation of its results. As a second step, those analyses develop LCIs for other plants, combining them in order to obtain an average inventory, based on installed power acting as a weighting factor.

Only one of the found studies went beyond this, using two LCI's for different dams installed on the same river, in order to establish an average between them, representing an improvement over the present situation (Swedish Environemntal Management Council 2002a, 2002b). However, criteria used to perform the average were not deeply described. Studies performed for hydropower generation systems have the tendency to follow the LCA's attributional approach. This focus can be justified due to the fact that it generally makes part of broader projects focused on electricity database development.

3.2 Function and Functional Unit

Even though some hydropower plants use their reservoirs to purposes other than electricity generation, all the LCA consider this as the single function of an energy production unit. Because of this, no allocation was made on any case.

Taking into account both requirements—the electricity generation as the only function of the product system, and the attributional approach of LCA—the Functional Unit (FU) of an LCA study for hydropower generation should necessarily match "*to generate a certain amount of energy*". In general, the amount of energy produced by the system is expressed either in Megawatt–hour (MWh), or Mega joule (MJ). It seems in fact that MWh is preferable because it is more familiar in electricity studies.

3.3 Product System Boundaries Definition

Boundary definition is probably the most controversial step in the development of a hydropower LCI, generally divided into two separate definitions: time boundary and spatial boundary.

Regarding *time boundary*, hydropower LCI's must consider the process that occurs within the power plant, and also the capital investments in materials, and the energy required to build and to operate the dam. This approach is recommended since the construction phase is claimed to be the most relevant source of environmental load.

Hydropower dams are expected to "last forever", if proper maintenance practices would be applied. So, construction impacts should be distributed to all the energy produced by the plant during its whole operative lifetime. Thus, the definition of how long the plant will last, or the amount of energy generated by the system, is decisive in terms of *scope definition*. This question is, maybe, the most controversial aspect of hydropower LCA. The solution commonly adopted by the examined studies was to establish a fixed value.

Originally, this value was determined as being 60 years (Brännstrom-Norberg et al.1996); however, after a review of the time horizons of systems under operation, a 100 years of operation has been considered more adequate (Swedish Environemntal Management Council 2002a, b, 1999; IEA 1998, 2000). As a requirement, all environmental burdens associated with maintenance activities must be included.

Another important definition on this kind of studies refers to the stages to be considered in the whole life cycle. In general, the studies used to separate it on construction and operation of the power plants. Demolition was not considered on any case, and mainly, because it is recognized that the dismantling of a dam could lead to severe environmental impacts, once the local ecosystems are adapted to the lacustrine situation. Thus, hydropower dismantling is not included, except on special cases when public security is endangered.

The *spatial boundaries* of hydropower LCIs are the same for the consulted studies. The refinement of boundaries is considered a key element on hydropower LCI development because of the large amount of mass and energy flows exchanged between the product system and the environment, especially during the construction stage of the project. In fact, all the considered studies had started their LCA screening adopting Vattenfall's boundary definition (Brännstrom-Norberg et al. 1996).

According to that procedure, the initial boundaries included material and energy flows from: the operation of civil work machines; extraction and/or production of construction materials (cement, sand, aggregate, rocks, steel, and diesel); and materials transportation. It were also included within the boundaries the environmental aspects associated with the production and the transportation of the materials necessary to manufacture the most important equipment (as steel, copper, lubricant, etc.), and also aspects related to civil workers transportation, greenhouse gases (GHG) emissions from the flooded biomass, and energy consumption during the operation stage.

Figure 1 illustrates the resultant boundary definition.

Besides, equipment, fuels, and materials life cycles should also be included (Krewitt 2001; Virtanen and Lubkert 1992; Frischknetcht 2001; Curran 2001). Capital goods—e.g., the turbine—are admitted into the boundaries. Generally, they are permanently allocated into the system. On the other hand, equipment temporarily used—such as lifting cranes—should not be included (USEPA 2001; Frischknetcht 2001).

It is also important to consider the environmental and the economic costs of maintenance and replacement of the electromechanical equipment, since the time horizon for the plant is higher than that defined for the equipment. This insertion is even more important due to the general agreement of not including the dismantling of the plant within the boundary; this is in accord with the hypothesis that the equipment should be able to operate in perfect conditions at the end of the time horizon, as recommended in the literature (Virtanen and Lubkert 1992; Setterwall 2001).

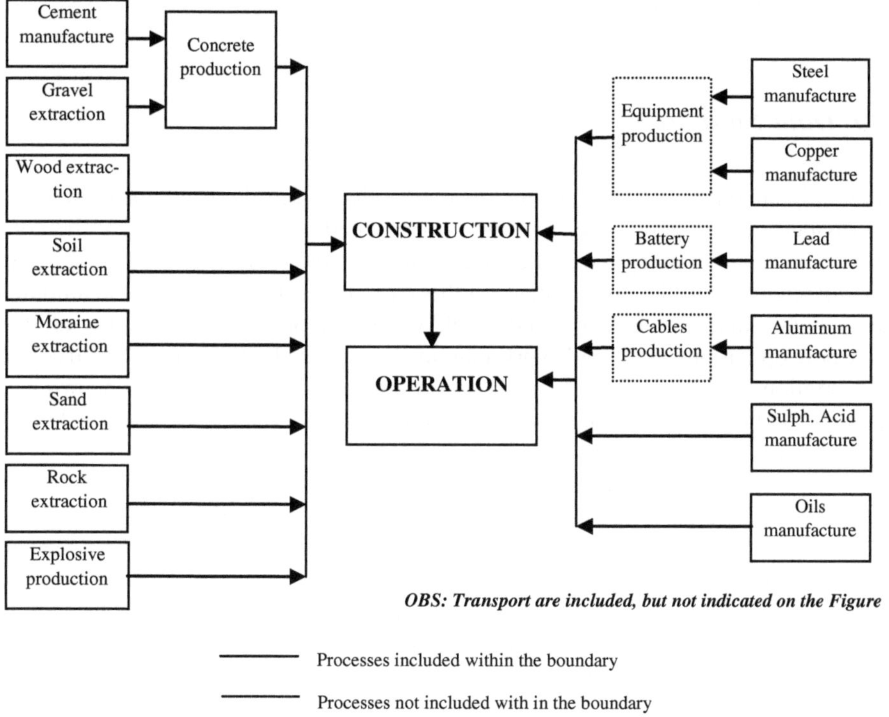

Fig. 1 Vattenfall's boundary definition [adapted from (Brännstrom-Norberg et al. 1996)]

For many authors (OECD 1992; Krewitt 2001; Frischknetcht 2001), the life cycles of construction materials and civil works, as well as emissions of particulate matter from earthworks, and GHG emissions from flooded areas, should not be disregarded. So a complete data collection must be conducted in these life-cycles processes.

Other important aspect of hydropower modeling for LCI purposes is the allocation procedure. Even though in many cases it could be observed a multiplicity of usages for the hydropower reservoir—e.g., recreation, fishing, etc.—it is recommended not to adopt any allocation procedure for this case (Curran 2001).

An important alert regarding environmental impacts associated with land occupation and transformation in hydropower systems should be done. Notwithstanding social impacts of building a dam and the environmental impacts of flooding the same area, only primary flows of mass and energy tend to be considered for this class of LCI. In order to establish an indicator of these social and environmental impacts, it is suggested to include an accounting of areas with potential of occupation and transformation. However, the authors are fully aware that this strategy does not fulfill the need for a broad environmental impact study, which should evaluate the burdens of the hydropower generation such as people displacement, ecosystem modifications, agriculturally productive land loss, and so on. This approach will certainly be one of the main challenges for the LCA community in the near future.

3.4 Data Sources

International experience reveals that the data collection for LCI of hydropower can be divided in two steps. The first one is performed in order to quantify the consumptions and emissions—i.e., inputs and outputs flows—associated with building and maintenance of the power plant. In this case, it is usual to include the amounts of cement, sand, gravel, iron and steel, water, copper, diesel oil, plus all the transport operations, and civil work machines employed either in its construction or operation. This part of the data collection is usually conducted based on primary data, i.e., collecting information directly from the constructor of the dam and other manufacturers and suppliers.

In the second step, the life cycle of those materials, products and services is estimated. In the cases in which there are no databases available, or even if the existing information is not representative, complementary data should be collected in technical literature, and a treatment process of the same information must be performed in order to provide consistency and representativeness for the whole LCI. Access to this information may be difficult or even impossible and even with all these precautions one should take care to avoid incorporating uncertainties to the study.

In all the studies consulted in this literature review, primary data were used on the power plant demand for material and energy—e.g., consumptions of cement, rocks, steel, copper, etc. They were collected in collaboration with the plant contractor. However, data on the life cycle of these inputs were obtained on available LCA databases that do not always correspond to the construction reality. This seems to be a limitation that studies of this class invariably have to face. On lack of data, it is widely recommended to make conservative estimations based on consultation of experts and other adequate and pertinent sources.

4 Case Study: Itaipu Hydropower LCI

4.1 The Brazilian Electricity Matrix

Due to its geographic and geologic conditions, hydroelectricity was strongly emphasized in Brazil from the beginning of its social and industrial development. The existence of large rivers, most of them running on uninhabited areas, led the government to build some of the biggest electricity generation structures in the world. This prevalence is evidenced by the majority of hydropower installed capacity, in comparison with other electricity generation modals. Table 1 presents the Brazilian electricity matrix in 2000, the reference year for the study.

Besides its prevalence over other generation options, Brazilian hydropower plants have in their size another important characteristic. According to records of the Brazilian National Agency for Electricity (ANEEL) (ANEEL 2001), in 2001 the country possessed 130 hydropower plants with more than 30 MW. From this group, exactly twenty-four units counted with 1.0 GW or more of installed power. Among them, the Itaipu Power Plant is the biggest one, with 12.6 GW. Its expressive capability in terms of installed power makes Itaipu nowadays the largest power plant in the world.

In terms of energy production, Itaipu achieved 93.4 GWh in 2000. It is equivalent to burn 434,000 barrels of petroleum per day in a thermoelectric plant and amounted to 23.8 % of the total electricity consumption in Brazil in the same period, or approximately one-third of actual USA hydro-power generation (Itaipu Binacional 2009). The case study of Itaipu power plant was developed as a research project, which aimed to generate a first approximation of Brazilian LCA electricity database. In spite of having a large reservoir Itaipu shows a high efficiency in terms of material and energy use per energy unit produced, due to its

Table 1 Brazilian electricity Matrix 2000 [adapted from ANEEL (2001)]

Electricity generation modals	Number of power plants	Installed power (kW)	% of the total
Hydropower	134	62,063,752	81.7
Thermal	614	11,006,848	14.5
Wind	7	21,200	0.03
Nuclear	2	1,966,000	2.6
Small hydro	313	861,271	1.1
Total	1,070	75, 919,071	

magnitude—economic scale—favorable land relief and high capacity factor with a usage up to 75 %.

4.2 Itaipu Power Plant Description

Located on the Parana River, the seventh most voluminous river in the world, on the border between Brazil and Paraguay, the Itaipu Power Plant was planned during the 1960s, when these countries experienced fast economic growth, which demanded great amounts of electricity to the region. An international agreement was signed in 1966, and after a long period of planning and local condition evaluation, in 1974 a company named Itaipu Binacional was established, in order to manage the construction and operation of the power plant (Itaipu Binacional 1994). The civil works to construct the dam were started in 1975 and finished during 1986. However, the first turbine began to operate some years earlier, in 1984. Some civil works continued till 2004, when the two last turbines were installed—raising the system's capacity to 14.0 GW (Itaipu Binacional 2009).

Itaipu Power Plant takes advantage of the large flow of Parana River and of the local topography, which creates a lake of 1,350 km^2 of area, and volume of more than 29 km^3 resulting in a coefficient of 9.33 MW/km^2—higher than any other Brazilian hydropower plants of its class: Tucurui (1.75 MW/km^2), Furnas (0.90 MW/km^2), Sobradinho (0.25 MW/km^2), or Balbina (0.11 MW/km^2) (Itaipu Binacional 2009).

Dam's construction demanded important civil works as: excavation of the deviation channel (with dimensions of 2 km length, 150 m width, and 90 m depth through basalt rock); earthen and gravel dams (7.7 km of length, and 225 m of average height); and construction of concrete installations (main dam, spillway, power house, etc). To manage all those civil works, two construction sites were installed, on both sides of the river. Both units had centers for silage, gravel production, ice production, woodworks and ironworks, for concrete reinforcing, and concrete mixing and transport, among other services. The development also led to the establishment of a city of over 9,000 houses, lodgings, schools, clubs, markets and stores, services of entertainment, and hospitals, where more than 30,000 civil workers lived along 10 years (Itaipu Binacional 1994). All of these processes and capital goods were considered on the study, as part of the LCI. Detailed description of these structures and information about their environmental burdens can be obtained on the original study (Ribeiro 2003).

4.3 Research Development and Method

After a literature review on hydropower LCA, LCI of Itaipu power plant was developed. The project was carried out along three years and produced a master degree dissertation (Ribeiro 2003). Due to the limited human and financial resources, a literature review was carried out instead of screening procedures for the LCI's development. Initially, a broad search for information about power plants was performed. A visit to the Itaipu power plant made possible the collection of field data and contact with the technical staff from the time of the construction. A former consultant was interviewed, bringing a clear picture of the building site, but also quantitative data, and relevant information about the venture. From these procedures, it was possible to establish the scope of the study.

The project's second step comprised the application of data collection procedure. The activity was divided on two parts: obtaining information on consumptions of materials and energy; and, afterwards, data able to express the most important environmental burdens of the life-cycle processes of this material and energy inputs were estimated. The first of these steps used information previously collected, from specific literature on the dam. On the other hand, second step demanded a new literature review, from which data of the inputs life-cycle processes were collected from the time of the construction in the 1970s. Subsequently, both estimations were joined with the help of a software proper to support LCA studies. As the objective of the study was to develop a database, no impact assessment was planned. The complete set of hypothesis and assumptions, data, and calculations, results and conclusions, apart from the LCI itself, are available for consultation (Ribeiro and Silva 2010). Some methodological aspects were also previously published in a conference paper by the authors (Ribeiro and Silva 2004).

4.4 Life-Cycle Inventory of Itaipu Hydropower Plant

As general guideline, the development of the LCI for Itaipu Power Plant followed the ISO standards for LCA (ISO 1997, 1998). To this framework, all previous methodological recommendations were added . In order to fulfill the formal structure proposed by ISO, the study outcomes are presented according to the normative steps on the following.

4.4.1 Goal and Scope Definitions

In terms of goal definition, the study had intended to develop a representative LCI for hydropower generation in Brazil. From this effort, it was expected to fulfill a gap in LCA studies developed in the country, where international databases were used in order to express environmental burdens associated to electricity. LCA practitioners are the target readership of the study.

According to ISO statements, the first step in terms of scope definition embraces the establishment of the function and the functional unit for the study. As recommended by international experience, the function considered for the power plant is "the generation of electricity". In order to measure this function, a Functional Unit (FU) of "generation of 1 MWh of electricity" was selected.

Operationally, the process of boundary definition was a major challenge. The core of the product system is the Itaipu Power Plant itself. However, the product system's boundaries also encompassed processes involved in delivery of materials for the construction and operation of the plant—such as cement, steel, copper, diesel oil, and lubricants. Moreover, it should be highlighted that the LCI addresses the construction and operation of the plant, but not the dam dismantling. So, in terms of temporal boundaries, the first statement was the division of the power plant life-cycle into two phases: construction and operation. The construction phase started with the beginning of the civil works. The operation step lasts for a 100-year period—according to temporal coverage originally defined—counted from 1984. For estimation and data collection of environmental loads of consumed materials, the base year of 1977 was selected. The greatest part of the materials was acquired on that period.

The spatial boundaries of the product system should include not only the dam and its related structures, but also the residential and construction site, which demanded energy and materials to be built. However, after a first screening, it was possible to verify that the amount of consumed materials and civil works needed to construct the residential site was not significant in comparison with the whole civil work conducted, and thus it was excluded (Ribeiro 2003). Moreover, it should be considered that all of these residences and support structures were incorporated afterwards into the city. On the other hand, the construction site equipment and structures that were fully dedicated to the power plant construction had its environmental burdens included within the LCI. Apart from consumption and emissions occurred during the operation phase, it was decided to consider also the consumption of steel, concrete and electricity to build the dam, and their life cycles.

The definition regarding inclusion of other unit processes within the system's boundaries was much more complex. This adjustment was initially made by taking a boundary definition from international studies (Brännstrom-Norberg et al. 1996; Swedish Environemntal Management Council 2002a, b, 1999; Vattenfall 1999), which was modified through a screening process, based on an extensive literature review, complemented by a personal interview with the professionals who took part in Itaipu's construction.

Figure 2 presents the final boundary for the product system.

No *allocation procedure* was conducted, due to the consideration of only one product and one function for the dam. Also, considering that only the "Definition" and "Inventory" phases of LCA were conducted, no impact assessment criteria definition was made.

4.4.2 Life-Cycle Inventory

Data collection

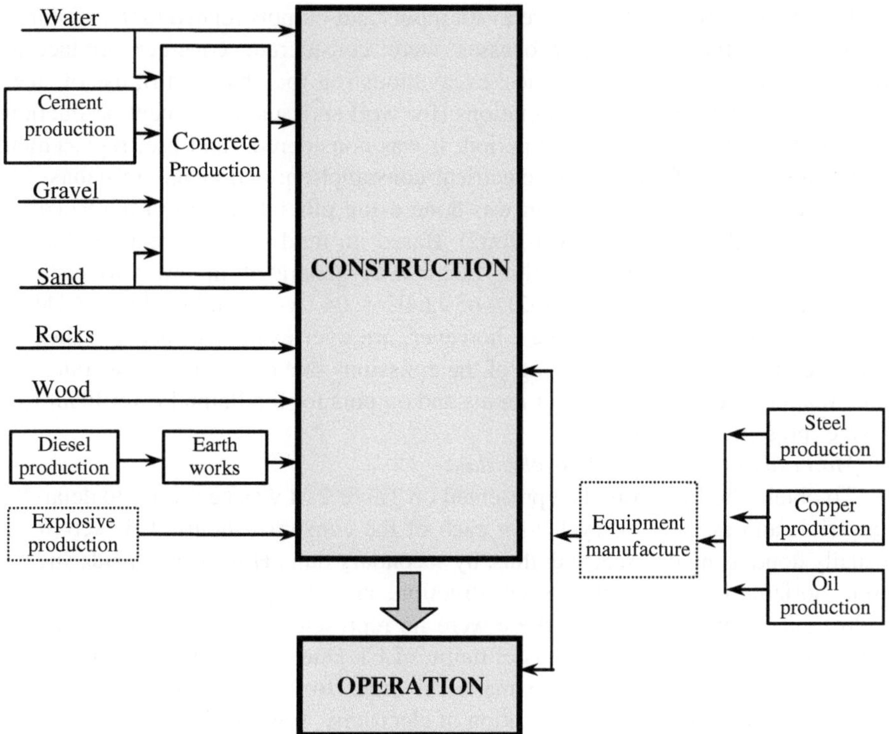

OBS: Transport operations were included, but are not indicated on the figure; processes represented with dotted lines are not included;

Fig. 2 Simplified product system boundary for Itaipu power plant LCI (Ribeiro 2003)

Itaipu's LCI data collection was performed in two sequential steps: Itaipu Power Plant data; and material and energy life-cycles data. Both are described in more details in the following paragraphs.

Itaipu Power Plant data

The first part of data collection was the adjustment of the product system model from international studies, to Itaipu's reality.

In order to obtain information for the essential considerations, personal interviews were conducted during a field trip to the dam, with some civil work specialists who had worked in Itaipu's project and construction. The information collected by interview was completed by an extensive literature review carried out on the first part of the study. This included power plant construction reports (from 1977 to 2003); lists of materials and equipment; original blueprints; reports from the different engineering companies that worked in the power plant project and construction; and technical articles published on a monthly civil engineering magazine (editions from 1976 to 1983). All of these data, together with the previous collected information, made it possible to conduct the necessary calculations, summarized in the annexes of the original study (Ribeiro 2003).

In order to estimate the most relevant inputs and outputs related to the venture's construction, the following processes were considered: concrete production; operation of construction machines; excavations (on rock and earth); use of steel; truck transportation; bus transportations (for workers); and equipment production. For the operation of a 100-year period, it was considered: land use; replacement and maintenance of equipment; electrical consumption; and GHG emissions.

The GHG emission estimation was done using official data from the Brazilian government (Santos 2000; MCT 2002). Based on field measurements performed on Itaipu's reservoir, an emission factor was generated in order to estimate emissions of carbon dioxide (CO2)—of 9.64E + 04 t/yr—, and methane (CH4)—of 1.18E + 04 t/yr. Those values, however, are over-estimated, because they do not take into account the decaying of the emissions rate over time. Data collection and treatment resulted in a list of inputs and outputs for the Itaipu Power Plant Life Cycle, presented in Table 2.

Materials and energy life-cycles data

To obtain the LCI from data presented on Table 2, it was necessary to detail the materials and energy life cycles for each of the consumed items. This activity is usually done using datasets and, thus, by secondary data. However, considering the time horizon for the venture's construction, and the peculiarities of Brazilian industrial system at that time, there were no representative databases for the different materials and energy inputs of Itaipu's LCI. Due to this limitation, additional data collection procedures were made, in order to estimate the environmental burdens of the processes of: generation of electricity; transportation (by truck, bus, train, and ship); and the production of cement, steel, copper, diesel oil, and lubricant oil.

For estimation of the life-cycle environmental burdens of the consumed materials, 1977 was considered to be the reference year for data collection. In this period, the greatest part of the materials was acquired. Data sources varied from primary data—from the factories that produced to Itaipu Power Plant—to international emission factors and industrial surveys of Brazilian industry, i.e., secondary data. The main data sources are summarized below:

- *Diesel and lubricant oil*: a previous adaptation from international LCA databases was used to obtain the environmental load of diesel fuel and lubricant;
- *Electricity*: the Brazilian electricity mix that prevailed in 1977 was modeled by an international LCI dataset for thermal generation;
- *Transport*: different transportation modals were used in the model: trucks of about 30 t load of capacity; buses; trains; petroleum and cargo ships. Emission factors were applied for road transportation considering 1977 technologies, and international datasets were used for ship displacements;
- *Cement*: after a literature review, it was concluded that the Brazilian cement production in 1977 could not be estimated using recent LCA databases. So, an independent data collection was conducted. It was based on data from manufacturers, technical reports, and sector analysis from official research agencies,

Table 2 Itaipu power plant life-cycle inputs and outputs [FU = 1 MWh) (adapted from (Ribeiro 2003)]

	Total (100 years)				Total (/FU)			
	Unit	Construction	Operation	Total	Unit	Construction	Operation	Total
Natural resource consumption								
Rock	m^3	3.20E+07	0.00E+00	3.20E+07	m^3/UF	3.59E−03	0.00E+00	3.59E−03
Earth	m^3	2.36E+07	0.00E+00	2.36E+07	m^3/UF	2.65E−03	0.00E+00	2.65E−03
Water	t	2.16E+06	0.00E+00	2.16E+06	t/UF	2.42E−04	0.00E+00	2.42E−04
Sand	t	3.68E+06	0.00E+00	3.68E+06	t/UF	4.12E−04	0.00E+00	4.12E−04
Wood	t	5.27E+03	0.00E+00	5.27E+03	t/UF	5.91E−07	0.00E+00	5.91E−07
Products consumption								
Electricity	GJ	9.50E+06	1.70E+03	9.50E+06	GJ/UF	1.07E−03	1.91E−07	1.07E−03
Cement	t	2.48E+06	0	2.48E+06	t/UF	2.78E−04	0.00E+00	2.78E−04
Ashes	t	3.00E+05	0	3.00E+05	t/UF	3.36E−05	0.00E+00	3.36E−05
Steel	t	7.97E+05	4.69E+05	1.27E+06	t/UF	8.94E−05	5.26E−05	1.42E−04
Copper	t	5.48E+03	1.63E+04	2.18E+04	t/UF	6.15E−07	1.83E−06	2.45E−06
Diesel	t	2.95E+05	0.00E+00	2.95E+05	t/UF	3.31E−05	0.00E+00	3.31E−05
Transf. oil	m^3	3.37E+03	0.00E+00	3.37E+03	m^3/UF	3.78E−07	0.00E+00	3.78E−07
Lubic. oil	m^3	1.50E+03	0.00E+00	1.50E+03	m^3/UF	1.68E−07	0.00E+00	1.68E−07
Transportation								
Truck	t.km	3.90E+09	0.00E+00	3.90E+09	t.km/UF	4.37E−01	0.00E+00	4.37E−01
Bus	km	1.50E+07	0.00E+00	1.50E+07	km/UF	1.68E−03	0.00E+00	1.68E−03
Air emissions								
Aldeides	t	3.03E+02	0	3.03E+02	t/UF	3.40E−08	0.00E+00	3.40E−08
CH_4	t	0.00E+00	1.18E+06	1.18E+06	t/UF	0.00E+00	1.32E−04	1.32E−04
CO	t	4.62E+03	0	4.62E+03	t/UF	5.18E−07	0.00E+00	5.18E−07
CO_2	t	9.26E+05	9.64E+06	1.06E+07	t/UF	1.04E−04	1.08E−03	1.19E−03
HC	t	8.45E+02	0	8.45E+02	t/UF	9.48E−08	0.00E+00	9.48E−08
PM	t	8.74E+02	0	8.74E+02	t/UF	9.80E−08	0.00E+00	9.80E−08
NOx	t	1.26E+04	0	1.26E+04	t/UF	1.41E−06	0.00E+00	1.41E−06
SOx	t	1.30E+03	0	1.30E+03	t/UF	1.46E−07	0.00E+00	1.46E−07
Non-material burdens								
Land use	km^2	0.00E+00	1.35E+03	1.35E+03	km^2/UF	0.00E+00	1.51E−07	1.51E−07

in order to produce an inventory minimally representative for cement production at that period;

- *Steel*: once again, none of the international steel production LCA databases were suitable for representing the Brazilian production in 1977, mainly due to the large use of charcoal. In this context and analogous to the cement case another independent LCI based again on data from manufacturers, technical reports and sector analysis from official research agencies was developed;
- *Copper*: all of its consumption was estimated using international databases, considering only imports from Chile.

Calculation procedures and data presentation

Once representative LCIs for Itaipu Power Plant were obtained, either a calculation procedure, or some adjustments were done in order to improve LCI's clarity. In an initial analysis, it was noticed that different terminologies were used for similar burdens. This required nomenclature adequacy, mainly for air emissions accountability: SO_2 emissions were expressed as SO_x; the same was done with NO and NO_2 emissions, consolidated as NO_x. Solid emissions, independently of the particles size were consolidated as "particulate matter". Besides, due to the different solid waste flows, all of them were indicated as "solid waste". Energy resources were expressed as energetic content through the High Heat Value (HHV). At last, consumption of earth and rock were converted from volumetric to mass units. Table 3 presents the complete consolidated LCI for Itaipu Power Plant.

Simplified Itaipu Power Plant LCI

Even though a complete LCI was the final objective of the study, an interpretative analysis of the dataset was conducted. It starts with an evaluation of the most significant environmental burdens of the LCI. Following the ISO standard (ISO 1998) a cutoff criteria was applied, in order to exclude material and energy consumptions and emissions with cumulative contribution of less than 1.0 %. This procedure generated a "Simplified LCI". It made possible to conclude that just six elementary flows—water, rock (basalt), earth, calcite, sand, and iron ore—were, in fact, significant. Regarding to energy, only four natural resources contributed with more than 1.0 % of the total: crude oil, coal, hydropower, and natural gas. In terms of atmospheric emissions CO_2, CH_4, CO, Particulate Matter, SO_x and NO_x contributed amounted together 99.9 % of the totals for this class. In order to complete the simplified LCI, "land use" was included, due to the importance of this parameter to the environmental and social impacts of hydropower. Table 4 presents the simplified LCI for Itaipu Power Plant, with environmental loads expressed per functional unit.

This simplified LCI was carried out according to clear and adequate methodological procedures and was based on international experiences (OEDC 1992; ORNL 1994; European Commission 1995; Brännstrom-Norberg et al. 1996; Swedish Management Council Environmental 1999, 2002a, 2002b; IEA 1998,

Table 3 Consolidated LCI for Itaipu power plant (Ribeiro 2003)

Environmental loads	Unit (/ UF = 1MWh)	Total	Construction	Operation
Material resources consumption				
Water	kg	8.90E+00	7.16E+00	1.74E+00
Air	kg	1.24E−05	7.86E−06	4.55E−06
Sand	kg	4.12E−01	4.12E−01	X
Clay	kg	8.34E−02	8.34E−02	4.15E−07
Basalt	kg	5.47E+00	5.47E+00	X
Bauxite	kg	4.00E−05	3.56E−05	4.36E−06
Calcite	kg	4.86E−01	4.70E−01	1.62E−02
Dolomite	kg	9.69E−04	6.14E−04	3.55E−04
Fluorspar	kg	6.63E−04	4.20E−04	2.43E−04
Gipsite	kg	8.33E−03	8.33E−03	X
Wood	kg	1.44E−01	9.13E−02	5.25E−02
Copper ore	kg	2.01E−03	5.47E−04	1.46E−03
Iron ore	kg	1.66E−01	1.05E−01	6.10E−02
Manganese ore	kg	1.27E−03	8.04E−04	4.65E−04
Quartzite	kg	2.54E−03	1.61E−03	9.31E−04
Salt	kg	1.90E−05	1.69E−05	2.07E−06
Iron scrap	kg	6.53E−02	4.14E−02	2.39E−02
Copper scrap	kg	2.93E−04	7.99E−05	2.13E−04
Earth	kg	3.05E+00	3.05E+00	X
Energy resources consumption				
Coal	MJ	1.65335	1.038186	0.615164
Energy (inesp.)	MJ	4.33E−02	3.86E−02	4.73E−03
Uranium energy	MJ	9.50E−04	8.46E−04	1.04E−04
Hydroelectricity	MJ	1.46E+00	1.36E+00	9.61E−02
Natural gas	MJ	0.285592	0.23958	0.045512
Crude oil	MJ	5.4753	4.7862	0.69368
Air emissions				
1.3 Butadiene	kg	1.40E−07	1.28E−07	1.17E−08
Aldehyde	kg	3.40E−05	3.40E−05	X
Ammonia	kg	2.16E−07	1.36E−07	8.06E−08
Benzene	kg	1.64E−05	1.60E−05	4.17E−07
Benzopyrene	kg	2.42E−09	1.52E−09	9.00E−10
CaO	Kg	1.21E−03	7.69E−04	4.45E−04
CH_4	kg	1.32E−01	4.50E−04	1.32E−01
Pb	kg	1.65E−11	1.04E−11	6.10E−12
CO	kg	1.12E−01	7.12E−02	4.08E−02
CO_2	kg	1.56E+00	4.45E−01	1.12E+00
VOC	kg	3.74E−04	2.75E−04	9.87E−05
Etene	kg	3.70E−05	2.34E−05	1.36E−05
F_2	kg	7.65E−08	4.80E−08	2.85E−08
FeO	kg	8.18E−04	5.18E−04	3.00E−04
Fluoretene	kg	2.42E−08	1.52E−08	9.00E−09
Fluoride	kg	1.72E−06	1.10E−06	6.19E−07
H_2	kg	1.83E−04	1.16E−04	6.70E−05

(continued)

Table 3 (continued)

Environmental loads	Unit (/ UF = 1MWh)	Total	Construction	Operation
H_2S	kg	1.10E−05	6.98E−06	4.07E−06
HCl	kg	4.97E−08	4.42E−08	5.42E−09
Hydrocarbons	kg	4.21E−04	3.74E−04	4.72E−05
Inespec.	Kg	1.03E−05	6.47E−06	3.84E−06
Particulate matter	kg	2.49E−02	1.96E−02	5.32E−03
Hg	kg	8.27E−15	5.22E−15	3.05E−15
Heavy metals	kg	9.50E−08	8.46E−08	1.04E−08
Methyl mercaptan	Kg	1.35E−09	1.20E−09	1.47E−10
N_2O	kg	5.89E−07	5.77E−07	1.21E−08
NO_x	kg	2.98E−03	2.71E−03	2.70E−04
Crude oil	kg	1.79E−05	1.60E−05	1.95E−06
SOx	kg	3.76E−03	2.10E−03	1.66E−03
Toluene	kg	1.05E−07	6.57E−08	3.90E−08
Xilene	kg	1.13E−07	7.07E−08	4.20E−08
Water discharges				
Acetic acid	kg	5.80E−03	3.68E−03	2.13E−03
Acetaldehyde	kg	8.09E−05	5.12E−05	2.96E−05
Acetone	kg	1.50E−04	9.52E−05	5.51E−05
Acid (H+)	kg	2.85E−06	2.54E−06	3.11E−07
Tar	kg	1.39E−02	8.79E−03	5.08E−03
Ammoniac	kg	5.19E−08	3.29E−08	1.90E−08
Lead	kg	4.03E−09	2.53E−09	1.50E−09
Cyanide	kg	3.18E−07	2.02E−07	1.17E−07
Cl-	kg	3.75E−06	3.34E−06	4.09E−07
Cu	kg	1.21E−09	7.58E−10	4.50E−10
$Chromium^{3+}$	kg	1.47E−09	9.32E−10	5.39E−10
COD	kg	2.46E−07	2.19E−07	2.68E−08
Phenol	kg	5.37E−07	3.41E−07	1.95E−07
Fe	kg	5.38E−06	3.41E−06	1.97E−06
Fluoride	kg	2.23E−06	1.41E−06	8.16E−07
H_2	kg	1.54E−07	1.37E−07	1.68E−08
Hexane	kg	5.36E−08	3.40E−08	1.97E−08
Hydrocarbons	kg	2.10E−06	1.82E−06	2.82E−07
General inorganic	kg	6.36E−03	1.73E−03	4.63E−03
Metalic ions	kg	5.00E−07	4.45E−07	5.45E−08
Manganese	kg	1.15E−07	7.28E−08	4.21E−08
Hg	kg	2.42E−10	1.52E−10	9.00E−11
Metanol	kg	2.10E−03	1.33E−03	7.71E−04
Methyl acetate	kg	1.85E−04	1.17E−04	6.78E−05
N total	kg	4.03E−08	2.53E−08	1.50E−08
NH_3	kg	4.76E−06	3.02E−06	1.75E−06
Nitrate	kg	4.75E−06	3.01E−06	1.74E−06
Oil	kg	1.34E−05	8.67E−06	4.69E−06
Dissolved organic	kg	1.90E−06	1.69E−06	2.07E−07
PAH	kg	4.03E−10	2.53E−10	1.50E−10

(continued)

Table 3 (continued)

Environmental loads	Unit (/ UF = 1MWh)	Total	Construction	Operation
Crude oil	kg	2.22E−06	1.98E−06	2.43E−07
Sodium	kg	1.88E−06	1.67E−06	2.05E−07
Dissolved solids	kg	6.30E−06	3.99E−06	2.31E−06
Suspended solids	kg	1.92E−05	1.21E−05	7.03E−06
Dissolved subst.	kg	9.50E−07	8.46E−07	1.04E−07
Suspended subst.	kg	6.65E−06	5.93E−06	7.26E−07
Sulfite	kg	3.56E−07	2.25E−07	1.30E−07
Zn	kg	1.78E−05	1.13E−05	6.53E−06
Other				
Solid waste	kg	3.10E−01	8.75E−02	2.23E−01
Heat loss (air)	MJ	1.24E−02	9.44E−03	2.91E−03
Heat loss (water)	MJ	6.35E−02	4.85E−02	1.50E−02
Land use	m^2	1.52E−01	6.64E−04	1.52E−01

2000; Carrignton 2000; Vattenfall 1999; USEPA 2001; Gagnon 2002. However, from detailed LCIA (ISO 1997), it would be possible to measure the quality losses imposed by the process of exclusion to the simplified version.

4.5 Results Discussion

4.5.1 Environmental Hotspots

An additional purpose of this study consisted to analyze the environmental hotspots of the Itaipu LCI. Figure 3 shows a bar graph with the contributions of the main processes in terms of environmental burdens.

Figure 3 evidences that the "*Reservoir filling*"—with emissions of CO_2 and CH_4, and land use; the '*Steel life-cycle*'—with water and energy consumption, and emissions of CO, particulate matter, SO_x and NO_x; '*Cement life-cycle*'—with emissions of CO_2 and particulate, and water and energy consumption; and the '*Operation of civil construction machines*'—with diesel consumption; and NOx emissions—are the most important contributors to the environmental profile of Itaipu.

4.5.2 Comparison with international results

A compilation of the LCI results from different power plants selected on literature showed a wide variability. This phenomenon seems to have different origins, mainly on methodological assumptions, and on differences on the constructive

Table 4 Simplified LCI for Itaipu power plant (UF = 1 MWh) (Ribeiro 2003)

	Unit (/FU)	Total	Construction	Operation (100-year)
Material resources consumption				
Water	kg	8.90E+00	7.16E+00	1.74E+00
Sand	kg	4.12E−01	4.12E−01	x
Basalt	kg	5.47E+00	5.47E+00	x
Calcite	kg	4.86E−01	4.70E−01	1.62E−02
Iron Ore	kg	1.66E−01	1.05E−01	6.10E−02
Earth	kg	3.05E+00	3.05E+00	x
Energetic resources consumption				
Coal	MJ	1.65335	1.038186	0.615164
Hydroelectricity	MJ	1.46E+00	1.36E+00	9.61E−02
Natural gas	MJ	0.285592	0.23958	0.045512
Oil	MJ	5.4753	4.7862	0.69368
Atmospheric emissions				
CH_4	kg	1.32E−01	4.50E−04	1.32E−01
CO	kg	1.12E−01	7.12E−02	4.08E−02
CO_2	kg	1.56E+00	4.45E−01	1.12E+00
Particulate	kg	2.49E−02	1.96E−02	5.32E − 03
NOx	kg	2.98E−03	2.71E−03	2.70E−04
SOx	kg	3.76E−03	2.10E−03	1.66E−03
Nonmaterial burdens				
Land use	m^2	1.52E−01	6.64E−04	1.52E−01

aspects—size, capability, age, technological level, construction material, and others.

In order to check Itaipu LCI values together with other hydropower studies, Table 5 presents a comparison with some international experiences.

These data evidence that even with a large variability between the different studies most of the results are inside the same order of magnitude. However, in the comparison of the results for each environmental burden, it is possible to identify a clear environmental performance advantage of Itaipu among the other hydropower plants. This prevalence can be explained by the economic scale of the venture—due to the great capacity factor, the plant generates a huge amount of energy for each unit of capital investment. This assumption must be taken with care, since the studies used in this comparison were developed with different objectives and methodologies and, most of them, were not full LCIs. The environmental gains of Itaipu are still more evident on the energy pay-back, parameter which measures the ratio between the energy output and input.

4.5.3 Time horizon sensitiveness

Time horizon is one of the important definitions of the study once the environmental burdens of construction were calculated proportionally to the energy

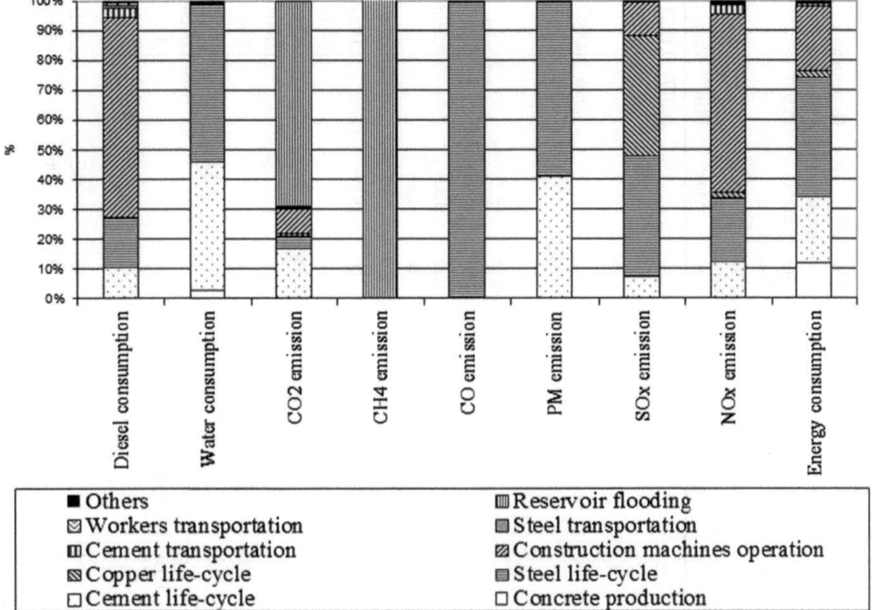

Fig. 3 Process contributions to the most important environmental burdens (Ribeiro and Silva 2010)

generated on the whole period. As mentioned before, this parameter was defined as 100 years. Nevertheless, some simulations were developed changing the time horizon to 60 years, and afterwards to 200 years.

The results of these tests not only had confirmed the high influence of the time horizon over the environmental profile but also indicated that if the LCI results do not vary so much, it was because the greatest part of the environmental loads comes from a life-cycle phase that is not time-dependent. The obvious conclusion was that the dam construction occurs on this phase.

4.5.4 Comments regarding the magnitude of the LCI results

A screening on the LCI results showed that the most part of the environmental burdens would not be significant, if a cutoff criteria had been applied. This procedure was not adopted on the study due to its objective and resources restrictions, but it is recommended on LCAs which intends to use this LCI as an electricity database.

A comparison with LCIs of other electric generation options pointed out that hydropower seems to present a better environmental performance than others, such as natural gas thermal plants, but one should consider that there are plenty of social

Table 5 LCI results comparison with international experience (Ribeiro 2003)

		ITAIPU LCI	(Gagnon et al. 2002)		(IEA 2000)	(IEA 2000)		(ORNL 1994)		(Vattenfall 1999)	(Swedish Environemntal Management Council 1999, 2002a, b)		(European Commission 1995)	
			Min	Max		Min	Max	Min	Max		Min	Max	Min	Max
Steel	kg/ FU	0,142		–	0,056		0,866		0,3	0,08			0,077	0,1221
Structural steel	kg/ FU	0,054					0,111		0,028					0,0275
Cement	kg/ FU	0,278		–	0,556					0,13				
Concrete	m^3/ FU	0,0014			0,002	0,0017	0,0044		0,002				0,006	0,0087
Copper ore	kg/ FU	0,0024		–			0,0003		1E-04		0,0025			
Iron ore	kg/ FU	0,166		0,06						0,17	0,15	0,3		
Crude oil	kg/ FU	0,12		0,11						–	0,07	0,23		
Sand	kg/ FU	0,412		3,62						–	3,25	13,8		0,0808
Calcite	kg/ FU	0,486		1,12						–	0,44	0,56		
Rock	kg/ FU	5,47		–						–	35,9	53,5		0,014
Earth	kg/ FU	3,05		0,12						–	20,8	30,7		0,1937
Diesel oil	kg/ FU	0,057			0,083									0,0972
Water	kg/ FU	8,9		12,04						–	2,13	2,29		
NOx emissions	kg/ FU	0,003		0,005	0,042	0,003	0,0448		0,023	0,0059				0,0061
SO_x emissions	kg/ FU	0,0037	0,007	0,011	0,1	0,005	0,0161	0,001	0,008	0,0015				0,0006
CO emissions	kg/ FU	0,112		0,005						0,0046				0,002
MP emissions	kg/ FU	0,0249		0,008	0,005		0,0031		0,002	0,0007				0,0004
CO_2 emissions	kg/ FU	1,56	0,01	1,59	5,9	1	5,1	0,01	2,48	0,71				0,31
GHG emissions	kg CO_2eq/ FU	4,33	15	–		48	2	2		–	4	11	0,93	
Land Use	km^2/ FU	1,52E-07	1,52E-04	–		1,52E-05	2,00E-06	1,00E-06		3,58E-07	3,59E-07	1,08E-06		3,55E-07
Energy Pay-Back	MJ/ MJ	403,6	267	–		260	48	205		–				

and environmental burdens and impacts that are not included on this study, like people displacement, ecosystem substitution, etc., impacts that certainly have high social and environmental importance.

4.5.5 Most relevant limitations of the study

A detailed analysis of the final version of Itaipu LCI reveals that the most important limitation of the study is related to the LCA itself as an environmental evaluation tool for hydropower projects, mainly in the Brazilian condition. As stated earlier in this document, the LCI process was not expected to cover all the environmental or social impacts of hydropower—e.g. impacts on population displacement; local climate; ecosystem modifications; hydrology and so on—for which international impact assessment studies and methodologies are available. The scope of the present LCI is limited to the environmental loads associated to the life cycle of the power plant in terms of exchanges of mass and energy with the environment, and the impacts caused by land occupation and transformation.

Regarding to data quality, the most significant uncertainties are related to GHG emission. It was assumed that GHG emissions over the whole area of the reservoir would be the same during all the life time of the venture. This assumption, taken in previous studies (Coltro 2003), is surely a rough estimation; on the other hand, until the moment, there are no reliable models able to estimate such losses consistently with the conditions in which they occur. Still in this same approach, some other suggestions to improve data quality could be:

- Update the estimation of the total energy production during the 100-year operation period, from the start of the operation of Itaipu in 1984 up to the date of the study;
- Conduct a more detailed data collection on the use of chemical products for power plant maintenance;
- Better modeling the air emissions from civil work machines. The calculations were performed using USEPA emission factors, which were applied to the total amount of consumed fuel. The power of the engines was used as criterion for the allocation. Furthermore, it was assumed that all machines had worked during the same period of time;
- Seek for more homogenous sources of data.

Finally, other relevant limitation was the impossibility to carry on another data collection in order to refine the boundaries of product system. Some processes were excluded from the boundaries based only on the international experience, which not always correspond to the Brazilian situation. So, processes like electromechanical equipment fabrication, explosives production and use, other materials life cycles, and so on, were not investigated.

5 Conclusion and Recommendations

The Itaipu LCI case study brought some information on scope definition and inventory development that could be useful to future LCA practitioners. This topic draws conclusions and presents some recommendations.

Regarding the proposed objective, it can be considered that the study achieved the objective to develop a hydropower LCI representative of Brazilian conditions. This result is useful as a basis for the extrapolation of the whole Brazilian grid, following international practice. Nevertheless, the values presented must be taken with care, given that each hydropower scheme is different from another. Moreover, it should be noted that the evaluation of environmental and social impacts was not conducted in this study, given the original objective developing a LCI.

After evaluating the results, it is possible to verify that hydropower LCI results are very sensitive to the time horizon definition, which leads the authors to conclude that the huge contributions for environmental load of hydropower, derives from the construction phase, confirming a general assumption. Also, it was possible to find that the most important sources of environmental burdens are as follows: reservoir filling; steel life-cycle; cement life-cycle; and the operation of civil work machines.

Regarding the results, it was possible to conclude that some level of cutoff criteria should be applied, in order to ensure that environmental burdens that are not representative are excluded from the LCI, avoiding misunderstandings. Moreover, an extra care must be taken on data collection, in order to ensure quality in terms of consistency and confidence between the different data sources.

A first conclusion in terms of the magnitude of the values refers to the relevance of the time horizon and, more than this, the mathematical evidence of the construction phase predominance on the life cycle, as a major source of environmental load to the LCI. Another important conclusion on this topic is the verification of the position of Itaipu results within the range of international values for hydropower life-cycle consumption and emissions as very favorable, which indicates, with high probability, the inverse proportion between the magnitude of the hydropower plant and the environmental load per unit of generated electricity.

In the absence of the impact analysis step of the LCA methodology, a discussion based on hot spots was performed in order to verify the contribution of each part of the product system to the whole LCI, as already described.

The analysis—based on Itaipu results and in international experience—reveals that the most important unit processes of the product system are the following: reservoir filling, steel life cycle, cement life cycle and civil work machines operation, besides the energy consumption on the building site operation, specifically on concrete production. In future studies, these processes are not recommended to be excluded, mainly when considering the electromechanical equipment manufacture, maintenance and substitution within 100 years of operation.

A last observation regarding process is that compared with the civil works machine operation, the unit process of transports included on the materials life cycle had proved to be not significant. This conclusion must be observed carefully, owing to the huge amount of civil work on Itaipu case, which should not be directly repeated in other hydropower schemes.

The initiative was considered successful once it opens the perspective to make more precise Brazilian LCIs, which are actually on development. It is believed that the availability of representative datasets for Brazilian conditions will improve readiness, quality, and accuracy to LCA studies, all of them, indispensable conditions to incorporate this important tool into the Brazilian environmental management community.

5.1 General Guidelines for the Elaboration of a Hydropower LCI

From the conclusion obtained from the Itaipu case study, it was possible to propose some guidelines for LCI to be performed hereafter over the hydropower domain. The most remarkable recommendations are as follows:

- Use "generated electricity" as the product system function. Do not include other uses of the reservoir as function of the product system, except on the case the reservoir had a previous usage other than generate energy;
- Establish the Functional Unit (FU) using the electricity unit of measure;
- Adopt the period of 100 years as a standard time horizon. In this frame, maintenance operation required to keep the plant operative along the period must be considered;
- Focus the attention on data collection on the processes of: civil machines operations; steel, cement, diesel, and copper life-cycles; materials used by the electromechanical equipment; transportation of raw materials and employees; land occupation and transformation; and GHG emissions from the reservoir filling;
- Develop the LCI with procedures that search for homogeneity of criteria on data collection;
- Conduct procedures as Muster Pedigree and Monte Carlo Analysis, in order to estimate uncertainties of the LCI. The use of LCIA can be an important kick off on this kind of analysis; and
- Develop the LCI with a multidisciplinary team of practitioners, with task-specific focus and frequent meetings to share impressions, discuss results and debate hypothesis to be used.

The recommendations above must be taken carefully, considering that each case study differs from the others.

The realization of Itaipu case study resulted not only on the first Brazilian electricity LCA database developed with primary data, but also gave unaffordable experience on inventory development, essential to the future of LCA on the country.

Acknowledgments The authors are grateful to the Itaipu Binacional, which contributed with large amounts of information and provided a field trip, indispensable to support the local evaluation and for conducting the interviews.

References

ANEEL—Agencia Nacional de Energia Eletrica (2001) Corporative website. Available at: www.aneel.gov.br. Consulted on 18 Dec 2001

Brännstrom-Norberg BM et al (1996) Life cycle assessment for vattenfaĺs electricity generation—summary report. Vattenfall AB, Stockholm

Carrignton G (2000) A study of the Lake Chelan hydroelectric project based on life-cycle stressor-effects assessment. SCS - Scientific Certification Systems, Lake Cheelan

Coltro L, Garcia EEC, Queiroz GC (2003) Life cycle inventory for electric system in Brazil. Int J LCA. doi:10.1007/BF02978921

Curran MA (2001) Summary of feedback on issues. In: International workshop on electricity data for life cycle inventories, Sylvatica, Cincinnati

European Commission (1995) ExternE—externalities of energy, vol 6: wind and hydro. Office for Official Publications of the European Communities, Luxemburg

Frischknetcht R (2001) Feedback on background paper. In: International workshop on electricity data for life cycle inventories, Sylvatica, Cincinnati

Gagnon L, Bélanger C, Uchyama Y (2002) Life-cycle assessment of electricity generation options: the status of research in year 2001. Energy Police. doi: 10.1016/S0301-4215(02)00088-5

IEA—International Energy Agency (1998) Benign energy? The environmental implications of renewables. OECD/IEA, Paris

IEA—International Energy Agency (2000) Implementing agreement for hydropower technologies and programs. OECD/IEA, Paris

ISO- International Organization for Standardization (1997) Environmental management—life cycle assessment—principles and framework—ISO 14.040. ISO, Geneva

ISO- International Organization for Standardization (1998) Environmental management—life cycle assessment—goal and scope definition and inventory analysis—ISO 14.041. ISO, Geneva

Itaipu Binacional (1994) Itaipu—hydroelectric project—engineering features. Itaipu Binacional, Foz do Iguaçu

Itaipu Binacional (2009) Corporative website. Available at: www.itaipu.gov.br, Consulted at: 06 Mar 2009

Krewitt W (2001) Feedback on background paper. In: International workshop on electricity data for life cycle inventories, Sylvatica, Cincinnati

MCT—Ministério da Ciência e Tecnologia (2002) Emissões de dióxido de carbono e de metano pelos reservatórios hidrelétricos brasileiros. Reference report of the "Primeiro Inventário Brasileiro de Emissões Antrópicas de Gases de Efeito Estufa", MCT, Brasilia

OECD—Organizatiom for Economic Cooperation and Development (1992) Highlights: synthesis of the issue. In: Expert workshop on life-cycle analysis of energy systems, methods and experiences. OECD, Paris

ORNL—Oak ridge national laboratory (1994) Estimating externalities of the hydro fuel cycles. ORNL, Oak Ridge

Ribeiro FM (2003) Inventário de ciclo de vida da geração hidrelétrica no Brasil- Usina de Itaipu: primeira aproximação, vol 2. PIPGE/USP, Sao Paulo. Available at: http://www.teses.usp.br/teses/disponiveis/86/86131/tde-23082004-123349/publico/MestradoFlavio.pdf

Ribeiro FM, Silva GA (2004) Hydropower life-cycle inventories: some methodological considerations based on a Brazilian case study. Global conference on sustainable product development and life-cycle engineering, Berlin

Ribeiro FM, Silva GA (2010) Life-cycle inventory for hydroelectric generation: a Brazilian case study. J Cle Pro. doi:10.1016/j.jclepro.2009.09.006

Santos MA (2000) Inventário de emissão de gases de efeito estufa derivadas de hidrelétricas. PPE—UFRJ, Rio de Janeiro

Setterwall C (2001) Input to the workshop. In: International workshop on electricity data for life cycle inventories, Sylvatica, Cincinnati

Swedish Environemntal Management Council (1999) Certified environmental product declaration (EPD)—hydropower electricity from the Lule river. Swedish Environmental Management Council, Stockholm

Swedish Environemntal Management Council (2002a) Summary of Vattenfall AB's certified environmental product declaration (EPD) of electricity from the river Lule. Swedish environmental management council, Stockholm

Swedish Environemntal Management Council (2002b) Summary of Vattenfall AB's certified environmental product declaration (EPD) of electricity from the river Ume alv. Swedish Environmental Management Council, Stockholm

USEPA—United States Environmental Agency (2001) Report on the international workshop on electricity data for life cycle inventories. USEPA, Cincinnati

Vattenfall AB (1999) Vattenfalĺs life cycle studies of electricity. Vattenfall AB, Stockholm

Virtanen Y, Lubkert B (1992) Life-cycle inventories and evaluations of energy systems and technologies. In: Expert workshop on life-cycle analysis of energy systems, methods and experiences, OECD, Paris

A Comparison of Life Cycle Assessment Studies of Different Biofuels

Dheeraj Rathore, Deepak Pant and Anoop Singh

Abstract The intensive increase of biofuel demand has pushed the researchers to find a sustainable biofuel production system. LCA is the most accepted tool to assess the sustainability of biofuel production systems. The functional unit, scope, system boundary, reference system, data source, and allocation are the most important steps of an LCA study. Variations in these steps between studies affect the results significantly. Previous studies have shown that different biofuel feedstocks have different environmental burden hot spots, which refer to elevated greenhouse gas (GHG) emissions associated with a specific life cycle stage or facility process. The present chapter is an effort to compare various LCA studies on different biofuels. The well-to-wheel (cradle-to-grave) system is recommended for the assessment of biofuels production system. An LCA study of biofuels can demonstrate their sustainability and can guide the policy makers in adopting the policies for their promotions.

1 Introduction

Biofuels are plant-derived energy sources that can either be burnt directly for heat or converted to a liquid fuel such as ethanol, biodiesel, biogas, biohydrogen (Davis et al. 2009; Nigam and Singh 2011). The global biofuel sector has grown

D. Rathore
Department of Conservation Biology, School of Biological Science, College of Natural and Mathematical Sciences, University of Dodoma, Dodoma, Tanzania
e-mail: rathoredheeraj5@gmail.com

D. Pant
Separation and Conversion Technology, Flemish Institute for Technological Research (VITO), Mol, Belgium
e-mail: deepak.pant@vito.be; pantonline@gmail.com

A. Singh (✉)
Department of Scientific and Industrial Research (DSIR), Ministry of Science and Technology, Technology Bhawan, New Mehrauli Road, New Delhi 110016, India
e-mail: apsinghenv@gmail.com

A. Singh et al. (eds.), *Life Cycle Assessment of Renewable Energy Sources*,
Green Energy and Technology, DOI: 10.1007/978-1-4471-5364-1_12,

considerably in the recent years, driven primarily by concerns about fossil fuel prices and availability. Large-scale biofuel industries are being promoted to decrease reliance on petroleum in response to an abrupt rise in oil prices and to develop transportation fuels that reduce greenhouse gas (GHG) emissions compared to conventional fuel (IPCC 2007). This growing interest in biofuels is a means of "modernizing" biomass use and providing greater access to clean liquid fuels while helping to address energy costs, energy security, and global warming concerns associated with petroleum fuels. Industrial use of biofuels, particularly in North America and Latin America, has been expanding over the past century (Fernandes et al. 2007). However, the energetic use of biomass also causes impacts on climate change and, furthermore, different environmental issues arise, such as land-use and agricultural emissions, acidification, and eutrophication (Emmenegger et al. 2012; Dressler et al. 2012). Therefore, the environmental and climate benefits of bioenergies must be verified according to life cycle assessment (LCA) methods (ISO 14040 2006; ISO 14044 2006) to make them a sustainable energy source.

The environmental performance of products and processes has become a key issue, which influences some companies to investigate ways to minimize their effects on the environment. Many companies have found it advantageous to explore ways of moving beyond compliance using pollution prevention strategies and environmental management systems to improve their environmental performance. One such tool is LCA. This concept considers the entire life cycle of a product (Curran 1996). Life cycle assessment is a tool for assessing the environmental impacts of a product, process, or service from design to disposal, i.e., across its entire lifecycle, a so-called cradle-to-grave approach. The impacts may be beneficial or adverse depending on a variety of factors most of which has been discussed in great detail in the previous chapters. These impacts are sometimes referred to as the "environmental footprint" of a product or service. The results of an LCA study depend on several factors, e.g., consideration of system boundaries, functional unit, data sources, impact categories, allocation. This chapter is an effort to compare different LCA studies of biofuels to highlight the main unresolved problems in performing an LCA study for biofuel production systems.

2 Role of LCA in Improvement of Biofuels Production System

Modern bioenergy can be a mechanism for economic development, enabling local communities to secure the energy they need, with farmers earning additional income and achieving greater price stability for their production (UNEP/GRID-Arendal 2011). Cultivation of the energy crops has raised concerns due to their high consumption of conventional fuels, fertilizers, and pesticides, their impacts on ecosystems and competition for arable land with food crops. Safeguards are needed and special emphasis should be given to options that help mitigate risks and create positive effects and co-benefits (UNEP/GRID-Arendal 2011). Responding to these challenges effectively requires a life cycle perspective of the biofuel production pathway/system. Since biofuels are considered a major alternative for the future energy demands, several LCA studies were carried out for the enhancement of biofuel production system (Muys and Quizano 2002; Kim and Dale 2009; Chiaramonti and Racchia 2010; Dressler et al. 2012). If biofuels are to become a major alternative to petroleum, it has to be both environmentally and economically advantageous. LCAs of these transitions will require much stronger integration between economists and systems engineers to address what happens during the transition phase when large-scale changes occur in many components of a complex, market driven, technological system (McKone et al. 2011). To achieve the target as per EC Directive 2009/28/EC (EC 2008), i.e., GHG savings of 60 % by 2020, selection of feedstock for considering local factors and land utilization, process technology, and consumption perspective are major steps to be considered under LCA for improvement in production system. LCA studies conducted in the recent past reported the process phases that can be improved by advancing the technology to consider a product as biofuel according to European Directive 2009/28/EC (Watson et al. 1996; Kaltschmitt et al. 1997; CONCAWE 2004; Larson et al. 2006; Larson 2006; Korres et al. 2010). A generalized scheme for LCA of biofuel production is presented in Fig. 1.

By the LCA study of energy crops, Emmenegger et al. (2011) concluded that producing biofuels can reduce the fossil fuel use and GHG emissions when compared to a fossil reference. The focus on GHG emissions of the main regulatory schemes neglects other relevant environmental impacts and may provide the wrong incentives. Thus, water consumption may become a major concern, offsetting the benefits of biofuel use with respect to climate change. McKone et al. (2011) explained the following seven grand challenges that must be confronted to enable LCA to effectively evaluate the environmental footprint of biofuel alternatives.

(a) understanding farmers, feedstock options, and land use
(b) predicting biofuel production technologies and practices
(c) characterizing tailpipe emissions and their health consequences
(d) incorporating spatial heterogeneity in inventories and assessments
(e) accounting for time in impact assessments
(f) assessing transitions as well as end states
(g) confronting uncertainty and variability

Dressler et al. (2012) conducted LCA study of biogas from maize at three different sites and find a variation in results due to local factor suggesting consideration of local and regional factors before selecting energy crops. In a study with biofuel from grass, Korres et al. (2010) consider that agronomy and digester use are the biggest issues for controlling the GHG savings.

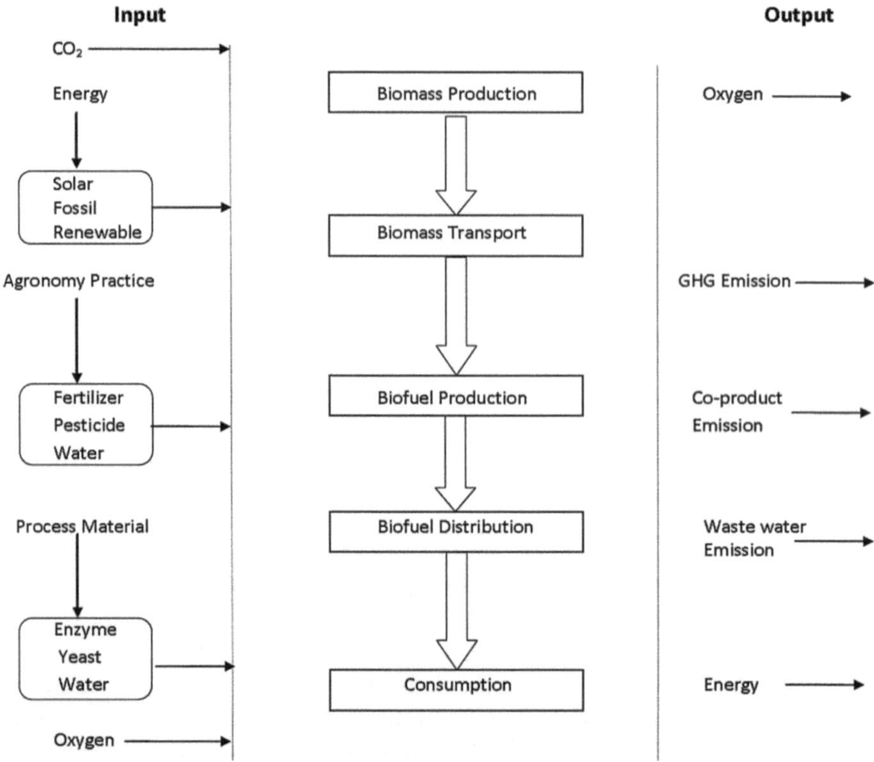

Fig. 1 A generalized scheme for LCA of biofuel production

3 Comparison of Different LCA Studies of Biofuels

Bioethanol and biodiesel are the most commonly produced biofuels, and currently these are derived mainly from food crops such as maize, soya, and sugarcane. Biofuels derived from food crops are known as first-generation biofuels. New technologies in advanced stages of development will allow alternative feedstocks to be used for bioenergy production and are known as second-generation and third-generation biofuels (IEA 2008; Maltitz et al. 2009; Nigam and Singh 2011, Singh et al. 2011). Over 200 feedstocks have been listed for the biofuel family. Use of biofuel over the fossil fuel requires a critical assessment for actual benefit from it. Various LCA studies showed variable results with different energy crop and products (Davis et al. 2009). A comparison of several LCA studies conducted by different researchers focusing on different biofuel for different purpose is presented in Table 1.

Huo et al. (2009) analyzed four different biofuels scenarios, produced from soybean oil. It was identified that allocation methods for coproducts and avoided emissions are critical to the outcome of the study. Additionally, it was also pointed

Table 1 Comparison of life cycle assessment studies of different biofuels

Feedstock	Product	System adopted	Functional unit	System boundary	Reference system	Environment Potential	Country	Reference
Rapeseed	Biodiesel	Field to wheel	1 km travelled by bus	Defined	Conventional diesel	56 % GHG savings	Italy	Finco et al. (2012)
Grass	Bio-methane	Cradle to grave	g CO_2 equivalent (CO_2e) MJ^{-1} energy replaced	Well defined	Fossil diesel	54–75 % GHG saving	Ireland	Korres et al. (2010)
Maize	Biogas	Cradle to grave	1 kg of fresh matter of maize and 1kWh of electricity	Well defined	Fossil fuel	GHG emission 0.179–0.058 kg CO_2eq./kWhe	Germany	Dressler et al. (2012)
Pongamia pinnata	Biodiesel	Field to wheel	1 MJ of energy	Well defined	Diesel	CO_2 sequestration 1.0–1.5	India	Chandrashekar et al. (2012)
Switchgrass	Ethanol	Cradle to grave	g CO_2 equivalent (CO_2e) MJ^{-1} energy replaced	Defined as scope of the study	Coal	114 % GHG saving	USA	Adler et al. (2007)
Reed canarygrass	Ethanol	Cradle to grave	g CO_2 equivalent (CO_2e) MJ^{-1} energy replaced	Defined as scope of the study	Coal	84 % GHG saving	USA	Adler et al. (2007)
Hybrid poplar	Ethanol	Cradle to grave	g CO_2 equivalent (CO_2e) MJ^{-1} energy replaced	Defined as scope of the study	Coal	117 % GHG saving	USA	Adler et al. (2007)
Corn-soybean	Ethanol	Cradle to grave	g CO_2 equivalent (CO_2e) MJ^{-1} energy replaced	Defined as scope of the study	Coal	38–41 % GHG saving	USA	Adler et al. (2007)
Jatropha	Biodiesel	Well to Tank	1 MJ of JME	Defined	Fossil Diesel	72 % GHG saving	Ivory Coast and Mali	Ndong et al. (2009)
Switchgrass, Cynara, Giant reed and Miscanthus	Biomass	Cradle to farm gate	Per unit energy/per unit land	Defined as scope of the study	Production of conventional crops	50–60 % less impact	Italy	Monti et al. (2009)
Corn stover	Bioethanol	Energy product to gate	Not defined	Well defined	Gasoline, A hypothetical case of pure ethanol	Reduction in GWP	The Netherlands	Luo et al. (2009)

(continued)

Table 1 (continued)

Feedstock	Product	System adopted	Functional unit	System boundary	Reference system	Environment Potential	Country	Reference
Switchgrass and corn stover	Ethanol	Cradle to wheel	Per km	Defined	Low-sulfur reformulated gasoline	Up to 70 % lower GHG emissions	Canada	Spatari et al. (2005)
Household and biodegradable municipal waste	Ethanol	Cradle to grave	MJ of fuel equivalent	Defined	Gasoline	Up to 92.5 % GHG emission saving	UK	Stichnothe and Azapagic (2009)
Corn stover	Ethanol	Cradle to grave	1 ha/1 km	Defined	Gasoline	Reduction of 267 g CO_2/km	USA	Sheehan et al. (2004)
Blue-green Algae	Ethanol	Cradle to grave	g CO_2e/MJ	Defined	Gasoline	67 and 87 % reductions in the carbon footprint	USA	Luo et al. 2010
Microalgae	Biodiesel	Cradle to grave	Combustion of 1 MJ biodiesel	Defined	First-generation biodiesel and oil diesel	Significantly decrease environmental impacts	France	Lardon et al. 2009
Potato steam peels	Hydrogen	Cradle to grave	g CO_2 per kilogram of hydrogen produced	Defined	Potato steam peels directly for animal fodder	Reduction in greenhouse gas emissions	USA	Djomo et al. 2008
Microalgae	Biodiesel	Cradle to grave	g CO_2e/MJ	Defined	Fossil diesel	About 80 % lower GWP	UK	Stephenson et al. 2010
Microalgae	Biodiesel	Well to pump	1,000 MJ	Defined	Fossil diesel	Up to 45 % emission	USA	Sander and Murthy 2010
Rapeseed	Biodiesel	Cradle to grave	1 person kilometer	Defined	Conventional Petrol	Shift in environmental problem	Argentina	Emmenegger et al. 2011
Corn, soybean	Bioethanol and Biodiesel	Cradle to grave	1 ha of arable land	Defined	Gasoline	Reduction in GHGs Increase in acidification and eutrophication	USA	Kim and Dale 2005
Rapeseed, oil palm, jatropha	Hydrotreated vegetable oil (biodiesel)	Cradle to grave	1 kWh energy out	Defined	Conventional diesel	About half the GWP	USA	Arvidsson et al. 2011

out that by using displacement approach, all four soybean-based fuels can achieve a modest to significant reduction in well-to-wheel GHG emissions (64–174 %) versus petroleum-based fuels. In this study, Huo and co-worker concluded that the method used to calculate coproduct credits is a crucial issue in biofuel LCA that should be carefully addressed and extensive efforts must be made to identify the most reasonable approach for dealing with the coproducts of biofuel production system. Finco et al. (2012) conducted an LCA study on rapeseed and reported a 56 % less CO_2 equivalent GHG emission from the rapeseed biodiesel than diesel. However, this study does not include negative impact caused by land use particularly from the use of N fertilizer. N_2O emissions, a by-product of N fertilization in agriculture, as one responsible factor to enhanced GHG emissions compared to consumption of fossil fuels (Crutzen et al. 2008) and can overrule the benefit of biofuel. Halleux et al. (2008) conducted a detail comparative LCA between ethanol from sugar beet and methyl ester from rapeseed and concluded an advantage of rapeseed over sugar beet biofuel in terms of total environment impact and GHG emission. Table 1 is explaining the environmental potentiality of various feedstock biofuels over reference fuel (i.e., mostly fossil diesel or fossil gasoline).

Result of Stucki et al. (2012) on LCA of biogas from different purchased substrates and energy crops viz. sugar beet, fodder beet, beet residues, maize silage, molasses, and glycerin shows that the environmental impacts of biogas from purchased substrates are in the same range than those from liquid biofuels. Chandrashekar et al. (2012) find 1.25 times negative global warming potential of *Pongamia pinnata* compared to fossil fuel and Jatropha biodiesel, and nil acidification and eutrophication potential. However, variations in the LCA result are also observed by the differences in selection of scope, system boundary, and other phases of LCA (Table 1). These issues were reviewed in detailed by Reap et al. (2008a, b) and Singh et al. (2010).

The life cycle stages can have harmful effects or benefits of different environmental, economical, and social dimensions. Therefore, an assessment of the complete fuel chains from different perspectives is of crucial importance in order to achieve sustainable biofuels (Markevicius et al. 2010). Comprehensive LCA of biofuels illustrating environmental benefits and impacts can be a tool for policy decisions and for technology development.

Current disagreements about the performance of biofuels rest on different approaches and assumptions used by the investigators (Farrell et al. 2006). The use of different input data, functional units, allocation methods, reference systems and other assumptions complicates comparisons of LCA bioenergy studies and uncertainties and use of specific local factors for indirect effects (e.g., land-use change and N-based soil emissions) may give rise to wide ranges of final results (Cherubini and Strømman 2011). The system choice for comparing different biofuels must be identical because different systems could results improper results, e.g., the choice of passenger car, the efficiency and emissions of EURO V and EURO III varied a lot, so different passenger car, bus, and other transportation vehicles could not be identical to compare different biofuels. The system boundaries of different biofuels also need to be identical, as inclusion and exclusion of coproduct use changed the whole results of the study. Liska and Cassman (2008) revealed that the prediction of emerging biofuel system's performance can pose additional challenges for LCA due to insufficient data of commercial-scale feedstock production and conversion systems. LCA of biofuel systems is currently depending on laboratory- or pilot-scale data. Extrapolation of these laboratory-/pilot-scale results to commercial-scale deployment must be made with caution because of multiple unknowns that introduce significant uncertainty in the estimation of life cycle energy efficiencies and GHG emissions (Liska and Cassman 2008). Standardized LCA methods and agreement on the most relevant metrics for assessing different biofuel systems are essential to forge a consensus in the scientific community, industrialist, and local people. That would help advance public policy initiatives to encourage development of commercial-scale biofuel industries.

There are two issues with regard to standardization. The first is choosing the appropriate metric for the goal of the assessment, and the second is the appropriate analysis framework to support the selected metric. Standardization procedure for regulatory LCA metrics for GHG and energy balances of biofuel systems is summarized by Liska and Cassman (2008) and presented in Table 2. The LCA quantifies the potential benefits and environmental impacts of biofuels but existing methods limit direct comparison of different processes within the biofuel production system and between different biofuel production systems due to inconsistencies in performance metrics, system boundaries, and available data. Therefore, the standardization of LCA methods, metrics, and tools are critically needed to evaluate biofuel production systems for estimating the net GHG mitigation of an individual biofuel production system.

Table 2 Standardization procedure for regulatory LCA metrics for GHG and energy balances of biofuel systems (adapted from Liska and Cassman 2008)

LCA element	Standardization procedure
Biofuel system boundaries	Explicit definition of system components and metrics for each component and the entire system
Input parameters	Evaluate variability, justify which are considered constant or variable, use most recent and directly measured values where possible
Crop production system	Most appropriate county, state, or regional data depending on the most appropriate scale and data availability for the biorefinery facility under evaluation
Coproduct credits	Based on representative coproduct use for the facility
Soil carbon emissions balance	Based on measured changes in soil, if not available, an estimated by appropriate ecosystem models
Nitrous oxide emissions	Based on measured emissions, if not available, use estimated by IPCC guidelines
Land-use change indirect GHG emissions	Estimated using an appropriate global econometric model depending on accepted national or international standards for allocating these effects

4 Key Issues

Life cycle assessment is carried out in phases (ISO 14044 2006; European Commission 2010a, b), and different phases of LCA are presented in Fig. 2. Various key issues in a LCA system of any process to product such as biofuel are scope and functional unit, reference system, system boundary, data source, allocation, inventory analysis, impact assessment, and sensitivity analysis (Singh et al. 2010; Askham 2012).

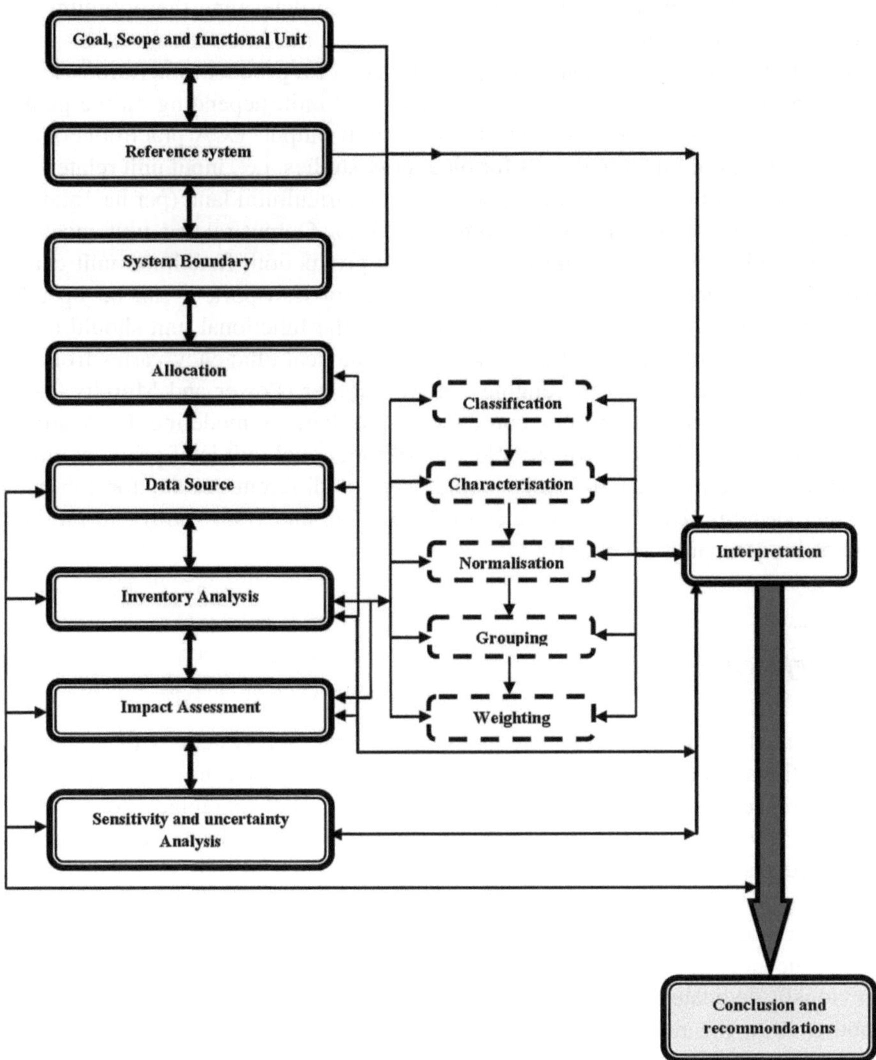

Fig. 2 Various phases of life cycle assessment

4.1 Scope and Functional Unit

First step of a LCA system requires a well-defined scope of the study, which should be compatible to the goal. Functional unit sets the scale for comparison of two or more products, provides a reference to which the input and output data are normalized, and harmonizes the establishment of the inventory (Jensen et al. 1997). The main goal for LCA of biofuels is to evaluate the environmental impacts of the system under examination and to quantify the ecological benefits from the replacement of the reference system basically conventional fossil fuels. It may also provide a tool for policy makers and consumers to determine the optimum eco-friendly fuel (Singh et al. 2010).

Functional unit is the "quantified performance of a product system for use as a reference unit" (ISO 14044 2006). The functional unit, depending on the goal of the study, must be expressed in terms of per unit output. LCA practitioners consider four types of functional units for bioenergy studies, i.e., input unit related (per tone biomass), output unit related (per MJ), unit agricultural land (per ha), and unit time (per year) (Cherubini and Strømman 2011). Output-related unit most frequently used in bioenergy studies. For energy production, functional unit can be expressed as "per kWh energy produced" and for transport, it can be "per km distance travelled" basis. For transport services, the functional unit should not be expressed in "unit energy at fuel tank"; as mechanical efficiency varies from one fuel to another and from one engine type to another (Power and Murphy 2009). Scale, if not properly chosen, could be a problem in modeling LCA studies (Addiscott 2005). Thus, adequate selection of functional unit is of prime emphasis because different functional units could lead to different results for the same product systems (Hischier and Reichart 2003; Kim and Dale 2006) and products cannot be compared accurately.

4.2 Reference System

System analysis is possible by comparing the biofuel system with a targeted (conventional) reference system. The goal of the study determines the choice of the reference system (e.g., whether biofuel is intended to replace conventional transport fuel or coal for electricity or wood pellets for heat). The choice of reference system influences the results of LCA study; therefore, it is important to choose an identical reference system to the conventional system (Singh and Olsen 2011). In most biofuel studies, reference system is limited to a fossil fuel system. It should be noticed that when production of feedstocks for bioenergy uses land previously dedicated to other purposes or when the same feedstock is used for another task, the reference system should include an alternative land-use or an alternative biomass use, respectively (Cherubini and Strømman 2011). In fact, fossil-derived electricity can be assumed to be produced from oil, natural gas, coal, or other sources, all of which having different GHG emission factors (Cherubini and Strømman 2011). The impact of different reference system can be observed in the study conducted by Pettersson and Harvey 2010), where GHG emission savings of bioelectricity production from black liquor are estimated using electricity coming from different fossil sources as reference. The Renewable Energy Directive (EC 2008) requires a 60 % savings in GHG emissions as compared to the fossil fuel it replaces to allow the biofuel to be used for national renewable energy targets after 2017. Thus, a detailed description and impact analysis of the reference system is crucial as well as mandatory for comparing the results of biofuel LCA (Singh et al. 2010).

4.3 System Boundary

On the basis of goal and scope, initial boundaries of the system are determined. Davis et al. (2009) concluded that different system boundaries among various studies of biofuel production from biomass have caused considerable variation in LCA estimates since they vary not only according to start and end points (e.g., well to tank and well to wheel) but also over space and time in a way that can dramatically affect energy and GHG balances.

Many researchers use the "well-to-tank" system boundary to compare environmental impact of biofuels with fossil fuels (Luo et al. 2009; Monti et al. 2009), while others use "well-to-wheel" or "cradle-to-grave" system (Sheehan et al. 2004; Spatari et al. 2005; Power and Murphy 2009; Stichnothe and Azapagic 2009; Korres et al. 2010).

The risk of improper boundaries selection include that LCA results may either not reflect reality well enough and lead to incorrect interpretations and comparisons (Graedel 1998; Lee et al. 1995) or provide the perception to the decision maker that it does not excogitate actual results and thus lower the confidence level of policy maker in making decisions based on the results (Reap et al. 2008a). Inconsistency of system boundaries in LCA analysis of biofuel through omission of the production of various inputs (e.g., enzymes which is used to degrade cellulosic feedstock, fertilizer, pesticides, lime), and utilization of bioethanol (Luo et al. 2009; Gnansounou et al. 2009) could cause a significant variation on the outcome of the analysis. A recent example of such problem can be observed in the debate surrounding the energy balance of ethanol where criteria for the selection of boundaries (like the inclusion of corn-based ethanol coproducts or energy from combustion of lignin in cellulosic ethanol) are strong enough to change the results significantly (Farrell et al. 2006; Hammerschlag 2006). A uniform and clear determination of system boundaries is necessary to accurately estimate the possible environmental impacts including GHG emissions in LCA comparisons between biofuels and conventional fuels (Farrell et al. 2006).

4.4 Data Source and Quality

The use of fixed databases such as ecoinvent, Edu DB, Xergi, NOVAOL srl for conducting an LCA study of bioenergy is not enough because the available databases do not have all processes required for LCA study of bioenergy. Monti et al. (2009) also realized that available databases were generic for specific agricultural problems during conducting the LCA of four potential energy crops (i.e., giant reed, miscanthus, switchgrass, and *Cynara cardunculus* or *Artichoke thistle*) in comparison with conventional wheat/maize rotation and clarify that external data from scientific literature should be obtained for life cycle inventory (LCI) enhancement and accurate representation of the system.

In a survey of approaches to improve reliability Björklund (2002) identifies the main types of uncertainty due to data quality, e.g., badly measured data/inaccurate data, data gaps, unrepresentative (proxy) data, model uncertainty, and uncertainty about LCA methodological choices. Standardized LCA databases are sought to reduce the burdens of data collection (UNEP 2003). There are few established, standardized, or consistent ways to assess and maintain data quality (Vigon and Jensen 1995). Data can become outdated, compiled at different times corresponding to different materials produced over broadly different time periods (Jensen et al. 1997), could be due to technology shift, new invention, etc. LCI data may be unrepresentative because it could be taken from similar but not identical processes (Björklund 2002). In general, the literature tends to agree that data for life cycle inventories are not widely available nor of high quality (Ayres 1995; Ehrenfeld 1997; Owens 1997), due to that during inventory analysis data with gaps are sometimes ignored, assumed, or estimated (Graedel 1998; Lent 2003), and LCA practitioners may extrapolate data based on limited data sets (Owens 1997). Such assumptions and/or extrapolation resulted inappropriate interpretation and/or huge uncertainty for decision makers.

4.5 Allocation

Allocation is the process of assigning to each of the functions of a multiple-function system only those environmental burdens associated with that function (Azapagic and Clift 1999). Allocation can be done on the basis of mass, volume, energy or carbon content or economic value of the coproducts if the inputs and outputs of the system should be partitioned between different products or functions based on physical relationships, i.e., they shall reflect the way in which the inputs and outputs are changed by quantitative changes in the products or functions delivered by the system (SAIC 2006). It is recommended that allocation, if possible, should be avoided (ISO 14044 2006) through subdivision of processes, if possible, or system expansion. Allocation on a mass basis relates products and coproducts using a physical property that is easy to interpret (Singh et al. 2010), although some researchers argued that it cannot be an accurate measure of energy functions (Malça and Freire 2006; Shapouri et al. 2002) and it is not a measure of environmental impacts also. When physical properties alone cannot be established or used, allocation may be based on the economic value of the products although price variation, subsidies, and market interferences could imply difficulties in its implementation (Luo et al. 2009).

In a study of soybean-derived biodiesel, Huo et al. (2009) compared five approaches to address the coproduct issues for various coproducts including protein products (such as soy meal), industrial feedstock (such as glycerin), and energy products (such as propane fuel mix and heavy oils). These five approaches includes the displacement approach, an energy-based allocation approach, a market-value based allocation approach, hybrid approach I, which employs both the displacement method (for soy meal and glycerin) and the allocation method (for other energy coproducts) and hybrid approach II, which is exactly like hybrid approach I except that it addresses soy meal with a market-value-based allocation method. The results of the displacement approach are influenced significantly by the extent of the energy and carbon intensity of the products chosen to be displaced and argued that soy meal displacement could introduce uncertainties because soy meal can displace many kinds of fodder and each fodder could have different energy and carbon intensities. Huo and coworker suggested that when the choice is between the displacement method and the allocation method, the displacement method tends to be chosen if the uncertainties and difficulties associated with it are solved, because it can reflect the energy use and emissions actually saved as a result of the coproducts replacing other equivalent products. They also pointed out that "energy-value-based allocation method is a favorable choice for a system in which the value of all the primary product and coproducts can be determined on the basis of their energy content, such as the production processes of renewable fuels. If a non-energy coproduct is involved and there are difficulties associated with using the displacement approach, the market-value-based allocation method could be an acceptable choice, although the fluctuation of prices could affect the results." Huo et al. (2009) concluded that the integration of displacement method and allocation method (hybrid approaches) could be the most reasonable choice of allocation method for every coproduct. The results of the two hybrid approaches were very close in terms of GHG emissions, indicating that the uncertainty associated with using soy meal to displace soybeans would be in an acceptable range. Reap et al. (2008b) observe that allocation failures hide or exaggerate burdens associated with a product system, effectively biasing all downstream results with an artifact of the analysis.

A number of scientific literatures are available which addresses the allocation issue in LCA and describe the alternative approaches to allocation (Frischknecht 2000; Wang et al. 2004; Curran 2007; Luo et al. 2009). Wang (2005) showed significant impact on overall energy and emission results of alternative allocation methods for corn ethanol LCA, ranging from benefits relative to petroleum of 16–52 % in the case when the ethanol is made by a wet milling process. In another study, Fergusson (2003) also found somewhat smaller (but nevertheless significant) range in GHG results for biofuels when different coproduct allocation methods are used. The expansion of system for use of coproducts within the system is recommended for biofuel production system. If allocation cannot be avoided, then allocation could be done on the basis of carbon content of all products as the target of biofuel production is to minimize the GHG emission and the mass/volume of products is not a precise measure of energy/emission and economic value is fluctuating with the market.

4.6 Inventory Analysis

A LCI is a process of quantifying energy and raw material requirements, environmental pollution for the entire life cycle of a product, process, or activity (SAIC 2006). The inventory analysis requires data on the physical inputs and outputs of the processes of the product system, regarding product flows as well as elementary flows (Singh and Olsen 2011). The main issue of inventory analysis includes data collection and estimations, validation of data and relating data to the specific processes within the system boundaries. After the initial data collection, of which the source should be clearly declared, the system boundaries can be refined as a result of decisions on exclusion of subsystems, exclusion of material flows or inclusion of new unit processes. The validation of data as a mean of data quality improvement or the need for supplementary data would improve the outcome of the analysis (Jensen et al. 1997). The inventory analysis requires very extensive data. The outcome of the study totally depends on the availability and quality of the datasets. So that, there is a great need to collection of standardized data, especially for background processes (Singh and Olsen 2011).

4.7 Impact Assessment

Impact assessment establishes a relationship between the product or process and its potential impacts on human health, environment, and sources depletion (SAIC 2006). ISO developed a standard for conducting an impact assessment entitled ISO 14042, LCIA (ISO 1998). Life cycle impact assessment (LCIA) is structured in classification, characterization, normalization, and weighting. The first three steps are mandatory steps for the determination of impact categories, which corresponds to an important environmental problem (e.g., eutrophication, depletion of non-renewable energy resources, and ozone depletion) (Singh and Olsen 2011). There is no standardized list of impact categories (IFEU 2000). Guinée et al. (2002) has tabulated most of the impact categories in the "Handbook of LCA." The main problems faced during LCIA result from the need to connect the right burdens with the right impacts at the correct time and place (Reap et al. 2008b), in this regard, impact category selection is the most important step which can influence results significantly.

Finnveden (2000) noted the slightly different impact category lists that have been proposed by different organizations. The lack of standardization of some impact categories is demonstrated in the recent debate as to whether certain impact categories such as soil salinity, desiccation, and erosion should be their own category or part of another category such as land-use impact and freshwater depletion (Jolliet et al. 2004). McKone et al. (2011) pointed out a key challenge for applying LCA to a broadly distributed system (e.g., biofuels) is to rationally select appropriate spatial and temporal scales for different impact categories without adding unnecessary complexity and data management challenges as significant geographical and temporal variability among locations over time could influence not only the health impacts of air pollutant emissions, but also soil carbon impacts and water demand consequences, among other factors. McKone and co-worker suggested that accurate assessments must not only capture spatial and variation at appropriate scales (from global to farm-level), but also provide a process to aggregate spatial variability into impact metrics that can be applied at all geographical scales. The selection of midpoint or end point (damage) impact categories is another potential result affecting criteria for both the level of confidence or relevance for decision making on the basis of LCA study results (Reap et al. 2008b). End point categories are less comprehensive and have much higher levels of uncertainty than the better defined midpoint categories (UNEP 2003), and midpoint categories, on the other hand, are harder to interpret because they do not deal directly with an end point associated with an area of protection (Udo de Haes et al. 2002) that may be more relevant for decision making (UNEP 2003).

The International Program on Chemical Safety (WHO 2006) proposed four tiers, ranging from the use of default assumptions to sophisticated probabilistic assessment to address uncertainty in risk assessment:

Tier 0: Default assumptions; single value of result

Tier 1: Qualitative but systematic identification and characterization of uncertainties

Tier 2: Quantitative evaluation of uncertainty making use of bounding values, interval analysis, and sensitivity analysis

Tier 3: Probabilistic assessments with single or multiple outcome distributions reflecting uncertainty and variability.

Cherubini and Strømman (2011) reviewed several biofuel LCA studies and found that very few studies (about 9 %) included land-use category in their impact assessment. This is an important indicator particularly for bioenergy systems based on dedicated crops or forest resources, since land use may lead to substantial impacts, especially on biodiversity and on soil quality. This is mainly due to the fact that there is no widely accepted methodology for including land-use impacts in LCA, despite some recent efforts (Dubreuil et al. 2007; Koellner and Scholz 2008; Scholz 2007). Cherubini and Strømman (2011) also stated that for the same reason, none of the reviewed studies included in the assessment the potential impact of bioenergy on biodiversity, despite an existing accurate methodology (Michelsen 2008).

Tokunaga et al. (2012) concluded that by ignoring emissions associated with land-use change, significant emissions savings could achieve via biofuel use, ranging from 10 to 80 % reductions than fossil fuel emissions. The land-use changes could significantly increase life cycle emissions, while byproduct credits could significantly reduce life cycle emissions. Emmenegger et al. (2011) reported that the use of marginal arid land for cultivation reduces land-use impacts but induces a higher demand for irrigation, which finally compensates for the environmental benefits. Emmenegger and co-worker concluded that changing from petrol to biofuels results in a shift of environmental burdens from fossil fuel resource depletion to ecosystem quality damages.

4.8 Sensitivity Analysis

The key purpose of sensitivity analysis is to identify and focus on key data and assumptions that have the most influence on a result. It can be used to simplify data collection and analysis without compromising the robustness of a result or to identify crucial data that must be thoroughly investigated. According to IFEU (2000), the sensitivity analysis can typically be carried out in three ways, i.e., data uncertainty analysis, different system boundaries, and different life cycle comparisons. The identification of lower and upper values of the process parameters could introduce subjectivity to the analysis and will reflect better on the characteristics of the parameter analyzed (Fukushima and Chen 2009).

Reap and co-workers summarize their opinions about severity and solution adequacy using a simple ordinal scale (Table 3). "Each number represents a

Table 3 Problems in LCA qualitatively rated by severity and adequacy of current solutions (1, minimal severity while 5, severe; 1, problem solved while 5, problem largely unaddressed) (adapted from Reap et al. 2008b)

Problem	Severity	Solution adequacy
Functional unit definition	4	3
Boundary selection	4	3
Alternative scenario considerations	1	5
Allocation	5	3
Negligible contribution criteria	3	3
Local technical uniqueness	2	2
Impact category selection	3	3
Spatial variation	5	3
Local environmental uniqueness	5	3
Dynamics of the environment	3	4
Time horizons	2	3
Weighting and valuation	4	2
Uncertainty in the decision process	3	3
Data availability and quality	5	3

qualitative estimate. Severity represents a combination of problem magnitude, likelihood of occurrence, and chances of detecting the error should it occur. For instance, spatial variations can lead to multiple order of magnitude differences in characterization factors for commonly used impact categories such as acidification. Solution adequacy integrates capacity to address the discussed problem and difficulty of using available solutions." (Reap et al. 2008b).

McKone et al. (2011) indicated that in developing and applying LCA to assess the environmental sustainability of transportation fuels, LCA practitioners commonly address the climate forcing, other pollutant emissions and impacts, water-resource impacts, land-use changes, nutrient needs, human and ecological health impacts, and other external costs. McKone and co-worker suggested that LCA practitioners may also consider social impacts and economic factors for more accurate sustainability assessment of transportation fuel.

5 Conclusion

The most critical issue for the development of biofuel support policies includes environmental and social sustainability of biofuel production and use. The LCA methodology is most acceptable tool for the estimation of the impact of biofuel chains, even in quantitative terms, which ultimately reflects the sustainability of biofuels. Conducting LCA of bioenergy production systems is challenging task because it attempts to combine disparate quantities in ways that require considerable explanation and interpretation as well requires large amounts of practical information. The biofuel LCA studies must have cradle-to-grave approach and function unit should be unit energy utilization as conversion efficiency varies greatly.

References

Addiscott TM (2005) Nitrate, agriculture and the environment. CABI Publishing, Oxfordshire, UK, pp 66–67

Adler PR, Grosso SJD, Parton WJ (2007) Life-cycle assessment of net greenhouse-gas flux for bioenergy cropping systems. Ecol Appl 17(3):675–691

Arvidsson R, Persson S, Froling M, Svanstromb M (2011) Life cycle assessment of hydrotreated vegetable oil from rape, oil palm and Jatropha. J Cleaner Prod 19:129–137

Askham C (2012) REACH and LCA—methodological approaches and challenges. Int J Life Cycle Assess 17:43–57

Ayres RU (1995) Life cycle analysis: a critique. Resour Conserv Recycl 14:199–223

Azapagic A, Clift R (1999) Allocation of environmental burdens in co-product systems: product-related burdens. Int J Life Cycle Assess 4(6):357–369

Björklund AE (2002) Survey of approaches to improve reliability in LCA. Int J Life Cycle Assess 7:64–72

Chandrashekar LA, Mahesh NS, Gowda B, Hall W (2012) Life cycle assessment of biodiesel production from pongamia oil in rural Karnataka. Agric Eng Int: CIGR J 14(3):67–77

Cherubini F, Strømman AH (2011) Life cycle assessment of bioenergy systems: state of the art and future challenges. Bioresour Technol 102:437–451

Chiaramonti D, Recchia L (2010) Is life cycle assessment (LCA) a suitable method for quantitative CO2 saving estimations? The impact of field input on the LCA results for a pure vegetable oil chain. Biomass Bioenerg 34:787–797

IPCC, Climate Change (2007) Mitigation of climate change. Working group III contribution to the Intergovernmental panel on climate change fourth assessment report. Cambridge University Press, Cambridge

CONCAWE: CONservation of clean air and water in Europe (Oil companies' European association for environment, health and safety in refining and distribution) (2004) Joint Research Centre of the EU Commission, and European Council for Automotive R&D. Well-to-wheels analysis of future automotive fuels and powertrains in the European context, version 1b, Jan 2004. http://ies.jrc.cec.eu.int/Download/eh

Crutzen PJ, Mosier AR, Smith KA, Winiwarter W (2008) N2O release from agro-biofuel production negates global warming reduction by replacing fossil fuels. Atmos Chem Phys 8:389–395

Curran MA (ed) (1996) Environmental life cycle assessment. ISBN 0-07-015063-X, McGraw-Hill

Curran MA (2007) Co-product and input allocation. Approaches for creating life cycle inventory data. A literature review. Int J Life Cycle Assess 12(1):65–78

Davis SC, Anderson-Teixeira KJ, DeLucia EH (2009) Life-cycle analysis and the ecology of biofuels. Trends Plant Sci 14(3):140–146

Djomo SN, Humbert S, Blumberga D (2008) Life cycle assessment of hydrogen produced from potato steam peels. Int J Hydrogen Energy 33:3067–3072

Dressler D, Loewen A, Nelles M (2012) Life cycle assessment of the supply and use of bioenergy: impact of regional factors on biogas production. Int J Life Cycle Assess 17:1104–1115

Dubreuil A, Gaillard G, Müller-Wenk R (2007) Key elements in a framework for land use impact assessment within LCA. Int J Life Cycle Assess 12:5–15

EC (2008) Commission of the European communities. Proposal for a directive of the European parliament and of the council on the promotion of the use of renewable sources. COM 19 final, 2008/0016 (COD)

Ehrenfeld J (1997) The importance of LCAs—warts and all. J Ind Ecol 1:41–49

Emmenegger MF, Pfister S, Koehler A, Giovanetti LDe, Arena AP, Zah R (2011) Taking into account water use impacts in the LCA of biofuels: an Argentinean case study. Int J Life Cycle Assess 16:869–877

Emmenegger MF, Stucki M, Hermle S (2012) LCA of energetic biomass utilization: actual projects and new developments—23 April 2012, Berne, Switzerland. Int J Life Cycle Assess 17:1142–1147

European Commission (2010a) IUCLID dataset substance ID/CAS no.:71-43-2, Benzene. http://ecb.jrc.ec.europa.eu/iuclid-datasheet/71432.pdf. Accessed 11 July 2011

European Commission (2010b) International reference life cycle data system (ILCD) handbook—general guide for life cycle assessment—detailed guidance. EUR 24708 EN. European commission, joint research centre, institute for environment and sustainability, 1st edn. Publications Office of the European Union, Luxemburg

Farrell AE, Pelvin RJ, Turner BT, Jones AD, O'Hare M, Kammen DM (2006) Ethanol can contribute to energy and environmental goals. Science 311:506–508

Fergusson M (2003) Expert paper on the global impacts of road transport biofuels: a contribution to the government's analysis, national society for clean air (NSCA), cleaner transport forum, and the Institute for European Environmental Policy, p 25

Fernandes SD, Trautmann NM, Streets DG, Roden CA, Bond TC (2007) Global biofuel use, 1850–2000. Global biogeochem cycles 21, GB2019 DOI:10.1029/2006GB002836

Finco A, Bentivoglio D, Rasetti M, Padella M, Cortesi D, Polla P (2012) Sustainability of rapeseed biodiesel using life cycle assessment. Presentation at the international association of agricultural economists (IAAE) triennial conference, Foz do Iguaçu, Brazil, 18–24 Aug 2012

Finnveden G (2000) On the limitations of life cycle assessment and environmental systems analysis tools in general. Int J Life Cycle Assess 5:229–238

Frischknecht R (2000) Allocation in life cycle inventory analysis for joint production. Int J Life Cycle Assess 5(2):85–95

Fukushima Y, Chen SP (2009) A decision support tool for modifications in crop cultivation method based on life cycle assessment: a case study on greenhouse gas emission reduction in Taiwanese sugarcane cultivation. Int J Life Cycle Assess 14:639–655

Gnansounou E, Dauriat A, Villegas J, Panichelli L (2009) Life cycle assessment of biofuels: energy and greenhouse gas balances. Bioresour Technol 100(21):4919–4930

Graedel TE (1998) Streamlined life-cycle assessment. Prentice Hall, Upper Saddle River, p 310

Guinée JB, Gorrée M, Heijungs R, Huppes G, Kleijn R, de Koning A, van Oers L, Wegener Sleeswijk A, Suh S, Udo de Haes HA, de Bruijn H, van Duin R, Huijbregts MAJ (2002) Handbook on life cycle assessment: operational guide to the ISO standards. Kluwer Academic, Dordrecht

Halleux H, Lassaux S, Renzoni R, Germain A (2008) Comparative life cycle assessment of two biofuels. Ethanol from sugar beet and rapeseed methyl ester. Int J Life Cycle Assess 13(3):184–190

Hammerschlag R (2006) Ethanol's energy return on investment: a survey of the literature 1990–present. Environ Sci Technol 40:1744–1750

Hischier R, Reichart I (2003) Multifunctional electronic media-traditional media: the problem of an adequate functional unit. Int J Life Cycle Assess 8:201–208

Huo H, Wang M, Bloyd C, Putsche V (2009) Life-cycle assessment of energy use and greenhouse gas emissions of soybean-derived biodiesel and renewable fuels. Environ Sci Technol 43:750–756

IEA (International Energy Agency) (2008) Energy technology perspectives: scenarios and strategies to 2050. OECD/IEA, Paris

IFEU (2000) Bioenergy for Europe: which ones fit best?–a comparative analysis for the community. Institute for Energy and Environmental Research, Heidelberg

ISO 14040 (2006) ISO Norm 14040:2006. Life cycle assessment: principles and framework. Environmental management. International Organisation for Standardisation, Geneva

ISO 14044 (2006) Environmental management—life cycle assessment—requirements and guidelines

ISO (1998) ISO 14041: environmental management—life cycle assessment—goal and scope definition and inventory analysis. ISO 14041:1998(E). International Standards Organization

Jensen AA, Hoffman L, Møller BT, Schmidt A, Christiansen K, Elkington J, van Dijk F (1997) Life-cycle assessment (LCA)—a guide to approaches, experiences and information sources. Environmental issues series no. 6. European Environment Agency, Copenhagen

Jolliet O, Mueller-Wenk R, Bare J, Brent A, Goedkoop M, Heijungs R, Itsubo N, Peña C, Pennington DW, Potting J, Rebitzer G, Stewart M, Udo de Haes HA, Weidema B (2004) The LCIA midpoint-damage framework of the UNEP/SETAC life cycle initiative. Int J Life Cycle Assess 9:394–404

Kaltschmitt M, Reingardt GA, Stelzer T (1997) Life cycle analysis of biofuels under different environmental aspects. Biomass Bioenergy 12(2):121–134

Kim S, Dale BE (2005) Life cycle assessment of various cropping systems utilized for producing biofuels: bioethanol and biodiesel. Biomass Bioenerg 29:426–439

Kim S, Dale BE (2006) Ethanol fuels: E10 or E85—life cycle perspectives. Int J Life Cycle Assess 11:117–121

Kim S, Dale BE (2009) Regional variations in greenhouse gas emissions of biobased products in the United States—corn based ethanol and soybean oil. Int J Life Cycle Assess 14:540–546

Koellner T, Scholz R (2008) Assessment of land use impacts on the natural environment. Int J Life Cycle Assess 13:32–48

Korres NE, Singh A, Nizami AS, Murphy JD (2010) Is grass biomethane a sustainable transport biofuel? Biofuels Bioprod Bioref 4:310–325

Lardon L, Helias A, Sialve B, Steyer JP, Bernard O (2009) Life-cycle assessment of biodiesel production from microalgae. Environ Sci Technol 43(17):6475–6481

Larson ED (2006) A review of life-cycle analysis studies on liquid biofuel systems for the transport sector. Energy Sus Dev 10(2):109–126

Larson ED, Williams RH, Jin H (2006) Fuels and electricity from biomass with CO2 capture and storage, prepared for 8th international conference on greenhouse gas control technologies, Trondheim, Norway, June 2006

Lee JJ, O'Callaghan P, Allen D (1995) Critical review of life cycle analysis and assessment techniques and their application to commercial activities. Resour Conserv Recycl 13:37–56

Lent T (2003) Toxic data bias and the challenges of using LCA in the design community, presented at GreenBuild 2003, Pittsburg, PA

Liska AJ, Cassman KG (2008) Towards standardization of life-cycle metrics for biofuels: greenhouse gas emissions mitigation and net energy yield. J Biobased Materials Bioenerg 2:187–203

Luo L, van der Voet E, Huppes G, Udo de Haes HA (2009) Allocation issues in LCA methodology: a case study of corn stover-based fuel ethanol. Int J Life Cycle Assess 14:529–539

Luo D, Hu Z, Choi DG, Thomas VM, Realff MJ, Chance RR (2010) Life cycle energy and greenhouse gas emissions for an ethanol production process based on blue-green algae. Environ Sci Technol 44:8670–8677

Malça J, Freire F (2006) Renewability and life-cycle energy efficiency of bioethanol and bio-ethyl tertiary butyl ether (bioETBE): assessing the implications of allocation. Energy 31:3362–3380

Maltitz G, von Haywood L, Mapako M, Brent A (2009) Analysis of opportunities for biofuel production in sub-Saharan Africa. CIFOR

Markevicius A, Katinas V, Perednis E, Tamasauskiene M (2010) Trends and sustainability criteria of the production and use of liquid biofuels. Renew Sustain Energy Rev 14:3226–3231

McKone TE, Nazaroff WW, Berck P, Auffhammer M, Lipman T, Torn MS, Masanet E, Lobscheid A, Santero N, Mishra U, Barrett A, Bomberg M, Fingerman K, Scown C, Strogen B, Horvath A (2011) Grand challenges for life-cycle assessment of biofuels. Environ Sci Technol 45:1751–1756

Michelsen O (2008) Assessment of land use impact on biodiversity. Int J Life Cycle Assess 13:22–31

Monti A, Fazio S, Venturi G (2009) Cradle-to-farm gate life cycle assessment in perennial energy crops. Eur J Agron 31:77–84

Muys B, García Quijano J (2002) A new method for land use impact assessment in LCA based on ecosystem exergy concept. Internal report. Laboratory for forest, and landscape research, Leuven, Belgium. http://www.biw.kuleuven.be/lbh/lbnl/forecoman/pdf/land%20use%20method4.pdf. Accessed 9 Jan 2012

Ndong R, Montrejaud-Vignoles M, Girons OS, Gabrielle B, Pirot R, Domergue M, Sablayrolles C (2009) Life cycle assessment of biofuels from Jatropha curcas in West Africa: a field study. Glob Change Biol Bioenergy 1(3):197–210. ISSN 1757-1707

Nigam PS, Singh A (2011) Production of liquid biofuels from renewable resources. Prog Energ Combust Sci 37:52–68

Owens JW (1997) Life-cycle assessment in relation to risk assessment: an evolving perspective. Risk Anal 17:359–365

Pettersson K, Harvey S (2010) CO_2 emission balances for different black liquor gasification biorefinery concepts for production of electricity or second-generation liquid biofuels. Energy 35:1101–1106

Power N, Murphy JD (2009) Which is the preferable transport fuel on a greenhouse gas basis: biomethane or ethanol? Biomass Bioenergy 33:1403–1412

Reap J, Roman F, Duncan S, Bras B (2008a) A survey of unresolved problems in life cycle assessment. Part 1: goal and scope and inventory analysis. Int J Life Cycle Assess 13:290–300

Reap J, Roman F, Duncan S, Bras B (2008b) A survey of unresolved problems in life cycle assessment. Part 2: impact assessment and interpretation. Int J Life Cycle Assess 13:374–388

SAIC (2006) Life cycle assessment: principles and practice. Scientific applications international corporation (SAIC), report no. EPA/600/R-06/060. National Risk Management Research Laboratory, Office of Research and Development, US Environmental Protection Agency, Cincinnati, Ohio

Sander K, Murthy GS (2010) Life cycle analysis of algae biodiesel. Int J Life Cycle Assess 15:704–714

Scholz R (2007) Assessment of land use impacts on the natural environment. Part 1: an analytical framework for pure land occupation and land use change. Int J Life Cycle Assess 12:16–23

Shapouri H, Duffield J, Wang M (2002) The energy balance of corn ethanol: an update. Agricultural economic report no. 813. US Department of Agriculture

Sheehan J, Aden A, Paustian K, Killian K, Brenner J, Walsh M, Nelson R (2004) Energy and environmental aspects of using corn stover for fuel ethanol. J Ind Ecol 7:117–146

Singh A, Olsen SI (2011) Critical analysis of biochemical conversion, sustainability and life cycle assessment of algal biofuels. Appl Energy 88:3548–3555

Singh A, Pant D, Korres NE, Nizami AS, Prasad S, Murphy JD (2010) Key issues in life cycle assessment of ethanol production from lignocellulosic biomass: challenges and perspectives. Bioreso Tech 1001:5003–5012

Singh A, Olsen SI, Nigam PS (2011) A viable technology to generate third generation biofuel. J Chem Tech Biotech 86(11):1349–1353

Spatari S, Zhang Y, MacLean HL (2005) Life cycle assessment of switchgrass- and corn stover-derived ethanol-fueled automobiles. Environ Sci Technol 39(24):9750–9758

Stephenson AL, Kazamia E, Dennis JS, Howe CJ, Scott SA, Smith AG (2010) Life-cycle assessment of potential algal biodiesel pro-duction in the United Kingdom: a comparison of raceways and air-lift tubular bioreactors. Energy Fuels 24:4062–4077

Stichnothe H, Azapagic A (2009) Bioethanol from waste: life cycle estimation of the greenhouse gas saving potential. Resour Conserv Recycl 53(11):624–630

Stucki M, Jungbluth N, Leuenberger M (2012) Life cycle assessment of biogas production from different substrates—Schlussbericht. Auftragnehmer: ESU-Services Ltd., Publikation 290514. Bundesamt für Energie BFE, Berne, Switzerland, Accessed from Int J Life Cycle Assess 17:1142–1147

Tokunaga K, Bernstein P, Coleman C, Konan D, Makini E, NASSERI I (2012) Life cycle analysis of biofuels implementation in Hawai'i. The economic research organization. University of Hawaii at Manoa, Hawaii

Udo de Haes HA, Finnveden G, Goedkoop M, Hauschild M, Hertwich EG, Hofstetter P, Jolliet O, Klopffer W, Krewitt W, Lindeijer E, Mueller-Wenk R, Olsen SI, Pennington DW, Potting J, Steen B (eds) (2002) Life-cycle impact assessment: striving towards best practice. Society of Environmental Toxicology and Chemistry (SETAC), Pensacola

UNEP (2003) Evaluation of environmental impacts in life cycle assessment, United Nations environment programme, division of technology, industry and economics (DTIE), production and consumption unit, Paris

UNEP/GRID-Arendal (2011) http://www.unep.org/climatechange/mitigation/Portals/93/documents/Bioenergy/VBG_Ebook.pdf. ecsase on 30 Dec 2012

Vigon BW, Jensen AA (1995) Life cycle assessment: data quality and databases practitioner survey. J Clean Prod 3:135–141

Wang M (2005) Energy and greenhouse emission impacts of fuel ethanol, presentation at the DOE/EC biorefinery workshop, Washington, DC, 21 July

Wang M, Lee H, Molburg J (2004) Allocation of energy use in petroleum refineries to petroleum products. Int J Life Cycle Assess 9(1):34–44

Watson RT, Zinyowera MC, Moss RH, Dokken DJ (eds) (1996) Climate change 1995: impacts, adaptations and mitigation of climate change. Cambridge University Press, Cambridge

WHO (World Health Organization), International Program on Chemical Safety (IPCS) (2006) Guidance document on characterizing and communicating uncertainty of exposure assessment. World Health Organization, Geneva, Switzerland, Nov 2006

Index

A

Agriculture, 14, 22, 27–29, 37–39, 50–54, 59, 61, 134, 275

Algae, 41, 42, 59, 146, 148, 152

Allocation method, 1, 2, 6, 53, 109, 272, 275, 281

Anaerobic digestion, 14, 38, 41, 43, 49, 61, 63, 79–81, 86, 87, 89, 162, 169, 172, 178–180, 185, 192

Assumptions, 6, 28, 33, 58–60, 65, 69, 107, 149, 175, 203, 204, 211, 213, 216, 219, 232, 236, 251, 259, 275, 280, 283, 284

Attributional LCA, 5, 6

B

Biodiesel, 19, 20, 26, 29, 40, 84, 86, 95–97, 99, 103–107, 109, 110, 115–118, 120, 125, 131, 145–149, 152, 159–162, 269, 272, 281

Bioenergy, 5, 13–33, 38–42, 50, 61, 63, 79–83, 146, 270, 272, 275, 278, 280, 283

Bioethanol, 14, 15, 20, 29, 38, 40, 41, 44, 46, 48, 51, 52, 64, 66, 68, 84, 272, 279

Biofuels, 2, 4, 7, 9, 14, 18–21, 23, 25, 29, 33, 39–41, 44, 48, 49, 51, 55, 60, 61, 63, 65, 68, 69, 80, 82, 84, 110, 131–135, 146, 180, 269–272, 275, 276, 278, 279, 282, 284, 285

Biogas, 38, 40–43, 51, 52, 61–63, 81, 87, 89, 90, 103, 137, 138, 140, 142, 176–179, 186, 269, 271, 275

Biomass, 5, 8, 13, 14, 16, 18, 23–26, 28, 29, 31–33, 41, 42, 44, 58, 65, 68, 69, 79–81, 84, 85, 89–91, 95, 98, 103 , 122, 131, 133, 134, 145–149, 152, 162, 181, 182, 218, 228, 232, 233, 235, 247, 270, 278

Bioreactor, 63, 145–147, 179, 180, 182–184, 186, 189, 192

C

Carbon, 5, 6, 8, 13–16, 18, 20, 22, 24–26, 31, 33, 37, 38, 43, 51, 59, 81, 83, 85, 86, 90, 101, 103, 105, 119, 131, 132, 134, 146, 178, 181, 200, 212, 232, 254, 280–283

Cassava, 131, 135–142

CO_2, 14, 22, 23, 25, 26, 31, 37, 41, 43, 56, 59, 62, 81, 83–85, 89–101, 105, 119, 122, 124, 125, 136, 138, 140, 145, 146, 148, 152, 161, 162, 178, 201, 206, 221, 226, 230, 232, 254, 256, 275

Combined heat and power (CHP), 81, 103, 115, 179

Consequential LCA, 5, 6, 13, 15, 17

Conventional energy sources, 206

Cradle to farm gate , 58

Cradle to grave , 5, 79, 91, 149, 270, 285

Cradle to wheel , 274

Crude palm oil (CPO), 99, 102

Cultivation, 7, 23, 26, 51, 86, 95, 96, 99, 100, 107, 109, 110, 115, 123, 132, 134, 136–139, 141, 142, 145–149, 152, 162, 177, 270, 284

Cumulative energy demand (CED), 213, 216–218

D

Damage assessment, 152

Data collection, 20, 28, 54, 56, 63, 110, 243, 248, 251–254, 263–265, 280, 282, 284

Data variability, 15, 16, 20, 28

Digestate, 49, 54, 61, 64, 81, 87, 179

Digester configuration, 90

A. Singh et al. (eds.), *Life Cycle Assessment of Renewable Energy Sources*, Green Energy and Technology, DOI: 10.1007/978-1-4471-5364-1,

E

Eco-indicator assessment method, 214
Ecoinvent, 28, 107, 110, 119, 149, 185, 203, 205, 212–216, 219, 220, 224, 226–228, 230, 234
Egalitarian, 211, 213, 215, 223, 233, 235
Electrical energy, 50, 57, 63, 170, 173, 175, 180, 181, 184, 185, 217
Electricity, 1, 14, 17, 20, 21, 28, 29, 33, 48, 52, 58, 68, 81, 86, 103, 107, 108, 115–117, 135, 138, 139, 141, 148, 152, 162, 173, 175–177, 179, 186, 187, 195, 196, 200–202, 204–206, 208, 212, 213, 217, 218, 226–228, 230, 232–236, 241–245, 249–252, 254, 261, 264–266, 278
Emissions, 1, 2, 6, 8, 14, 18–20, 25, 26, 29, 33, 37, 38, 41, 47, 49, 51, 53–56, 58, 59, 60, 63, 64, 79, 80, 84, 96, 100, 101, 104–107, 110, 117–119, 121–126, 132, 134, 135, 138, 142, 201, 205, 206, 233, 242, 247, 248, 252, 256, 269–272, 275, 281, 283, 284
End point, 279
Energy analysis, 115, 172, 173, 180, 187, 201
Energy efficiency, 69, 84, 175, 211, 213, 217, 220, 228, 229, 235
Energy indices, 56
Energy payback time (EPT), 169, 172, 173, 213, 220, 224, 235
Energy return on investment (EROI), 169, 171
Environmental hot spots, 259
Environmental impacts, 3, 5, 7, 9, 17, 32, 47, 49, 50, 53, 54, 58, 61, 63, 66, 69, 80, 95–97, 102–104, 107, 108, 110, 117–121, 123–126, 132, 133, 140, 142, 147, 203, 205, 245, 246, 248, 270, 271, 275, 276, 278, 279, 281

F

Feedstock, 9, 14, 16–21, 25, 26, 28, 30, 31, 33, 37–41, 44, 48, 49, 52, 61, 63–66, 80, 81, 87, 91, 96, 105, 107, 131, 132, 134, 135, 141, 178, 180, 271, 272, 275, 276, 278, 279, 281
Field to wheel , 273
Fossil energy requirement (FER), 213, 235
Foundation, 197–199, 203–205
Framework, 6, 47, 69, 106, 172, 276
Functional unit, 5, 6, 13, 15, 17, 19–21, 27, 33, 47, 52, 53, 58, 61, 63, 66, 82, 83, 107, 109, 145, 149, 202, 245, 251, 252, 256, 265, 269, 270, 275, 277, 278

G

GHG emissions, 7, 14, 15, 21, 23, 25, 26, 31, 33, 49–51, 53, 56, 58, 59, 62, 68, 82, 84, 85, 88, 90, 101, 132, 133, 162, 213, 221, 226, 230, 232, 235, 245, 248, 254, 263, 265, 271, 275, 279, 281
GHG savings, 41, 85, 86, 89, 90, 162, 271
Global energy requirement (GER), 173, 185, 186
Global warming potential (GWP), 83, 100, 136, 141, 161, 213, 216, 221, 222, 226, 230, 234, 235
Greenhouse gas (GHG), 1, 37, 53, 79, 80, 101, 132, 242, 269, 270

H

Harvesting, 23, 54, 98, 99, 122, 136, 138, 141, 145, 147–149, 152, 159, 162, 232
Heating energy, 181, 182
Hierarchist, 211, 213, 215, 223, 233, 235
Hydrogen, 15, 39, 43, 169, 172, 178–180, 182, 186, 269
Hydropower, 109, 241–246, 248–251, 256, 260, 261, 263–265

I

Impact assessment, 8, 13, 15, 16, 30–33, 48, 58–60, 82, 99, 135, 138, 152, 205, 213, 235, 243, 244, 251, 252, 263, 277, 282
Impact category, 17, 30, 59, 84, 110, 118, 119, 159, 206, 282
Individualist, 211, 213, 215, 223, 224, 233–235
Interpretation, 47, 53, 60, 69, 82, 120, 135, 229, 230, 279, 280, 285
Inventory analysis, 13, 15, 20, 47, 60, 82, 277, 280, 282
Intergovernmental Panel on Climate Change (IPCC), 22, 26, 119, 134, 229, 232
ISO 14040, 3, 47, 52, 54, 69, 106, 107, 133

L

Labour, 51, 55, 134, 136, 138, 169, 173, 175, 177, 185, 186, 189
Land use change, 19, 41, 51, 134, 142, 232, 275, 284
LCA framework, 15, 22, 34, 52
Life cycle assessment (LCA), 1, 2, 13, 38, 46, 79, 80, 82, 91, 96, 131, 133, 145, 146, 211, 270

Life cycle Impact assessment (LCIA), 31, 53, 58, 118, 211, 212, 282
life cycle inventories (LCI), 241, 242
Lifetime energy production (LEO), 218, 235
Lignocellulose, 30, 33, 37, 38, 41, 42, 44, 46, 61, 64, 66, 68, 79–81, 83–87, 89–91
Limitations, 110, 148, 199, 203, 263

M
Methane, 15, 19, 41, 43, 52, 59, 61, 83, 103, 110, 119, 135, 169, 178–180, 182, 186, 188, 254
Methodology, 4, 5, 15–17, 23, 29, 34., 47, 82, 96, 106, 110, 134, 135, 147, 172, 175, 177, 205, 211, 215, 230, 236, 242–244, 264, 283, 285
Microalgae, 146, 147
Midpoint, 136

N
Nacelle, 199, 200, 203, 205, 207
Nannochloropsis, 145–148, 152
2nd generation, 29, 40, 42, 96
Net energy balance (NEB), 131
Net energy ratio (NER), 131, 213
Normalization, 31, 119, 159, 215, 223, 282

O
Oil extraction, 99, 101, 145, 147–149, 151–162

P
Palm oil, 19, 20, 28, 29, 85, 95–97, 99–104, 107, 109, 116–118, 120, 123, 126
Photosynthesis, 15, 105, 116
Photovoltaic (PV) systems, 211
Power plant, 58, 197, 228, 235, 241, 246, 248–254, 256, 259, 263
Product system modelling, 242, 245, 253

R
3rd generation, 41, 96
Reference system, 2, 5, 6, 17, 18, 58, 63, 66, 68, 149, 269, 275, 277, 278
Renewable directive, 62, 80, 82
Renewable energy requirement (RER), 218
Renewable energy sources, 1, 2, 4, 6, 8, 9, 13, 33, 80, 131, 140, 142, 188, 196, 203, 218, 229, 241
Rotor, 199–201, 203, 205, 207, 208

S
Scope and goal, 2
Sensitivity analysis, 29, 64, 152, 161, 169, 277, 283, 284
1st generation, 30, 33, 40, 42, 44, 80, 96
Sugarcane molasses, 131, 135, 138, 140, 142
Sugars, 41, 44, 45, 68
Sustainability, 1, 2, 6, 8, 9, 14, 37, 38, 40, 44, 46, 51, 62, 65, 80, 84, 89, 95, 96, 131, 133, 142, 146, 147, 169, 171–175, 178, 180, 185, 187–189, 192, 211, 214, 216, 224, 229, 232, 235, 269, 285
Sustainability assessment, 7, 9, 15, 132–135, 216, 285
Sustainability evaluation, 9, 176, 186
Sustainable energy, 1, 6, 9, 41, 49, 52, 169, 180, 270
System boundaries, 2, 6, 13, 19–21, 33, 48, 54, 60, 61, 66, 69, 84, 134, 135, 149, 162, 146, 170, 275, 279, 282, 284
System limits, 202

T
Thermal energy loss, 181, 183
Total indirect energy spent (TIES), 175
Total net energy produced (TNEP), 175
Tower, 198–200, 203, 205, 207
Trans-esterification, 95, 103, 104, 107, 116, 124–126, 147
Transport, 1, 8, 19, 28, 37, 39, 41, 46, 61, 63, 65, 79–81, 84, 85, 89, 91, 145, 148, 149, 177, 184, 199, 222, 229, 250, 254, 278
Transportation, 4, 7, 14, 20, 38, 41, 61, 63, 64, 84, 95, 102, 108, 115–117, 121–123, 125, 126, 131, 134, 135, 137, 139, 148, 173, 176, 177, 202–204, 246, 254, 265, 270, 275, 285
Turbines, 195–201, 203, 205, 206, 208, 214, 233, 250

U
Useful energy, 116, 170, 172–174, 185, 186–189, 192

W
Well to pump, 274
Well to tank, 68, 162, 279
Wind energy, 196, 197, 200, 201, 203, 233

MIX
Papier aus verantwortungsvollen Quellen
Paper from responsible sources
FSC® C105338

If you have any concerns about our products,
you can contact us on
ProductSafety@springernature.com

In case Publisher is established outside the EU,
the EU authorized representative is:
Springer Nature Customer Service Center GmbH
Europaplatz 3, 69115 Heidelberg, Germany

Printed by Libri Plureos GmbH
in Hamburg, Germany